Maintainability

PRINCIPLES AND PRACTICES

Benjamin S. Blanchard, Jr.
General Dynamics, Electronics Division

E. Edward Lowery
General Dynamics, Electronics Division

McGRAW-HILL BOOK COMPANY

New York St. Louis San Francisco London
Sydney Toronto Mexico Panama

MAINTAINABILITY

 Library of Congress Catalog Card Number 69-12772

05855

1234567890 MAMM 754321069

We dedicate this book to our wives Dorothy and Beverly for their support, understanding, and encouragement, without which its development would not have been possible.

Foreword

The value of maintainability engineering in the design process and in support planning has been recognized for some time by leaders in industry and government. The advancement of the discipline and its application have been restricted, however, by the limited number of qualified, trained personnel to perform the required services. It has become evident that a great need exists for a comprehensive text for use in classroom instruction. The authors have responded to this need by providing a text that conveys both the basic principles of maintainability and methods of implementation. Each topic within the subject of study is treated so as to relate theory with method of application.

Much of the material has been applied in different form to training programs within the General Dynamics Corporation. It has been tested under the critical review of the students with favorable results. The basic material has also been used in teaching a graduate-level course in applied maintainability engineering at the Rochester Institute of Technology. In addition, a number of graduate students at the Red River Army Depot training center (in association with Texas A&M University) have used the original material as basis for their research studies. Information available from the industry and college programs has been evaluated by the authors and is reflected in this book.

The book is primarily intended for classroom use in training engineers for maintainability work in industry. It will also be of value to those people in government who are concerned with establishing program requirements. And, of course, the book will serve as an authoritative reference for those in the ranks of management who must direct and control a maintainability program.

John W. Snodgrass
Director of Product Assurance
General Dynamics, Electronics Division

Preface

Maintainability is a new discipline. No other area of management activity offers greater opportunity for development, creative application and reward for accomplishment. The problems of product support and maintenance of complex systems can be reduced—if not eliminated—by applying the principles of maintainability. The decisions made by the line manager in implementing a maintainability program have much to do with the competitive success or failure of his firm.

This book is an introduction to maintainability engineering. Its focus throughout is on the *principles* and *practices* of organization, planning, actuation, and control of a company maintainability program. The arrangement of the book was based on the assumption that the material would be used for classroom instruction. It is designed for use in courses at either the undergraduate or graduate level. The first chapter outlines the history of maintainability, relates the basic principles and concepts of the area of study, and defines the terms which are commonly used by maintainability engineers. The second chapter outlines the requirements and criteria for organization of maintainability within a company and specifies the tasks that are associated with a maintainability function. The remainder of the book deals with the treatment of topics within each defined area of work. Each chapter is a *building block* of information that logically develops the total concept of maintainability.

Although this is an introductory book, it covers a wide spectrum of product- and system-support problems. An attempt has been made to relate maintainability with the interacting effects of such factors as design engineering, training, human factors, reliability, technical documentation, spares provisioning, support-equipment definition, and operational requirements. This book is intended to develop an understanding of maintainability in relationship to these factors that, as a whole, form the ingredients for *system effectiveness.*

The material in this book is the product of our industry and military experiences over a period of 20 years. As such, it is not possible for us to acknowledge the individual contributions that have influenced our work. Some of the materials in the book have been used in different form in training programs, and the reactions of students have been helpful in defining and presenting the information for this text. We are greatly indebted to Mr. William C. Tinney, who made significant contributions in developing the chapters dealing with *quantitative factors* and *demonstrations.* He was helpful also by offering comments based on his teaching experiences with graduate students at the Rochester Institute of Technology. Special recognition is extended to the Electronics Division of the General Dynamics Corporation for their permission to use the document entitled "Organization for Maintainability" as the basis for this text.

Mr. Paul H. Hayden has done extensive research on fluid systems, and he has given us information and data to illustrate maintainability-design characteristics of equipment in this specialized area of engineering. The Chrysler Corporation has provided information on the maintainability features of automobiles which highlights the practical application of maintainability. And last, but not least, we owe a debt of gratitude to our secretaries Mrs. Mary Hint and Miss Lin Anderson for their assistance in the preparation of the text materials.

Benjamin S. Blanchard, Jr.
E. Edward Lowery

Contents

ONE
Introduction

Maintainability is a characteristic of equipment design and installation which is expressed in terms of ease and economy of maintenance, availability of the equipment, safety, and accuracy in the performance of maintenance actions.

The object of maintainability is to design and develop systems and equipment which can be maintained in the least time, at the least cost, and with a minimum expenditure of *support resources,* without adversely affecting the item performance or safety characteristics.

Support resources are (1) manpower utilization, (2) spare parts, (3) tools and test equipment, (4) services (contractor support), and (5) support facilities.

Maintainability is a discipline within the science of engineering. This discipline was developed to provide designers with a source of specialized skills and knowledge related to the support and maintenance of equipment and systems. The need for this service was brought about by the increased complexity, size, and quantity of items comprising a system, and by the attendant technical problems, increased support costs, and changes in technology. These conditions have forced the abandonment of the concept of design by one individual, and in its place the *design team* has emerged. Maintainability is one important element of the design team which (along with reliability, human engineering, components engineering, and others) provides the designer with specialized information and data for consideration in the definition of an equipment item or a system.

The realization of the objective of maintainability requires involvement in the total design process. This includes participation in advance planning, design development, production, installation, and activation of equipment or a system. Applied maintainability consists of four discrete work stages within each phase of the design process. These are (1) planning, (2) design application, (3) measurement, and (4) evaluation. In the planning stage, maintainability requirements are defined and translated into design criteria. Next, the criteria are applied to the design to establish the desired inherent functional and physical characteristics of the equipment or system. These design characteristics are then measured to verify quantitative and qualitative goals. And finally, the design and the results of the measurements are evaluated to assess areas for improvement. In this manner, the ultimate support requirements of the equipment or system are controlled to meet specific operational needs.

Maintainability is applicable to both commercial and military products. A commercial product containing maintainability features has sales appeal. This characteristic of design can also be the influencing factor in obtaining "repeat sales." A progressive company engaged in a competitive market is aware of these facts, and proper emphasis is placed on the conduct of its maintainability program. The success of its maintainability program contributes directly to the profit from sales. A company undertaking defense work reaps the same benefits from a maintainability program as the company that concentrates on commercial products. That is, it will receive more contracts, and therefore make more money, because it produces maintainable products.

The intent of maintainability, as defined and discussed above, is akin to good engineering and business practices. In this regard, maintainability is not new or different. What does give distinctive importance to maintainability is the *method* and *organization* of applying proven, well-founded engineering and management principles. A formulated concept of maintainability, therefore, is the composite result of understanding specific, selected principles, and how and why they are combined to attain certain detailed objectives.

The authors believe that a formal definition of maintainability would be of no value to the student. Instead, the student should establish a genuine feel for each topic covered in the rest of this book, wrestle with its problems, see the diversity of the subject, and formulate his own conclusions.

THE HISTORY OF MAINTAINABILITY

The concept of maintainability was formally recognized by the military services of the United States in 1954. The concept evolved from the results of the reliability programs conducted in the late 1940s and early 1950s, which indicated that 100 percent reliability of equipment was an unobtainable goal. Despite the fact that the reliability programs were effective in prolonging the life of equipment and systems, it became evident that maintenance requirements could not be eliminated. With the increased size and complexity of military systems, the maintenance costs for these systems approached one-third of all the operating costs. In addition, it was established that nearly one-third of all personnel were engaged in maintenance or support functions. Maintainability was conceived to deal with these problems of maintenance and support. An immediate goal was to reduce the costs of sustaining the military systems in existence at that time.

The initial step in the maintainability development process involved the establishment of definitions and the preparation of program specifications. From 1954 to 1959, maintainability requirements were primarily covered through special contractual exhibits or amendments. By 1959, formalized program specifications were available, one of the first being the Air Force's MIL-M-26512 (June, 1959). Subsequently, there have been a wide variety of specifications and documents developed to cover the subject. The requirements of these specifications have been applied to many types of equipment (pneumatic equipment, hydraulic systems, electrical/mechanical systems, electronics, etc.) designed and developed to satisfy varying operational requirements applicable to shipboard, aircraft, missile-vehicle, and ground environments.

The first maintainability specifications developed were limited to the inclusion of broad definitions and qualitative requirements. These specifications were later developed to incorporate refined definitions, specific maintainability-program tasks, and quantitative system requirements. Supplementing the specified program tasks were procedures governing methods for accomplishing maintainability prediction, analysis, test, and demonstration. These methods varied from one specification to another, resulting in significant differences in approach relative to the implementation of a maintainability program.

Early in 1966, an attempt was made to standardize (1) the approach to implementation of a maintainability program, (2) the techniques for accomplishing a prediction, and (3) the methods and procedures associated with maintainability demonstration. These goals were reflected through the release of two military standards (MIL-STD-470 and MIL-STD-471) and a military handbook (MIL-HDBK-472). The application of these documents has resulted in a better definition of maintainability, its parameters, and its methods of measurement.

During the period 1965 to 1968, maintainability was associated with the measurements of cost effectiveness, system effectiveness, and integrated logistics support. In this regard, the role of maintainability has assumed new proportions, but its basic purpose remains unchanged. This association has, however, brought more attention to the subject. This is evidenced by (1) the increased interest in the part of engineers and managers representing industry and government, (2) the establishment of numerous formal college and service-school programs on the subject of maintainability, (3) the adoption of the discipline by various engineering and management societies, and (4) the increase in presentation of technical papers on the subject at meetings. The association of maintainability with cost effectiveness, systems effectiveness, and other functions and disciplines will be discussed further in subsequent sections of this book.

The primary impetus for maintainability development is attributable to the Department of Defense and to military programs. It should be noted, however, that during the same period of time that the government was advancing the concept of maintainability, general industry was taking measures to counter similar problems of rising costs in support of their products. The basic principles applied by general industry were identical to those set forth in military specifications and standards, although they varied in application to suit particular needs. And, as may be expected, the terminology describing this work effort differed from one company to the next. The significant point is, however, that the need for maintainability was universal. And this need was brought about by the unique combination of events surrounding the engineering and technological advances of the 1950s and 1960s. It appears that the need for maintainability will remain in coming years—and perhaps increase—because of the trend in complexity of products and systems.

As indicated above, maintainability was developed over a period of years. It went through many stages of progressive development causing frequent changes in the meaning of the word and redefinition of the area of work. Initially, maintainability was associated solely with *design liaison*, limited to review of drawings and equipment to assess inherent characteristics. This effort was set apart from maintenance engineering, logistics, design engineering, and other activities. It was a satellite function with limited authority and responsibility. Gradually, the function extended into design services, systems planning, and systems analysis. And today, maintainability is a distinct organization in most companies and one that is fully integrated with all conventional line activities. As presently defined, the tasks performed by a maintainability organization are extensive and all inclusive. Since an entire chapter will be devoted to organizational structure and functions, we shall not treat the subject in greater detail here.

THE LANGUAGE OF MAINTAINABILITY

By way of introduction to the field of maintainability, it seems logical to direct some attention toward its language. The following terms and definitions were selected to provide the reader with the rudiments of the language. The listing was purposely limited to basic data since a complete glossary is not necessary to understand the material presented in this book. However, an understanding of the listed items is necessary to fully appreciate the discussions in subsequent chapters.

1. *Maintenance:* All actions necessary for retaining an item in, or restoring it to, a serviceable condition. Maintenance includes servicing, repair, modification, overhaul, inspection, and condition verification.

2. *Maintenance, preventive:* That maintenance performed to retain an item in satisfactory operational condition by providing system inspection, detection and prevention of incipient failures, overhaul, lubrication, calibration, etc.

3. *Maintenance, corrective:* That maintenance performed to restore an item to satisfactory condition after a malfunction has caused degradation of the item below the specified performance. The major tasks associated with corrective maintenance are:

PREPARATION—gathering tools; obtaining, setting up, and calibrating maintenance aids; warming up equipment; etc.

LOCALIZATION—determining the location of a failure to the extent possible without using accessory support equipment.

ISOLATION—determining the location of a failure through use of accessory support equipment.

DISASSEMBLY—equipment disassembly to the extent necessary to gain access to the item that is to be replaced.

INTERCHANGE—removing the defective item and installing the replacement.

REASSEMBLY—closing and reassembly of the equipment after the replacement has been made.

ALIGNMENT—performing any alignment, minimum tests, and/or adjustment made necessary by the repair action.

VERIFICATION CHECKOUT—performing the minimum checks or tests required to verify that the equipment has been restored to satisfactory performance.

4. *Mean-time-between-maintenance (MTBM):* Mean-time-between-maintenance includes both preventive and corrective-maintenance requirements. Relative to corrective maintenance, *MTBM* considers primary (random) failures, secondary (dependent) failures, quality and manufacturing defects, operator- and maintenance-induced failures, and others.

5. *Mean-time-between-replacements (MTBR):* Mean-time-between-equipment-item-replacements for preventive- and corrective-maintenance purposes. This factor forms the basis for spare/repair-parts determination. It should be noted that a maintenance action (represented by the *MTBM* factor) does not always result in an item replacement.

6. *Maintenance downtime:* That portion of downtime which can be attributed to preventive- and corrective-maintenance functions. Maintenance downtime may be expressed in a measure of central tendency (arithmetic mean, geometric mean, median, and mode). It may also be expressed in terms of a maximum value relative to a percentile point of distribution of downtime. Symbols of maintenance downtime are:

$\overline{M}_{ct}$—Mean active corrective-maintenance time (arithmetic mean). Equal to *MTTR*.

$\overline{M}_{pt}$—Mean active preventive-maintenance time (arithmetic mean).

$\overline{M}$—Mean maintenance time (function of $\overline{M}_{ct}$ and $\overline{M}_{pt}$).

MTTR—Mean-time-to-repair. Equal to $\overline{M}_{ct}$.

MDT—Maintenance downtime (total time during which an equipment item is not in condition to perform its intended function). *MDT* includes logistics time and waiting or administrative time.

$\widetilde{M}_{ct}$—Median active corrective-maintenance time.

Equal to *ERT*.

ERT—Equipment active repair time. Equal to $\widetilde{M}_{ct}$.

$\widetilde{M}_{pt}$—Median active preventive-maintenance time.

$MTTR_g$—Geometric mean-time-to-repair.

M_{max}—Maximum active corrective-maintenance time (95 percent confidence level).

Logistics or Supply Time—That portion of nonactive maintenance time during which maintenance is delayed solely because a needed item is not immediately available.

Wait or Administrative Time—That portion of nonactive maintenance time that is not included in logistics or supply time.

7. *Availability (inherent):* The probability that a system or equipment, when used under stated conditions, without consideration for any scheduled or preventive action, in an ideal support environment (i.e., available tools, spares, manpower, data, etc.), shall operate satisfactorily

at a given point in time. It excludes ready time, preventive-maintenance downtime, logistics time, and waiting or administrative downtime. It may be expressed as

$$A_i = \frac{MTBF}{MTBF + MTTR} \tag{1}$$

MTBF is a factor of reliability. It is the Mean-time-between-failures of the equipment or system.

8. *Availability (achieved):* The probability that a system or equipment, when used under stated conditions in an ideal support environment (i.e., available tools, spares, manpower, data, etc.), shall operate satisfactorily at a given point in time. It excludes logistics time and waiting or administrative downtime. It includes active preventive- and corrective-maintenance downtime. It may be expressed as

$$A_a = \frac{MTBM}{MTBM + \bar{M}} \tag{2}$$

9. *Availability (operational):* The probability that a system or equipment, when used under stated conditions in an actual operational environment, shall operate satisfactorily at a given point in time. It includes ready time, logistics time, and waiting or administrative downtime. It may be expressed as

$$A_o = \frac{MTBM + \text{ready time}}{(MTBM + \text{ready time}) + MDT} \tag{3}$$

10. *Cost effectiveness (C.E.):* The measure of a system or equipment in terms of a given level of cost. Cost includes initial acquisition costs (cost of conceptual studies, design, development, technical data, equipment fabrication and test, installation, and initial support) and utilization costs (annual operational costs for personnel, spares, support equipment and tools, facilities, and resources).

11. *Integrated logistics support (I.L.S.):* A management function providing the controls which help to assure that the customer will receive a system that will not only meet performance requirements, but one which can be expeditiously and economically supported throughout its programmed life cycle. I.L.S. accomplishes the integration of all support elements to include support equipment, spare/repair parts, personnel and training, technical data, facilities, and contractor maintenance support.

12. *Maintenance indices:* Primary expressions of quantitative measurements of equipment-maintenance characteristics.

MMH/OH—Maintenance man-hours per equipment or item operating hour.

MMM/OH—Maintenance man-minutes per equipment or item operating hour.

Cost/OH—Maintenance cost per equipment or item operating hour.

Others are *MTBM, MDT,* and achieved availability.

13. *Design criteria:* Narrative or quantitative statements that define the required functional and physical characteristics of an equipment item or system. (E.g., "self-test provisions shall be provided to verify the proper function of each plug-in assembly.")

14. *Design liaison:* The monitoring of design progress and the review of engineering data and equipment to assess maintainability characteristics of the design, and the interfacing activity with the design organization in order to influence the ultimate equipment-support characteristics.

15. *Design review:* The evaluation of drawings, sketches, mock-ups, assemblies, and other items which describe the equipment/system design. This evaluation is performed to assess potential and existing problems related to (1) the manufacture of the equipment, (2) its functional capability, and (3) the support of the equipment or system. The design review is normally conducted in a formal meeting which is attended by representatives from the various areas of interest (e.g., human engineering, maintainability, reliability, systems and design engineering, manufacturing engineering, etc.). As a result of this joint meeting, major trade-off decisions are made, and direction is issued which reflects these conclusions.

16. *Design services:* Providing design criteria, standards, procedures, policy statements, requirement sheets, and other data which will aid the designer in the definition of the equipment- or system-design characteristics. This function includes research and special studies to determine the feasibility of a design approach as it relates to maintainability.

17. *Demonstration:* The testing of an equipment item or system to prove that specified quantitative and qualitative maintainability characteristics are inherent in the design of the equipment or system.

18. *Maintainability analysis:* The sequential development and review of data—concurrent with, or preceding design development—to aid in describing the optimum design characteristics of the equipment or system. The elements considered in the review are (1) quantitative requirements, (2) support resources, (3) cost, (4) operational objectives, and (5) safety. The results of the review are translated into criteria which are applied to the design process.

19. *Maintainability engineering:* An organization that is associated with the functions of maintenance engineering, maintainability-design liaison, systems analysis, design services, support documentation, systems planning, safety engineering, and systems integration and test.

20. *Maintenance analysis:* The review of engineering drawings and equipment—concurrent with, or following design development—to vali-

date the maintainability-analysis data. This is a formal review with the findings recorded in a prescribed format. The scope of this review is identical to that of the maintainability analysis.

21. *Maintenance concept:* A narrative statement or illustration that defines the theoretical means of maintaining an equipment item or system. The statement or illustration relates the tasks that *should* be performed at each level of maintenance, the test equipment and tools that *should* be used in maintenance of the items, and the skill levels of the maintenance personnel that perform the identified tasks.

22. *Maintenance engineering:* An organization or function that is usually associated with the tasks of maintenance analysis, maintenance studies, maintenance-policy generation, support planning, and maintenance-procedures development. These tasks are usually included in the functions performed by the present-day maintainability organization.

23. *Maintenance policy:* A narrative statement or illustration that defines the practical means of maintaining an equipment item or system. This is an extension of the maintenance concept with due consideration for the inherent characteristics of the equipment/system design and other constraints or limitations.

24. *Systems analysis:* The review of system design and support requirements, the establishment of primary maintainability criteria, an incremental analysis of tasks and functions, association of tasks and functions with support needs, and the progressive review of design/support needs as the design process evolves.

25. *Trade-off:* A decision-making process. As related to this text, a trade-off is concerned with decisions between desired maintainability characteristics and cost, reliability- and maintainability-design characteristics, automatic-equipment use and manual-equipment use, and other factors associated with maintainability engineering.

26. *Maintenance level:* A division of maintenance tasks. As an example, primary USAF maintenance levels are classified as (1) organizational, (2) intermediate, and (3) depot. Certain defined tasks are performed at each level consistent with the capability of the associated organization and the provided facilities, and task allocations are also made with due consideration for the operational objectives of the system. Discrete maintenance levels are assigned within each primary level to clearly define the maintenance processing of each equipment item.

APPLIED MAINTAINABILITY

The concept of maintainability can best be understood by having knowledge of what is done to apply the underlying principles of the function. Since the spectrum of maintainability is broad, this topic will be pursued to great lengths in subsequent chapters. For purposes of introduction, however, a composite summary of what a maintainability organization does is outlined below.

1. Given specified maintainability qualitative and quantitative requirements by the customer, maintainability personnel interpret and translate these requirements into specific design criteria.

2. Given the customer's proposed operational concept and maintenance environments, maintainability personnel define the approach to the system/equipment design and logistics support.

3. During the design and development process, maintainability personnel provide direct assistance to the design engineers through (1) accomplishment of special studies, (2) participation in the selection of equipment, (3) accomplishment (on a sustaining basis) of maintainability prediction, and (4) participation in progressive formal and informal design reviews.

4. In support of the system/equipment design and development process, maintainability personnel produce various types of documentation. Such documentation not only constitutes an evaluation of the basic design, but provides data leading to the development of system-support requirements (e.g., anticipated scheduled and unscheduled maintenance requirements, support equipment and tools, personnel types and skill levels, spare and repair parts, facilities and resources). Examples of such data include:

- Maintenance-concept data
- Systems-analysis data (functional-flow diagrams)
- Maintainability-analysis data
- Trade-off–study reports
- Maintainability-prediction reports
- Support-equipment plan and illustration data
- Test-point data (TPD)
- Calibration-requirements summary (CRS)
- Design-review reports
- Maintenance-analysis data
- Maintainability-demonstration plan and report (including assessment reports)
- Maintainability-program reports (initial plan and status reports)

Documents such as the maintainability-prediction reports, the support-equipment plan, the design-review reports, the maintenance analysis, and the maintainability demonstration report directly reflect the basic system/equipment design. If the equipment is poorly designed from the standpoint of maintainability, the problems will be readily evident in the documentation. In this event, the documentation provides the basis for recommended corrective action. On the other hand, if the equipment design is good, the documentation will support this fact.

5. Maintainability personnel provide support in all logistics provisioning actions (e.g., support-equipment provisioning, spare/repair parts provisioning, etc.). Maintainability documents supplement the provisioning data.

6. As preproduction systems/equipment items are made available, maintainability personnel plan and implement demonstration testing to ensure that the equipment design meets the specified customer requirements.

PRIMARY CONSIDERATIONS

As a science, maintainability is founded on certain fundamental truths. To retain substance to the concept of maintainability, these truths must be observed in the study and application of the subject.

The primary considerations to be applied in all phases of maintainability work and study are:

1. The fulfillment of maintainability requirements cannot be accomplished by a single individual, a specific skill-level type, or by a single organizational entity. Maintainability in the design of an equipment item or system is the product of a joint effort by many organizations.
2. A maintainability organization is the catalyst in the joint effort by providing direction, coordination, and measurement of the maintainability program.
3. The responsibility for the maintainability characteristics of an equipment item or system lies with the designer.
4. Maintainability in equipment design can be specified, predicted, measured, and demonstrated. Prediction and measurement can be accomplished to varying degrees throughout all program phases.
5. As a design characteristic, maintainability features must be specified and incorporated in the system/equipment design concurrent with the item development. That is, maintainability is a part of design planning. It cannot be treated as a separate feature. Lack of consideration of maintainability features during initial design results in either costly product support throughout the equipment life cycle or the incorporation of costly product changes at some subsequent point in the development cycle.
6. Maintainability-design planning considers existing capabilities regarding facilities, test equipment, personnel manning of organizations, procedures, spares support, etc., in establishing system criteria. Theoretical maintainability principles may be waived to utilize existing capabilities.

7. Reliability, human engineering, and other effectiveness characteristics of a system or equipment are evaluated during all phases of maintainability-program implementation. A sacrifice of reliability, maintainability, or any other single characteristic is often necessary to obtain the best overall design configuration.

8. Equipment or systems comprising ideal maintainability-design characteristics may be prohibitive due to cost factors. *Applied* maintainability considers cost factors incurred during design and development of equipment, as well as cost factors associated with equipment support.

9. Maintainability in equipment design is not only beneficial to customer use in an operational environment, but is beneficial to the contractor relative to manufacturing and test operations. Good maintainability aids in saving costs in both areas.

10. Maintainability is planned to complement the operational requirements of a system. It is required, on occasion, to waive ideal maintainability characteristics in favor of meeting unusual or stringent operational objectives. *Applied* maintainability considers operational needs.

11. Safety provisions protecting personnel and equipment receive special consideration, and otherwise desirable design features are rejected to retain these provisions.

12. Maintenance is the consequence of design. Properly applied, maintainability forestalls the requirement of maintenance and alters the course of design to eliminate or reduce the effect.

COST EFFECTIVENESS

Cost effectiveness is a term which describes the relative value of a system. Its measure reflects the costs of purchase, installation, and maintenance of the system over its useful life, and the capability of the system to perform its intended function. Cost effectiveness is expressed as follows:

$$\text{Cost effectiveness (C.E.)} = \frac{\text{system effectiveness (S.E.)}}{\text{acquisition cost} + \text{utilization cost}} \tag{4}$$

System effectiveness is defined as "the probability that the system can successfully meet an operational demand within a given time when operated under specified conditions."

The *acquisition cost* of a system pertains to those funds expended to attain a specified performance (developmental costs), the purchase cost of the items that comprise the system (production costs), the cost of installing the system, and the initial investment to prepare the system for use (training costs, expenditures for publications, facilities, support equipment, etc.).

The *utilization cost* of a system refers to those funds that are required to maintain the system in a *ready* or *operational* condition during its planned life. This includes maintenance costs (expenditures for contractor services, cost of a sustained training program, funds for spares and labor, etc.) and general operating costs (staffing costs, expenditures for servicing and operating the system).

Please refer to Fig. 1-1.

Maintainability contributes to each of the three factors that determine cost effectiveness. This relationship will be discussed in detail in subse-

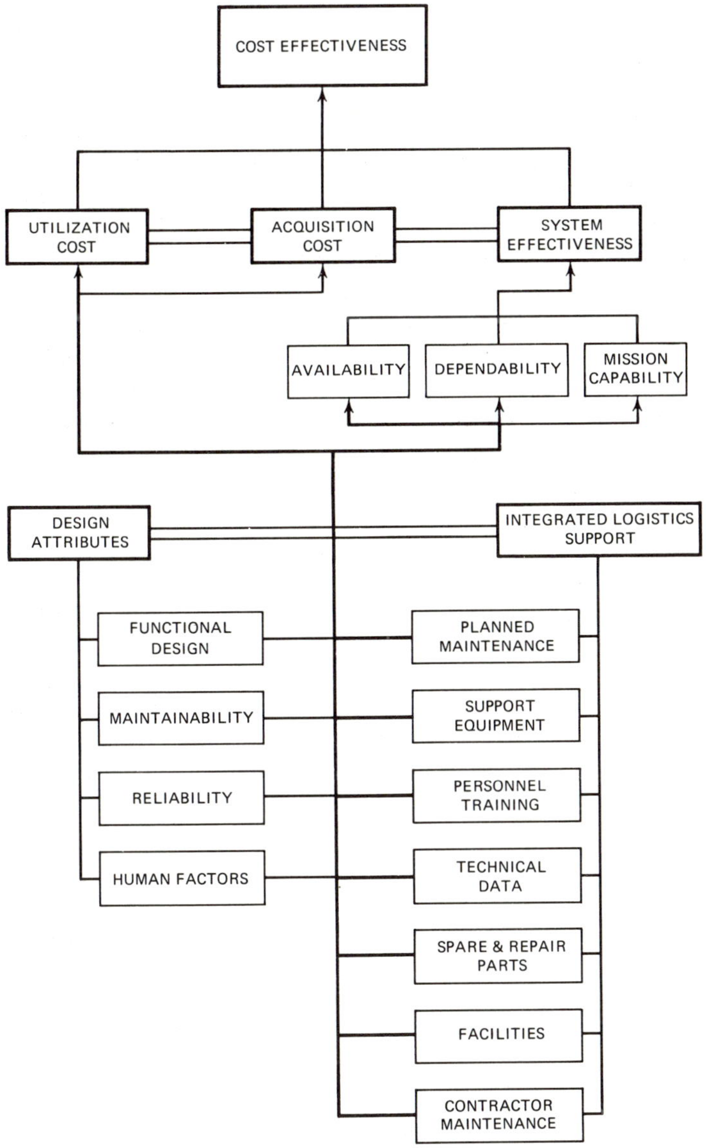

FIG. 1-1 The Ingredients of Cost Effectiveness.

quent paragraphs. To appreciate the role maintainability plays in defining the relative value of a system, it is necessary first to understand the function of system effectiveness, what it is comprised of, and how *integrated logistics support* influences its measurement.

System Effectiveness (S.E.)

As outlined earlier, system effectiveness is defined and expressed as a probability of system success. In practice, however, system effectiveness as a concept includes the engineering-management process concerned with the definition, control, and measurement of system performance. The functions or elements that are defined, controlled, measured, and related in the form of a probability differ in terminology from one organization to another. As an example, refer to the three concepts portrayed in Fig. 1-2. In any event, the object of the S.E. function is to determine:

1. Is the system ready to perform its function?
2. How well will it perform during a mission?
3. Will it produce the desired effects?

The engineering aspect of S.E. involves the technical assessment of the system to answer the above questions. The management aspect of S.E. is concerned with the control and measurement of the system development to attain the specified characteristics. Both use analytical prediction and modeling techniques to identify problem areas in a timely manner so that corrective action can be initiated. Hence, assurance is gained that full expectations of the system will be realized when it is placed in operation.

For purposes of discussion, the Air Force S.E. concept is adopted for use in this text.

Availability-operational. The elements that determine the probability that a system will be ready for use are (1) *MTBM*, (2) *MDT*, and (3) ready time. See Eq. (3). *MTBM* is a function of reliability and maintainability. *MDT* is a measure of maintainability, human factors, and logistics. Ready time is a function of reliability. The interaction of the functions that determine each element of system availability will become evident as the reader gains knowledge of the concept of maintainability.

Dependability. The elements that determine the probability of a successful performance of the system during its mission vary in each situation. As an example, the measurement of a ground-to-ground missile performance will differ from the considerations extended to a manned aircraft system. Safety may be a dominant factor in one case, and not applicable to another, etc. Generally speaking, however, the functions of maintainability, human factors, reliability, and logistics apply

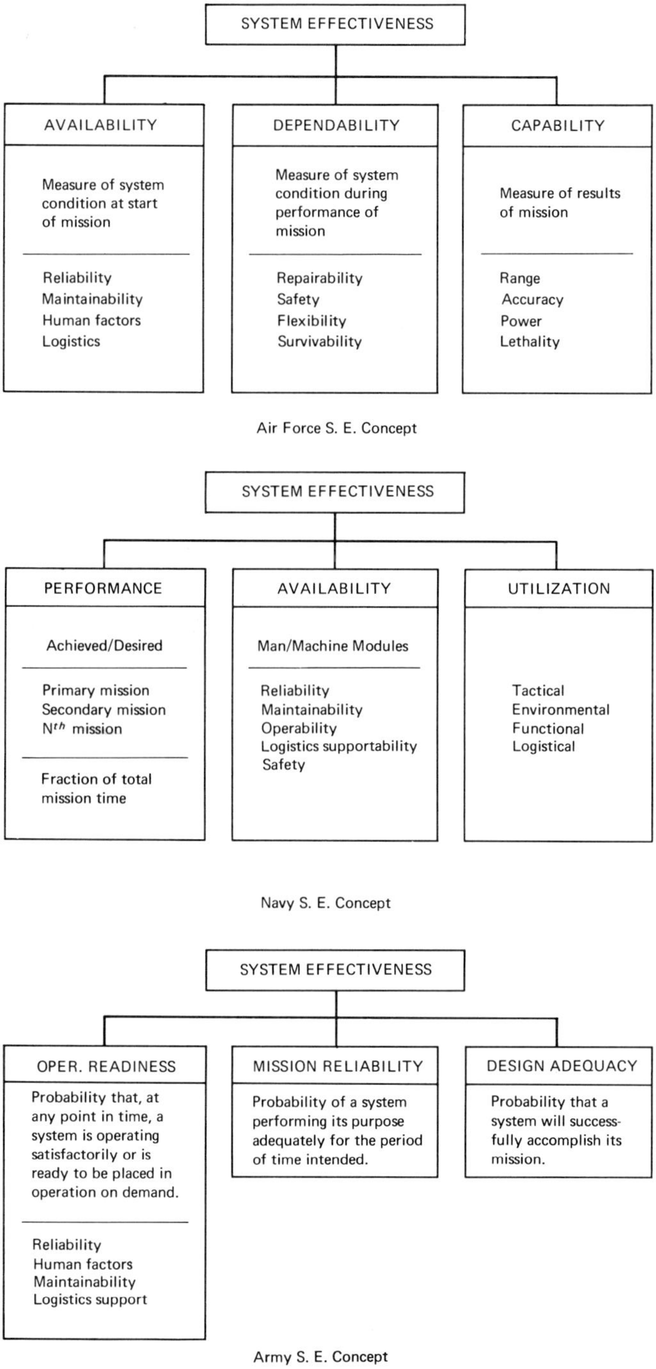

FIG. 1-2 Three Concepts of System Effectiveness.

when there is a man-machine interface activity during the performance of the mission. And, reliability is dominant in assessing the flexibility and survivability of a system that does not involve man in the conduct of the mission.

Capability. The measure of mission effectiveness also depends on the situation under analysis. As an example, the measurement of *accuracy* may be dependent to some extent on the involvement of man, in which case human factors is a function to be considered. Likewise, *range* capability may depend on the limits of human endurance. As a general rule, however, reliability is a dominant function in determining the mission capability of a system, with due consideration given to the physical functional characteristics of the system (e.g., weapon).

Integrated Logistics Support (I.L.S.)

Integrated logistics support is a management function. Its purpose is to assure that the system being designed and developed can be economically supported when placed in its operational environment. The object of I.L.S. is to fulfill this purpose with the least expenditure of funds.

I.L.S. is involved in all phases of the system-development cycle commencing with mission/system definition and extending through system operational deployment in the field (refer to Fig. 1-3). I.L.S. is a major consideration in:

1. The establishment of support-requirements criteria and the maintenance concept during the conceptual phase prior to commencing with end-item design and development.
2. The accomplishment of system analysis and trade-off studies where end-item and logistics support are considered on an equivalent basis in the system-selection process.
3. The accomplishment of design-analysis functions (e.g., maintainability/maintenance analysis) where design decisions are made which impact system support.
4. The timely provisioning and acquisition of items of support.
5. The integrated testing of end-item equipment along with support equipment, personnel, spares, technical data, and facilities.
6. The follow-on contractor maintenance support once systems or equipment have been delivered to the customer and operationally deployed.

The development of I.L.S. can be realized through three major steps to include: the identification of basic support requirements (system level); the engineering aspects associated with design analysis and the development of detailed support requirements; and the provisioning, acquisition, and follow-on support of the applicable system or equipment

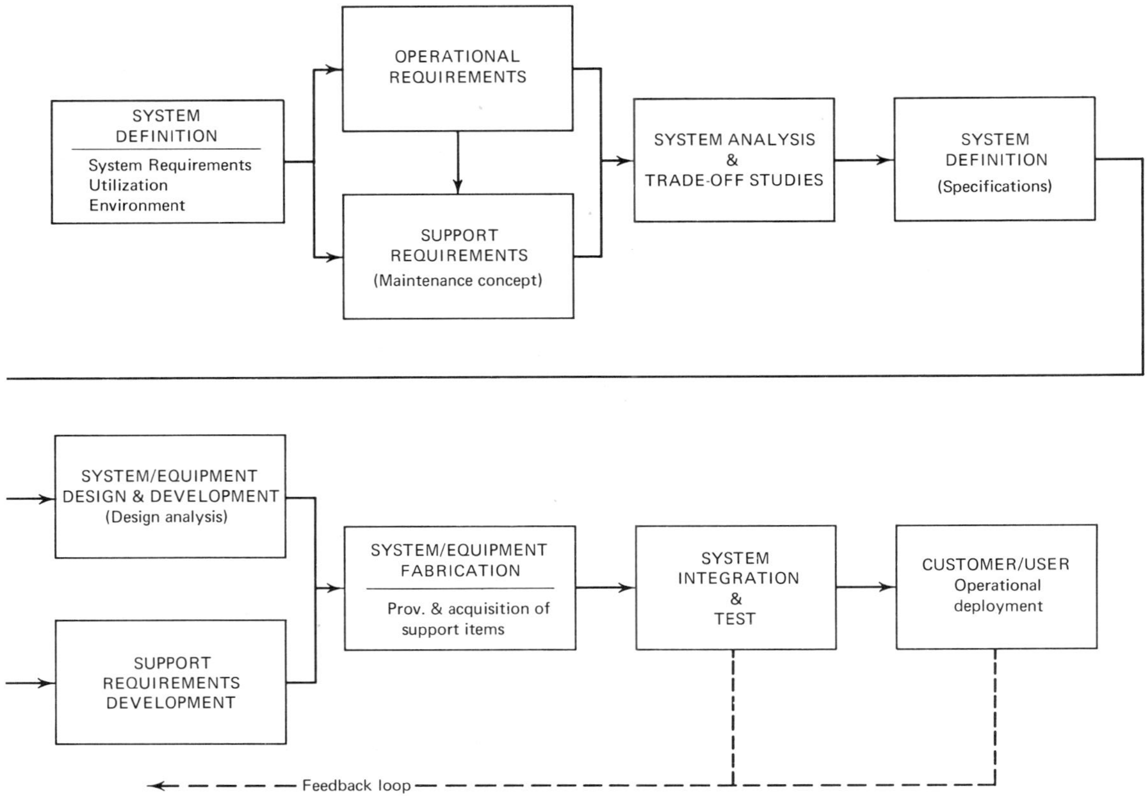

FIG. 1-3 The Basic System Development Cycle.

through operational deployment. The three steps are depicted in Fig. 1-4.

I.L.S. is concerned with the integration of all organizational activities that contribute to the support definition of a system. This involves the *coordination* of organizations such as (1) reliability engineering, (2) maintainability, (3) human factors, (4) support-equipment design, (5) technical publications, (6) spares provisioning, (7) field service, (8) training, and (9) facilities engineering. I.L.S. is normally performed by a staff of people associated with systems engineering or a program office. As such, the I.L.S. staff is usually an organizational entity responsible for the planning, implementation, evaluation, and control of all organizational activities associated with system support.

Maintainability is a key element of I.L.S. Maintainability characteristics in design (e.g., diagnostic provisions, packaging concepts) dictate methods of support, and the planned-maintenance (maintenance-analysis) aspect of maintainability serves as the key element in providing integrated support through a coordinated, centralized technical-data base for use in all subsequent logistic support processes. This technical-data base incorporates (1) the portrayal of decisions pertinent to detailed maintenance requirements, (2) the relationship of corrective- and preventive-maintenance tasks to maintenance levels, and (3) the designation of resource requirements with their estimated rates of consumption. Maintainability, in association with reliability and human factors, defines the basic support requirements of the system. This definition, in turn, establishes the parameters of work for other company functions (e.g., technical publications, spares provisioning).

I.L.S. and S.E. relationship. The I.L.S. functions, properly performed, serve to provide a support capability that is consistent with the system demands, and they control the quantitative elements that are used to measure system effectiveness.

An adequate support capability is assured for the system by having tools and test equipment, technical publications, spare parts, and facilities that complement the system operational and maintenance needs. Also, trained, qualified personnel are provided in a timely manner to operate and maintain the system. I.L.S. gives a measure of assurance that the system will be available for operation, and that it can successfully perform its mission. The required resources for operation and maintenance of the system are provided, and contingencies for mission success are eliminated.

I.L.S. contributes to the definition of the S.E. quantitative elements in the following manner:

1. Through *planned maintenance,* I.L.S. defines a support program that is compatible with the quantitative goals of the system.

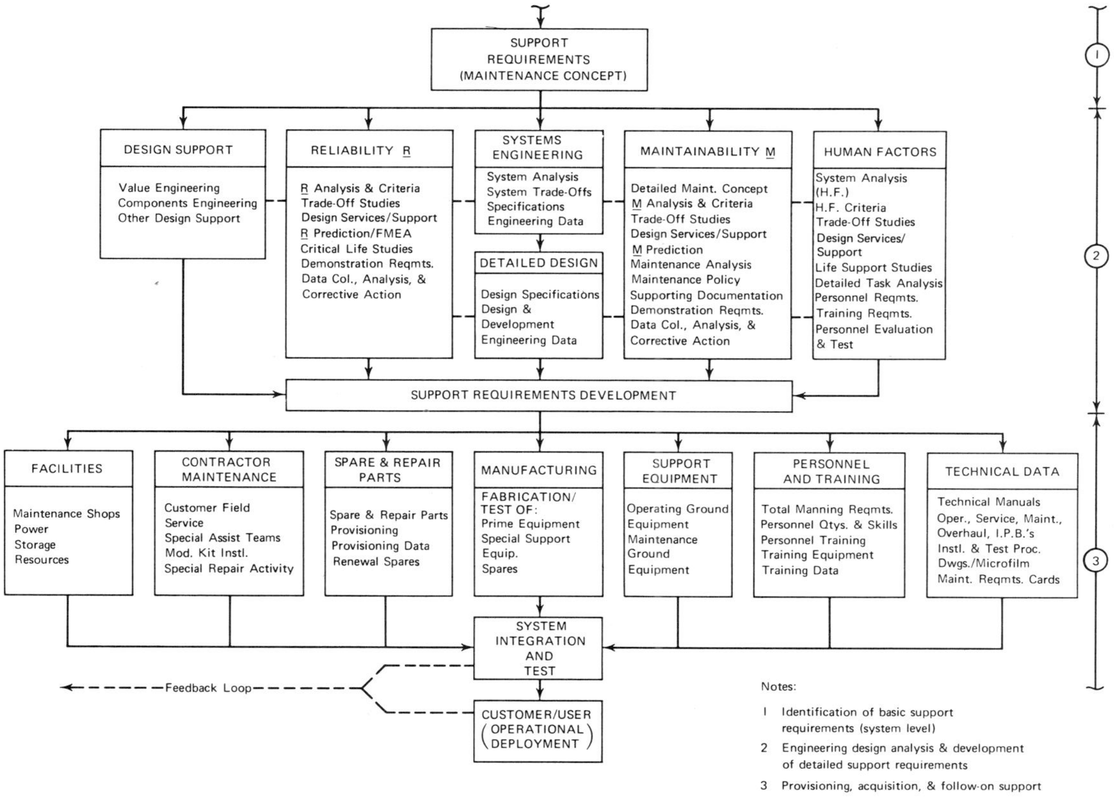

FIG. 1-4 Support Development Cycle.

2. Support equipment and tools are provided to accomplish maintenance tasks within the prescribed time limits. As an example, semiautomatic test equipment may be provided, instead of manual test equipment, in order to expedite certain tasks.

3. Technical documents are provided which relate maintenance-task information that is consistent with the planned maintenance criteria and which can be readily understood by maintenance personnel. This facilitates maintenance tasks and aids in keeping system downtime to a minimum.

4. Spare parts are allocated to each level of maintenance in accordance with the support plan. The availability of spare parts reduces to a minimum the logistics time in maintenance, and in turn expedites the maintenance task. Availability of the system is increased.

5. Facilities planning is necessary to decrease downtime of the system. As an example, power outlets must be available where certain tools and test equipment are expected to be used. Otherwise, time is lost in obtaining extension cords, etc. Facility considerations include lighting, heating, work-spare requirements, and storage needs, which all aid in facilitating the maintenance tasks of a system.

6. Planned maintenance considers the skill level of maintenance personnel and the training required to prepare these people to perform maintenance tasks. The allocation of tasks, the selection of equipment, the development of documentation, and facilities planning are all formulated with the maintenance man in mind. Motivation, skill, and knowledge on the part of the technician are major factors that influence or establish the downtime characteristics of the system. Availability of the system is increased when proper training is conducted, and when the system design and support characteristics are made compatible with the capabilities of the maintenance personnel.

7. During the activation phase of a program, it may be feasible to employ contractor maintenance personnel for a limited period of time. This will preclude the improper operation and maintenance of delivered equipment until the customer/user obtains a full organic capability. Improper utilization of the system equipment during the initial stages of a program can have a lasting effect on the reliability of the system. This, in turn, can significantly degrade the S.E. characteristics. I.L.S. considers the need of contractor support to ensure maximum S.E.

8. The definition of the system-design characteristics (physical and quantitative) is attained through the functions of reliability, maintainability, and human factors. Proper implementation of these functions as part of the I.L.S. program ensures compliance with the specified S.E. elements of *MDT*, *MTBM*, and ready time, which establishes the

availability of the system. As discussed previously, these functions also will likely impact the measurements of dependability and capability.

Maintainability and C.E. Relationship

Maintainability contributes to the minimum outlay of funds for system acquisition and utilization through attainment of objectives associated with maintainability design and support planning.

Maintainability design. Inherent system-maintainability characteristics provide for the following:

1. Accomplishment of all tasks in a minimum amount of time—with the least number of people
2. Accomplishment of tasks with a minimum amount of training of personnel
3. Minimum expenditure of spare parts
4. Least requirements in variety and quantity of tools and test equipment
5. Least facility requirements to maintain the system
6. Minimum requirement for contractor services
7. Minimum need of support documentation

These characteristics of design result in cost savings that enhance the cost effectiveness of the system.

Support planning. The maximum utilization of inherent system characteristics requires consistency in the development and maintenance of a support capability. The provisioning of spare parts, the training of personnel, and all other accompanying services and functions of a support program must complement the design of the system in order to realize maximum efficiency and economy in operation and support. The planned-maintenance aspect of maintainability provides direction to a support program. It assures that the products and services related to the functions of support-equipment design, training, technical publications, spares provisioning, facilities design, and contractor support are consistent with the needs of the system. In this manner, maintainability aids in keeping to a minimum the acquisition and utilization costs.

INTERACTING FUNCTIONS

The functions of reliability and human factors are inseparable from maintainability in defining and describing the inherent design and support characteristics of a system. The relationships of these disciplines are briefly discussed below to illustrate the required high degree of interaction to successfully implement a maintainability program.

Reliability

The objective of a reliability program is to extend the life of the equipment to the maximum degree possible with due consideration for cost and other pertinent factors. This is normally accomplished by selecting highly reliable components and assemblies for inclusion in the design of a major equipment item. This effort is supplemented with design studies, predictions, stress analyses, manufacturing assurance, and other tasks that aid in attaining the main objective of the function.

The measurement of reliability is expressed as *mean-time-between-failures* (*MTBF*). That is, under prescribed conditions the equipment item/system will function normally—on the average—for the duration of the time specified by the *MTBF* factor. It should be noted that the *MTBF* factor is predicted, and that it is based on anticipated primary and secondary failures due to the inherent characteristics of the design. The *MTBF* does not consider preventive-maintenance needs (e.g., lubrication, calibration), human error in operation or maintenance (induced faults), and the quality of workmanship in assembly.

The reliability *MTBF* factor is combined with the maintainability quantitative element *MTTR* (mean-time-to-repair) to predict the *inherent availability* of an equipment item or a system. In this regard, maintainability and reliability are directly related to each other. They are interdependent in describing inherent availability.

Maintainability and reliability are also related in other respects: (1) reliability predictions form the baseline for maintainability predictions; (2) failure-mode and effects-analysis data are used as references in the generation of maintenance logic diagrams; (3) maintainability considers the critical-life data supplied by reliability; and (4) both functions complement each other in providing design liaison and in rendering design services.

Human Factors

The objective of a human-factors program is to ensure compatibility between the system physical and functional design features and the *element of man* during operation, maintenance, and support of the system. This objective is accomplished by applying human-engineering criteria to the design of the equipment. This effort is supported by design studies, operator-task analyses, maintenance-task analyses, and test programs to assess the inherent human-factors characteristics of the system.

The primary measurement of human factors is expressed in terms of (1) safety and cost (staffing) and (2) system effectiveness (the influence on system availability, dependability, and capability).

The human-engineering characteristics of a system directly influence the *MTBM* and *MDT* quantitative factors of maintainability. As an example, human error in operation of the system induces a failure; therefore *MTBM* is affected. And, if maintenance tasks are difficult for the technician to perform, then *MDT* is affected. In this regard, human-engineering and maintainability-design features combine to aid in establishing the operational-availability characteristics of the system.

The two disciplines are also related in other respects: (1) maintainability- and maintenance-analysis reports form the baseline for the generation of the maintenance-task analyses; (2) manpower-requirement reports are considered in making maintenance policy decisions; (3) defined skill-level needs aid in establishing training requirements; (4) environmental needs are considered in facilities planning (lighting, heating, etc.); (5) the content of maintenance instructions is made consistent with the abilities and needs of the system personnel; and (6) maintainability and human-engineering design features reflect on each other.

QUESTIONS AND PROBLEMS

1. What are the terms used to express maintainability?
2. What are *support resources*?
3. Why is maintainability an important element of the design team?
4. Is maintainability concerned only with design and development of equipment?
5. Is maintainability applicable only to military products?
6. Why was maintainability conceived?
7. Is maintainability applied only to electronic equipment and systems?
8. The scope of maintainability has remained unchanged since its inception. True or false?
9. What is the significant difference between *MTBF* and *MTBM*?
10. A maintainability organization has the responsibility for fulfilling all maintainability requirements. True or false?
11. The designer has no responsibility for maintainability design of equipment. True or false?
12. It is feasible to incorporate maintainability features into the equipment design during any phase of a program. True or false?
13. Applied, effective maintainability considers other areas of interest or concern. Theoretical principles are often waived in the interest of reliability, cost, safety, etc. True or false?
14. Maintenance and maintainability have similar meaning. True or false?
15. What is *cost effectiveness*?
16. What is the formula for cost effectiveness?
17. What does S.E. mean?
18. What are the three factors that comprise the Air Force concept of system effectiveness?
19. What is the formula for operational availability?
20. Ready time is a function of reliability. True or false?
21. *MDT* is a measure of maintainability, human factors, and logistics. True or false?

22. What does I.L.S. mean?
23. What is I.L.S.?
24. Maintainability is the keystone of I.L.S. True or false?
25. Calculate A_o based on the following known data:
    a. *MTBM* = 200 hr
    b. Ready time = 50 hr
    c. *MDT* = 30 min
26. 100 percent utilization means full use of the equipment when it is available. If the *MTBM* of a system is 200 hr, and this represents 80 percent utilization, what is the ready time?

TWO

Organization[1]

As described in Chap. 1, maintainability was developed over an extended period of time, during which it took on added meaning. With this progressive change, maintainability as a function was formed into various organizational structures. At first, the function was performed by staff or group acti[illegible] with no association with the term maintainability. As tasks were a[illegible] specific organizational groups emerged. And in recent yea[illegible] ompanies relate the function to the unit, section, or departme[illegible] of organization. With this elevation in structure, the term *mai[illegible]ity* is widely used as an organizational title.

Because of [illegible] constant change in the meaning of maintainability during its development from an art to a science, a great difference still exists between companies in regard to the scope of work delegated to their maintainability organizations. This condition is improving, however, since an increasing number of companies are adopting standard maintainability tasks. Organizational structures are being altered accordingly.

[1] This chapter was prepared initially by E. Edward Lowery and Benjamin S. Blanchard and appeared in the document *Organization for Maintainability,* issued by the General Dynamics Corporation in June, 1965.

The purpose of this chapter is twofold. First, it is intended to convey the basic organizational principles involved in establishing or revamping a company maintainability activity. Second, the tasks outlined herein will serve to identify what a maintainability organization does, and thereby orient the reader for greater appreciation of the material in subsequent chapters.

DEFINITION OF BASIC NEEDS

The primary considerations involved in establishing a maintainability-engineering organization are not unlike those applied to any other situation. That is: the objectives must be known; work elements must be defined; common activities must be grouped; duties must be specified; and personnel must be assigned to accomplish the work.

For the company engaged in defense programs, primary maintainability requirements are set forth by military specification. In this event, the requirements must be reviewed, evaluated, and interpreted in terms of work tasks. The outlined work tasks are then related to other company functions resulting in an integrated management plan. The latter is necessary to avoid duplication of effort, to establish time phasing of products, and to ensure equitable assignment of responsibilities.

The non-defense-oriented company has to define its maintainability-service requirements through a study and evaluation of current service and support practices. Military-specification data can be utilized for comparison with study findings, but care must be exercised to correctly apply the data to a nondefense product-support program. Beyond this point, the development and processing of requirement data is the same for either type of company.

The identified requirements obtained from speci[illegible]s, or from an operations analysis, will likely fall within five categ[illegible]work. These are:

1. Maintainability system-design requirements
2. Maintainability equipment-design requirements
3. Maintenance-analysis requirements
4. Program-support requirements
5. Management (supervisory) requirements

Each category of work is outlined in the following paragraphs. It should be noted that the identified tasks are those expressed by most specifications and documents, and the fulfillment of these requirements necessitates the establishment of a complete maintainability capability. Some military contracts do not require this total coverage, and most certainly a nondefense company would not have need for a number of the defined functions.

Maintainability System-design Tasks

1. Review system technical requirements, mission requirements, the proposed operational-deployment concept, and the proposed maintenance and general-support concepts. Translate this information into basic system-design criteria.

2. Review system operational functional-flow diagrams. Define major system functional-interface requirements. Participate in system-analysis studies.

3. Establish system-maintenance concepts and generate maintenance functional-flow diagrams. Define major items required for maintenance functions.

4. Generate a maintainability analysis as part of the system-analysis effort. Accomplish a maintainability allocation/apportionment.

5. Participate in system trade-off studies. Generate maintainability data (e.g., maintenance time, frequency and cost data, support requirements, etc.) necessary for trade-off–study accomplishment.

6. Establish system-maintainability design criteria. Provide specific qualitative and quantitative information pertaining to individual equipment items.

7. Monitor system-design activity. Participate in systems-engineering planning sessions. Review system drawings (facility layouts, installation drawings, system functional flows, etc.) for inclusion of maintainability characteristics, and perform a drawing sign-off service. Participate in special system-engineering studies. Define equipment-item system data (accessibility limitations for maintenance of the item due to its proximity to other equipment; power or light restrictions; safety limitations, environment data, etc.). Submit maintainability recommendations.

8. Participate in system formal design–review sessions. Provide maintainability-support data (special reports, trade-off-study reports, recommendations, etc.).

9. Participate in the development of integrated logistics or material support plans (e.g., support-equipment plan, personnel and training plan, technical-publications plan, system-installation and checkout plan). Maintain coordination with logistics support activities.

10. Conduct system-maintenance orientation sessions for the benefit of other maintainability-engineering organization personnel.

11. Participate in system test and demonstration activities.

12. Review and evaluate maintainability prediction and analysis, maintenance-analysis data, and equipment assessment data to establish a system maintenance-profile record. Provide data for management re-

porting. Document critical problem areas for management dissemination.

Maintainability Equipment-design Tasks

1. Provide a maintainability-design liaison service for the engineering organizations. Station personnel in the engineering groups to participate in the design process and to function as consultants. Provide and interpret maintainability criteria.

2. Perform qualitative and quantitative maintainability analysis of equipment articles to ascertain the degree of maintainability characteristics required and to identify all significant maintenance and support needs. Participate in the selection of equipment, and perform detailed trade-off studies as required. Submit design recommendations.

3. Monitor design development. Ensure that the design includes the maintainability characteristics as defined in the article specification and that it meets the requirements of prescribed criteria. Ensure that the item can be maintained within the established allocated time period.

4. Review preliminary design drawings, lists of material, and documentation. Ensure that adequate maintainability coverage is included. Submit recommendations for inclusion in the final design.

5. Accomplish maintainability prediction utilizing design drawings, lists of material, and documentation. Compare prediction results against initial allocation, and submit recommendations for design improvement.

6. Perform a drawing/data sign-off service. Acknowledge the incorporation of maintainability in the design.

7. Initiate and conduct maintainability studies and design tests to obtain optimum solutions to maintenance problems (utilize design aids, special layouts, mock-ups, etc., as required).

8. Participate in formal design-review sessions involving individual equipment. Provide maintainability-support data (special reports, trade-off studies, recommendations, etc.).

9. Participate in review of design-change data. Determine impact on maintainability requirements.

10. Review and evaluate developed equipment relative to the inclusion of maintainability characteristics. Submit recommendations for design improvement.

11. Participate in maintainability-demonstration tests (actual test accomplishment).

12. Prepare and submit special reports covering maintainability problem areas (e.g., actual maintenance times exceed initial contracted requirements).

13. Coordinate maintainability-design activities with related organiza-

tions (i.e., reliability, human factors, value engineering, systems engineering, etc.).

Maintenance-analysis Tasks

1. Define all maintenance requirements (i.e., troubleshooting, repair, functional-test, inspection, servicing, overhaul, calibration) for all repairable items through review of engineering data. Record these requirements in established format to fully identify:
   - *a.* Maintenance tasks and task sequences for each echelon of maintenance (support tasks with logic diagrams)
   - *b.* Maintenance-task times and frequency
   - *c.* Personnel requirements (type, quantity, and skill level)
   - *d.* Support-equipment requirements (type, quantity, and utilization)
   - *e.* Tool requirements (type, quantity, and utilization)
   - *f.* Spare/repair-part requirements (type and usage rate)
   - *g.* Facility and resource requirements.
2. Evaluate defined maintenance-requirement data in regard to the article qualitative specification. Determine alternate methods of maintenance coverage; perform trade-offs; submit recommendations for product improvement.
3. Initiate documentation to obtain approval for purchase of special support equipment (identified during article-requirement analysis).
4. Develop a calibration-requirement summary (CRS).
5. Develop article-illustration layouts (item-illustration sheets).
6. Release maintenance-data reports to design engineering, technical publications, logistics, training, reliability, human factors, and other maintainability units.
7. Participate in spares-provisioning meetings.

Program-support Tasks

1. Provide personnel in support of proposal development (new business).
2. Establish general maintainability criteria for all programs.
3. Generate and maintain a current file on system- and maintainability-related specifications and reports.
4. Conduct maintainability training and indoctrination courses.
5. Establish general program policies and procedures.
6. Represent the maintainability organization at trade meetings. Function as focal point for all trade correspondence.
7. Conduct detailed requirement analysis for each program.
8. Generate detailed program plans.
9. Generate detailed program schedules and product schedules.

10. Review all maintainability products for completeness and conformity to specified requirements before their release from the maintainability organization (obtain the required level of signature).

11. Review program status at periodic intervals and advise supervision of variance from planned performance.

12. Prepare maintainability demonstration plans.

13. Prepare special reports and summaries to assist supervision in program-control functions.

14. Prepare program final report.

15. Perform subcontractor liaison functions.

16. Perform customer liaison functions (as delegated by supervision).

17. Provide EDP service (consultant and coordinator).

18. Provide general program-coordination services (departmental relationships).

Management (Supervisory) Tasks

1. Review program specifications and technical requirements; interpret the technical requirements in terms of manpower needs; and allocate manpower resources to accomplish the fulfillment of specific requirements.

2. Establish primary goals and schedules for the organization and subordinate activities regarding the development of technical products. Delegate responsibilities to the functional groups.

3. Monitor the progressive development of technical products. As required, interpret the requirements for subordinates.

4. Review technical products and reports for accuracy, quality, and completeness of content. Approve products prior to their release to other organizations.

5. Sustain technical planning for the organization through evaluation of periodic reports, through review of findings, and from information received through daily contacts. Review schedules or reallocate manpower resources as necessary to attain primary goals.

6. Function as maintainability-engineering consultant to management.

7. Make trade-off decisions regarding maintainability when significant costs are involved, or when a new concept is under consideration.

8. Advise or instruct subordinates in all technical aspects of program coverage as warranted to ensure efficient and effective performance of the functional groups.

9. Make periodic program technical presentations to management and the customer. Prepare written reports at periodic intervals as defined by contract.

10. Conduct special studies and prepare technical reports for presentation at trade meetings. Review specifications and standards at the request of government committees.

11. Perform economic analysis related to maintainability-program technical aspects. Prepare plans of alternate approaches. Summarize significant cost data for reporting to systems engineering, management, and the customer.

12. Coordinate all technical activity between subordinate functional groups.

13. Coordinate all technical activity between the maintainability organization and related organizations of the company pertinent to program requirements.

14. Perform administrative tasks related to the operation of the organization.

GROUPING OF TASKS

The grouping of maintainability tasks into an organization structure should be done with the following objectives in mind:

1. Each job within the group should contribute to the purpose of the group.
2. Each job within the group should require knowledge of about the same level.
3. Each job within the group should have defined responsibility.
4. Each job should be measurable to evaluate performance.

The group structure of the maintainability organization should be based on the following considerations:

1. The groups should be of nearly equal size.
2. The groups should be limited in size to afford control over job activities.
3. Communication between groups and the exchange of work products should be unrestricted.
4. Stability in organization should be sought despite changes in work requirements. (The need of various products and services at different stages of a program should be projected into organizational planning.)
5. The basic organizational structure should permit expansion in size and allow for growth, without causing complete reorganization if additional assignments were received.
6. The number of groups within the maintainability organization should be kept to a minimum.

Experience indicates that maintainability functions are dependent upon one another, and it is necessary to maintain very close liaison between group activities to minimize control problems. Therefore, special attention should be given to the grouping of functions, and care should be taken in delegating tasks to related departments. The tasks outlined in this chapter should be assigned to a specified maintainability organization to attain maximum efficiency in work performance.

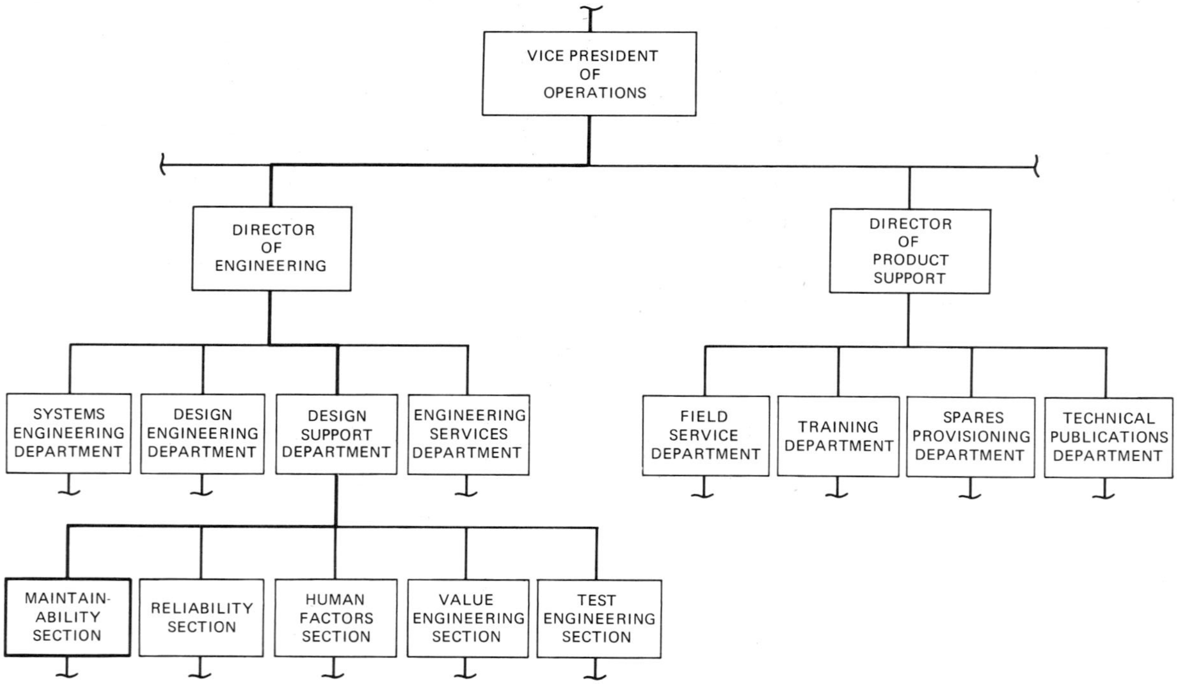

FIG. 2-1 Organization Alignment. Example No. 1.

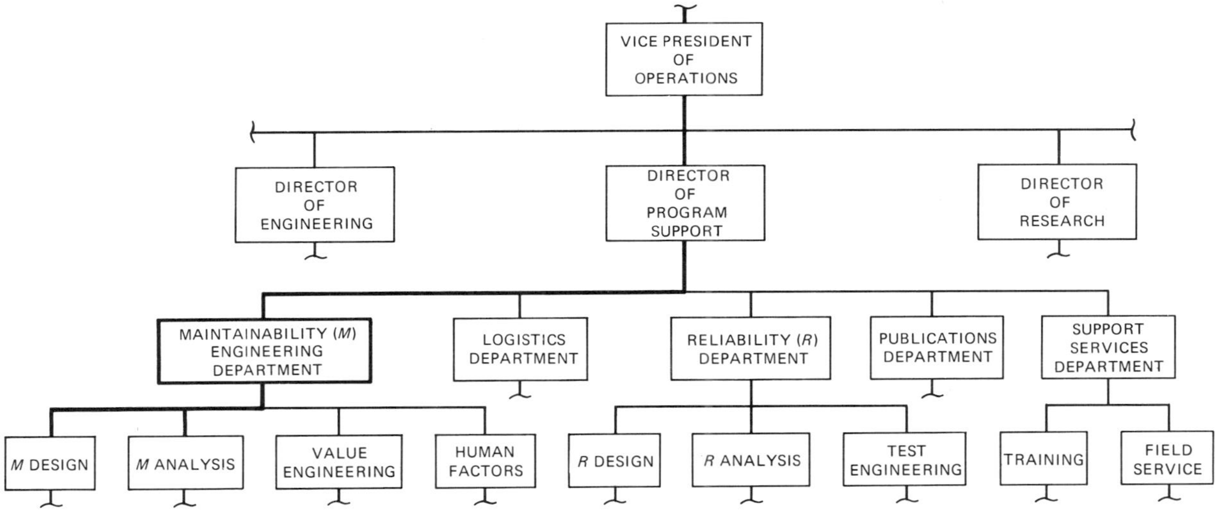

FIG. 2-2 Organization Alignment. Example No. 2.

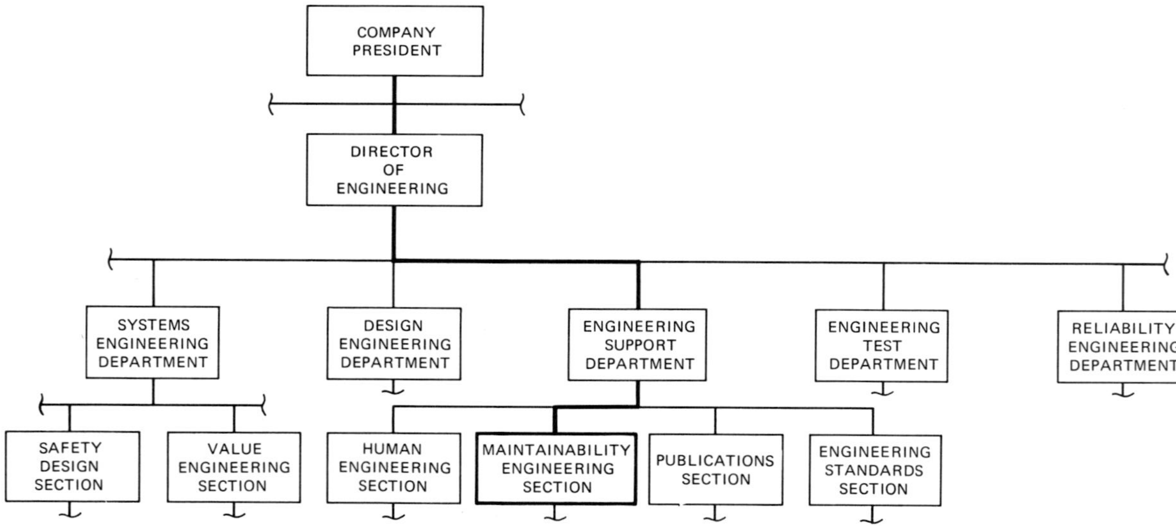

FIG. 2-3 Organization Alignment. Example No. 3.

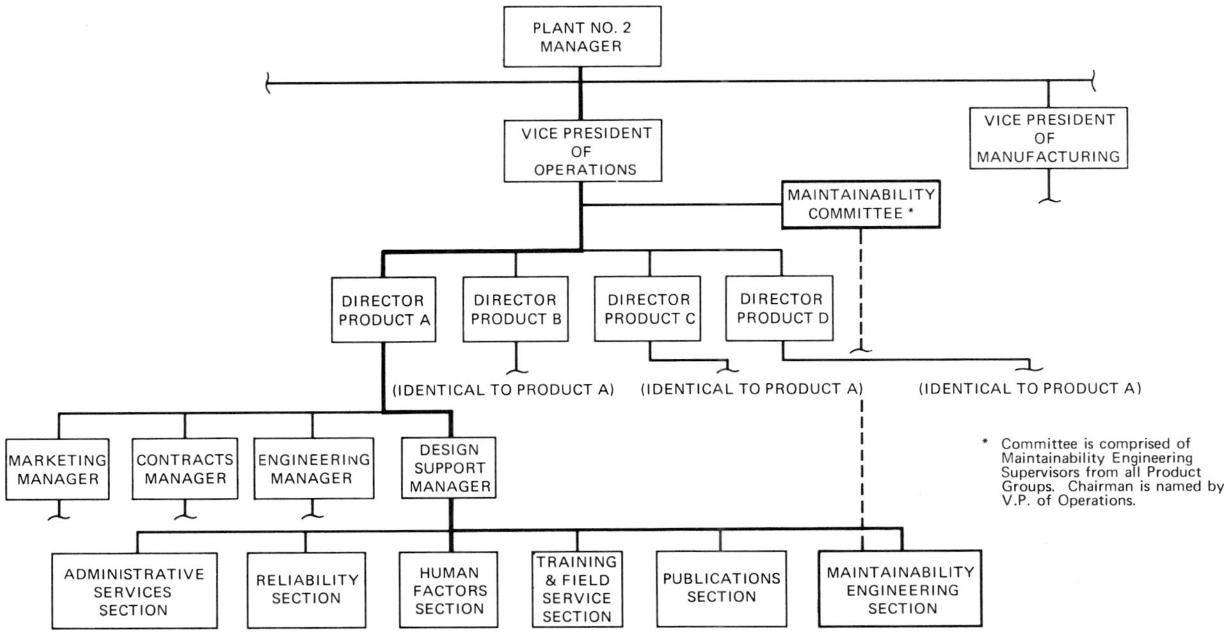

FIG. 2-4 Organization Alignment. Example No. 4.

MAINTAINABILITY AS PART OF THE COMPANY STRUCTURE[2]

The designation of a department to assume maintainability responsibilities should not be done in an arbitrary manner, as so often is the case. Instead, the selection of a department should be based on its ability to perform the function efficiently, and this can only be determined through the findings of a study. Through systematic review of existing departmental capabilities, it is possible to relate defined maintainability requirements and to evaluate the feasibility of integrating the projected maintainability functions with current departmental tasks.

The decision to select one department over another could be due to existing organizational relationships. That is, a high degree of compatibility does exist under present conditions, and the addition of maintainability would likely be received without opposition. In fact, it could further enhance the relationship. Unquestionably, factors of harmony and cooperation should be considered when placing the maintainability function in the company structure.

The need for division of responsibilities will also influence the decision of where to place the function. Aside from the aspect of equity in assignments, consideration must also be given to the size of an organization and its effect on the span of control.

The department manager who holds no interest in the subject of maintainability will probably do nothing to promote the function, and its placement in his department will likely result in failure. However, the presence of interest does not in itself qualify a person to manage a maintainability function. The person must also be competent in management matters, and should have knowledge of program-support requirements. The interest level of the department manager is also important for another reason. Relatively speaking, maintainability is a new field, and a high degree of interest and confidence must be evident to convince other managers of its importance to the company. This is necessary to ensure its adoption and application to the various areas of work. In brief, a certain amount of "selling" is required, and the manager who believes in what he is doing will transfer this conviction of purpose to others.

Additional factors to be considered in the placement of a maintainability organization are:

1. The maintainability organization must have status. The parent department should be on the same level as logistics, technical publica-

[2] A correlation of organization principles exists between the two documents *Why Organize for Maintainability?*, issued by the National Security Industrial Association, September, 1965, and *Organization for Maintainability*, issued by the General Dynamics Corporation, June, 1965.

tions, design engineering, systems engineering, training, and other company line departments.

2. The maintainability organization must be identified. The functions should not be hidden among other activities.

3. The maintainability organization must have authority as well as responsibility. It should be recognized by all levels of management.

4. Maintainability should have organizational ties with reliability, safety, human factors, value engineering, and test-group functions. All have common needs, and are interdependent on some aspects of work coverage.

5. The maintainability organization should be located near logistics, technical publications, systems engineering, design engineering, training, and the other groups of its parent organization. This is necessary due to the nature of maintainability where the functions are a part of the dynamics of daily work. Physical separation discourages the active participation necessary to fully implement maintainability requirements.

6. In large companies, provisions should be made to allocate space to maintainability engineers in the design-organization area of work. Although this is not an ideal arrangement, it does overcome some of the disadvantages of being totally separated from design activities.

One might conclude that the placement of maintainability should receive the same general considerations given to other activity assignments within the company structure.

Management Functions

Management's role in the organization of maintainability functions should consist of the following:

- Delegation of functions to departments
- Definition of functional relationships, departmental responsibilities, and delegated authority
- Definition of program objectives
- Establishment of control and reporting procedures
- Approval of the organizational structure of the maintainability activity within the department structure

Delegation of Functions. Management should clearly designate the placement of the maintainability organization within the company structure. Typical placements are illustrated in Figs. 2-1 through 2-4.

Functional Relationships, Responsibilities, and Authority. The information contained in the company integrated management plan will establish primary functional relationships, responsibilities, and departmental authority. However, this information should be restated in the

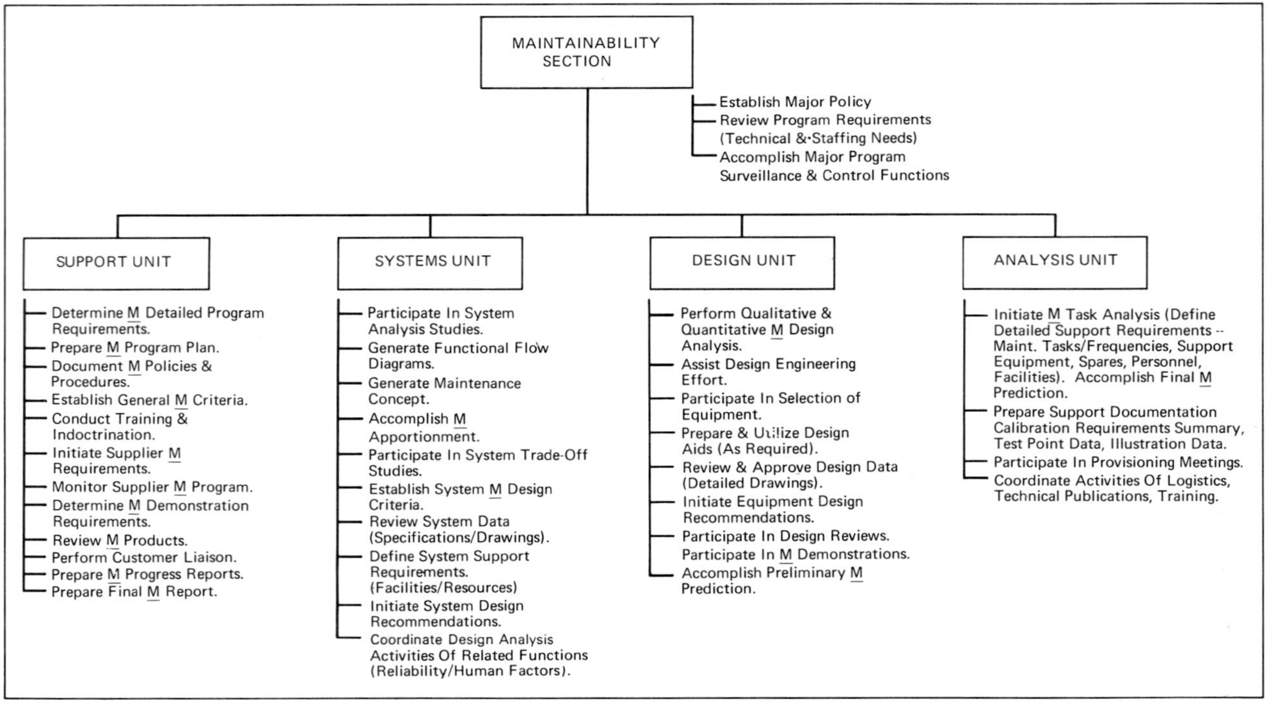

FIG. 2-5 Maintainability Organization. Example No. 1.

form of company directives (or operating procedures) to clearly relate the intent of management regarding program conduct.

Definition of Program Objectives. A statement of maintainability objectives should be generated by management to present milestones for use by the line departments for their planning purposes, and also to establish goals for the designated maintainability organization. A means of relating these objectives is in the form of a company maintainability plan or criteria document. In addition, a committee should be designated to review maintainability objectives periodically and to recommend to management changes in program scope.

Control and Reporting Procedures. The control and reporting procedures necessary for maintainability-program direction should be an integral part of those company directives that define relationships, responsibilities, and authority. In addition, management should receive periodic progress reports on each program to measure attainment of primary objectives. The company maintainability committee should also utilize the information in the periodic reports to formulate recommendations to management.

Approval of the Organizational Structure. The delegation of functions by management will in itself determine the placement of maintainability within the company structure, the result being that a department manager will be responsible for merging the functions into the existing department sections, or he will establish an organizational entity for

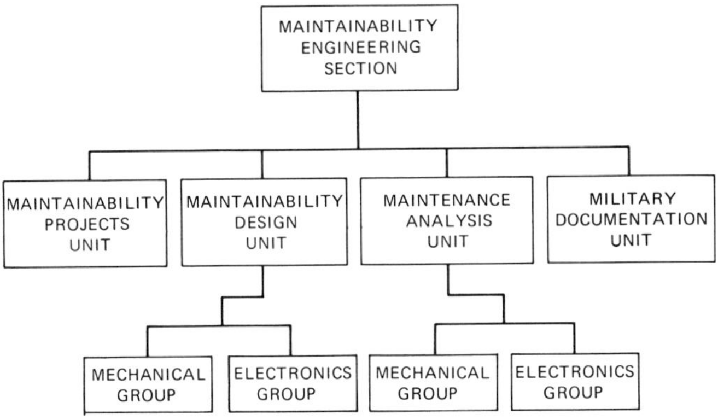

FIG. 2-6 Maintainability Organization. Example No. 2.

the sole purpose of performing maintainability tasks. Whatever approach the department manager takes, company management should review the organization alignment and pass judgment on whether or not the organization can function effectively. Without this critical review, the preceding management-planning data may prove to be of no value. Typical maintainability-engineering organization structures are illustrated in Figs. 2-5 through 2-7.

Types of Organizations

The types of maintainability organizations frequently appearing on company charts can be classified as *line, project,* or *staff*. Each is briefly discussed below.

Line Organization. This type of organization has total responsibility for the conduct of maintainability matters affecting the company product lines. It is a self-contained organization with single authority that provides services for all company design and support activities. An example of this type of organization is illustrated in Figs. 2-1 through 2-3. Experience indicates that this type is effective since responsibility can be defined and duplication of effort is eliminated. However, if the company has a great number of diversified product lines or is engaged in more than three major projects, then the maintainability-organization size becomes such that control is difficult. If this situation is encountered, the project-type organization is recommended.

Project Organization. This type of organization has total responsibility for the conduct of maintainability matters affecting a single company product line or a specific program contracted by the company. This type seems to be preferred by government contracting agencies in that activity and expenditures can be measured directly. The full resources of the organization are devoted to the program, and the recording and reporting of activity is distinct from other company functions. Figure

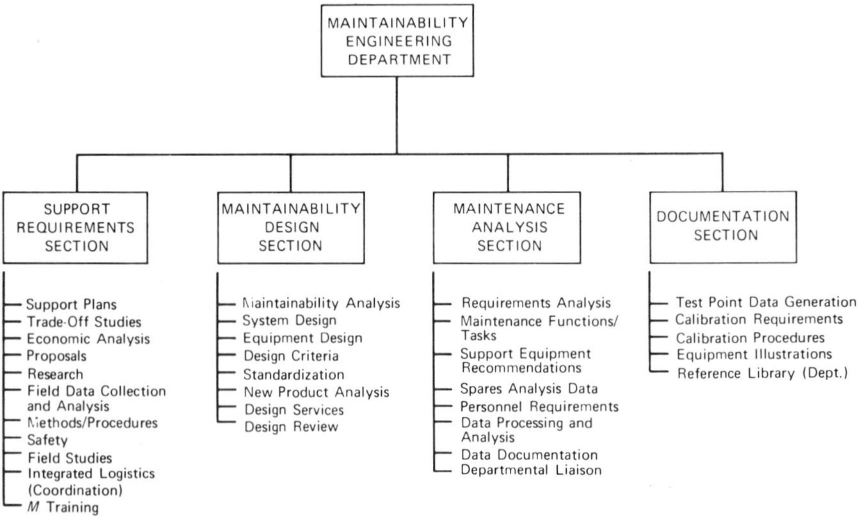

FIG. 2-7 Maintainability Organization. Example No. 3.

2-4 conveys this type of structure. The disadvantages of this type of organization are:

1. There is a duplication of services and activity between the various company maintainability organizations.

2. There is instability in the organization since phasing of the various programs is most difficult. The transfer of people from one program to another is possible, but they must be experienced in all phases of maintainability work to fully utilize their services. To retain this skill level, the average rate of pay is quite high for maintainability personnel of the company.

Staff Assignment. Some companies prefer to place a maintainability expert on the staff of the chief design engineer, or to assign him as a consultant to the director of engineering. Experience indicates that this is most ineffectual. A review of the basic principles and concepts of maintainability will reveal that this arrangement cannot possibly satisfy the function of a maintainability organization. The only visible purpose this arrangement serves is to indicate that the term maintainability is recognized by management.

Summary

It would be presumptuous to conclude that one type of organization is better than another under any conditions. The variables present in any one company defy attempts at generalization. We do know, however, that the principles discussed throughout this chapter must be adhered to for assurance of success over an extended period of time.

If the project-type organization (Fig. 2-4) is adopted, it is recommended that a company maintainability committee be established to:

1. Generate company maintainability standards
2. Resolve common problems
3. Exchange ideas
4. Communicate with customer agencies, other industry organizations, and societies
5. Serve as a company voice in dealings concerned with maintainability matters

PERSONNEL SELECTION AND PLACEMENT

Finding and placing qualified people is a very difficult assignment when organizing a maintainability activity. The main reason for this difficulty is that experienced people are not available in any great number and that the demands for their services are widespread. Therefore, the maintainability-organization surpervisor must, for the most part, develop the organization capability through selective recruiting of individuals who have related job experience, formal technical education, and the desire to perform maintainability work. This approach does, however, require an extensive amount of training, and this prevents a fast build-up of the organization capability. An alternate approach, and perhaps the best method, is to recruit a few experienced and qualified maintainability engineers. This instant capability, coupled with a training program for the inexperienced employees, provides a firm base for growth of the organization. Regardless of what method of staffing is adopted, the people must be reviewed on an individual basis and assigned to a specific job. That is, the organization will require a variety of skill levels, and the supervisor should make work assignments that are compatible with the individual's interest, ability, and background. This is accomplished in the conventional manner through a review of the employee's history, an interview, and the matching of the person's qualifications with the job requirements.

There seems to be a common belief among management personnel that a person classified as a maintainability engineer is capable of performing any task related to maintainability work. This general association of men to work efforts is very convenient and is a human response, but it is not supported by fact. To the contrary, each job in the maintainability organization will require the placement of an individual with particular characteristics, traits, and abilities. For instance, a person who requires social interaction with others, enjoys a fast work pace, and is emotionally stable will likely adjust to the job of being a customer coordinator. On the other hand, a person who enjoys working alone and does not become bored with details will probably adjust to maintainability-analysis work. Rare is the person who combines the characteristics, traits, and abilities to perform all maintainability jobs. Table 2-1 reveals the difference in personnel requirements between four work categories of the maintainability organization. A review of the ratings will indicate the necessary individual characteristics for placement in a specific work group. Although these indicators represent primary considerations only, they do highlight the differences in maintainability-engineer requirements throughout the organization.

There is no sure way of selecting the right person for a particular job. As indicated above, the supervisor must rely on his ability to properly assess the applicant's history, education, personal traits, and experience. And he must validate this information through techniques of interviewing. Special points to consider in this process of evaluation are:

1. The applicant who professes to have experience in all phases of maintainability is, likely, not proficient in any task. Most companies

have no rotation plan for training purposes, and the general practice has been to specialize within one work group. Likewise, a claim to having a "working knowledge" of all aspects of a particular specification is usually without foundation. However, either of these claims can be confirmed through a review of the applicant's employment history, with special attention given to the length of employment in a stated capacity.

2. An applicant with oustanding credentials of a profession other than maintainability should receive critical review to establish his reasons for seeking employment in maintainability work. Although there are sincere, dedicated men of this category, there are an equal number who have been failures in their own profession. It is unlikely that the latter will adjust to maintainability work.

3. Each applicant should be advised of what position he is being considered for, the tasks that he is expected to perform, and the maintainability products that he is expected to generate. The failure to

TABLE 2-1 Employee Placement Guide

REQUIREMENT	EVALUATION LEVELS/AREA OF WORK			
	SYSTEMS DESIGN	PROGRAM SUPPORT	EQUIPMENT DESIGN	MAINTENANCE ANALYSIS
Experience	High	High	Average	High
Education	High	Average	High	Average
Technical Ability	Average	Low	High	Average
Job Training	Low	High	Low	High
Learning Rate	Average	Average	High	High
Judgment	High	Average	Average	High
Precision	Low	Average	High	High
Analytical Ability	High	Low	Average	High
Intelligence (IQ)	High	Average	High	Average
Visualization	High	Low	High	Average
Ingenuity	High	Low	High	Average
Writing Ability	Average	High	Average	High
Verbal Communications	High	Average	High	Low
Adaptability	High	High	Average	Average
Persistence	Average	High	Average	High
Sociability	High	High	Average	Average
Diplomacy	High	High	High	Average
Optimism	High	Average	High	Average
Motivation	Low	Low	Average	High
Work Rate	High	Average	Low	Average
Group Cooperation	High	Average	Average	Average

have an understanding of where the new employee "fits" into the organization can deflate his expectations, lower his morale, and in turn adversely affect the performance of fellow employees.

4. It is often difficult for the zealous supervisor to assume the role of the listener, but the applicant should be given an opportunity to contribute to the interview. His opinions or comments should be sought to add information to that within the application blank or résumé.

5. Educational requirements will vary, depending on the area of work assignment and the position to be assumed by the employee. As a general rule, candidates for placement in maintainability system- and equipment-design groups should possess a B.S. degree in electrical engineering or mechanical engineering, as applicable to the company products. Candidates for the maintainability program-support and analysis groups should have attained an educational level equivalent to a B.S. degree, either by college credit accumulation or through actual job training and experience. However, stringent policy on educational requirements should be avoided. If the applicant is qualified for the job in all other respects, his level of formal education is a secondary consideration.

6. Jobs within each work group should be classified by levels of task complexity and degrees of responsibility to aid in the placement of employees. Job titles should be created to identify the position of the employee within the organization. A written job description should be utilized in associating job requirements with the employee's qualifications.

7. It is unlikely that a candidate will meet all expectations of the supervisor. A compromise will have to be made on some points of job qualification. However, these differences should not be so great that they cannot be bridged by counseling or training.

8. Maintainability-related experience is most commonly possessed by maintenance engineers, system engineers, technical writers, field service engineers, installation and test engineers, military service technicians, laboratory technicians, and design engineers.

9. The majority of candidates should have majored in engineering or physics during their period of college training. However, the establishment of a full maintainability capability will require some men who majored in mathematics, business, and English.

10. Personnel assigned maintainability system- and equipment-design tasks should have job titles of engineer or senior engineer.

MAINTAINABILITY TRAINING

The purpose of a company maintainability training program is twofold. The objectives are:

1. To train maintainability engineers in the proper performance of tasks
2. To acquaint other company personnel with the purpose and functions of the maintainability organization

The approach to fulfilling these objectives should be described in the maintainability-program plan. The training plan should include course outlines, schedules, attendance listing, course length, and identification of the instructors. In addition, the plan should include a list of training references and aids to be used in each course.

The major types of maintainability training courses are discussed in the following paragraphs.

Familiarization Course

A familiarization course should be conducted to inform all program-management personnel and supervisors of the purpose of the maintainability program, how the program will be implemented, and how maintainability products can be used effectively by their organizations. The length of the familiarization course should be limited to one hour, and it should be scheduled to coincide with the release of the maintainability-program plan.

Design-orientation Course

Design engineers should be briefed on maintainability criteria documents, maintenance policies, maintenance-design criteria, and program design standards. The use and interpretation of program quantitative data should be explained, as well as the services available to the engineer to aid in meeting the design goals. This course should be scheduled within a month after the familiarization course, and it should not exceed two hours in length. The design-orientation course should be supplemented with maintainability-information bulletins, which are distributed to the design engineers at random intervals throughout the length of the program.

Briefing of Support Personnel

Personnel assigned to logistics, technical publications, and human factors should receive detailed instruction on the interpretation and use of maintenance-analysis data as applicable to their area of work.

In addition to introductory material, the training session should include a discussion on the content of an analysis report. Emphasis should be placed on what data are contained in the analysis report, the purpose of each data entry, how they are applied to related products, and how they contribute to the system-support program. The briefing of program-support personnel should coincide with the release of the first maintainability-analysis–data report. The training session should not exceed two hours. A separate training period should be designated for each group of personnel (e.g., logistics), rather than having a joint training session.

Maintainability-engineer Training

Regardless of the experience level of the people comprising the maintainability organization, a certain amount of classroom instruction and on-the-job training (OJT) will be necessary to ensure efficient and effective performance of the organization. For instance, personnel of the analysis group must be familiar with the services performed by other groups of the organization to avoid duplication of effort. These functional interfaces must be clearly understood by each member of the organization to provide meaning to each job and to make certain that each man understands the scope of his job. This knowledge also gives the individual an appreciation of the work of others, and in general establishes a condition of esprit de corps.

On-the-job training is necessary to provide the individual guidance required by most men to orient their efforts toward full production. OJT overcomes the variance in the level of job understanding and increases the uniformity of maintainability products.

Classroom training of maintainability engineers should be conducted immediately upon establishing a maintainability organization. General program data should be related first, followed by instruction on specific job tasks, and concluded with instruction on special maintainability techniques and procedures (e.g., trade-offs). Classroom training should be completed, if possible, prior to the assignment of job responsibilities. In cases where this is not possible, an order of topic priority should be established wherein a maximum amount of job-related information is conveyed before tasks are assigned. The duration of classroom training will vary depending upon the experience level of the students, the complexity of program needs, and the amount of reliance placed on OJT to substitute for coverage of classroom topics. However, the average maintainability engineer should receive at least 40 hours of instruction before being charged with job responsibilities.

OJT will assume many forms. It includes that relationship between supervisor and employee where step-by-step guidance is issued and the employee's response is monitored, as well as that situation in which the employee is thrust into a new area of work for the purpose of job exposure. And, of course, it includes any condition between these two extremes. In either case, OJT is dynamic and continuous throughout the life of a program and directly contributes to fulfilling the objective of the maintainability organization.

QUESTIONS AND PROBLEMS

1. The basic approach to organizing a maintainability capability differs greatly from any other situation. True or false?
2. Maintainability requirements are identical for all companies. True or false?
3. By what means are maintainability requirements identified?
4. List the five common categories of maintainability work.
5. List three maintainability system-design tasks.
6. List three maintainability equipment-design tasks.
7. List two maintenance-analysis tasks.
8. The manager of a maintainability organization makes trade-off decisions. True or false?
9. There are no special considerations to be applied in the grouping of maintainability tasks. True or false?
10. Maintainability functions are usually dependent upon one another. True or false?
11. The maintainability organization can function apart from other company activities. True or false?
12. What are the three basic types of maintainability organizations? Draw an example of each.
13. A maintainability engineer is capable of performing any task related to maintainability work. True or false?
14. All people assigned to the maintainability organization should have had formal training in engineering subjects. True or false?
15. List the four major types of maintainability training courses.
16. What does OJT mean?

THREE

Program Implementation

Implementation refers to the proper sequencing and application of tasks to attain program objectives effectively. This involves the actuation of the maintainability functions in a predetermined manner to accomplish the tasks described in Chap. 2.

The demands placed on a company maintainability organization will change from one program phase to another. A number of the required tasks are nonrecurring. That is, each is performed only one time during the program. Other tasks are continuous or repetitive, but the majority of these change in scope of coverage as the program progresses. This fluctuation in task requirements imposes a need to continually adjust the number of personnel assigned to a specific program task. In addition, the assignment of personnel must be consistent with the skill-level needs of each task at any point in time. The method of program implementation, therefore, has a direct influence on the operating efficiency of the maintainability organization in terms of cost of operation and task performance.

The purpose of this chapter is to present (1) the time relationship of maintainability tasks to each program phase, (2) the skill requirements in association with each program phase, and (3) the methods of program control. An understanding of these principles should facilitate the implementation of a company maintainability program and result in the effective, efficient attainment of program objectives.

PROGRAM PHASES

A given program consists of a maximum of five primary phases of work effort. The phases are (1) conceptual, (2) proposal, (3) definition, (4) acquisition, and (5) operational. The program phases describe the stages of system/equipment evolution, namely, (1) the establishment of a need for the system or equipment, and basically what is required to fulfill the need (conceptual phase), (2) the proposed solution to fulfill the need in terms of equipment and program development (proposal phase), (3) the definition of the proposed equipment and program coverage to establish the feasibility and practicality of the proposed solution (definition phase), (4) the actual design, development, testing, and manufacturing of the system or equipment and the conduct of related program tasks (acquisition phase), and (5) the application of the system or equipment to fulfill the need as initially identified (operational phase). The typical order of program phases is illustrated by path *A* in Fig. 3-1.

As an element of a major program, maintainability is associated with each of the five phases. The tasks to be performed during each phase

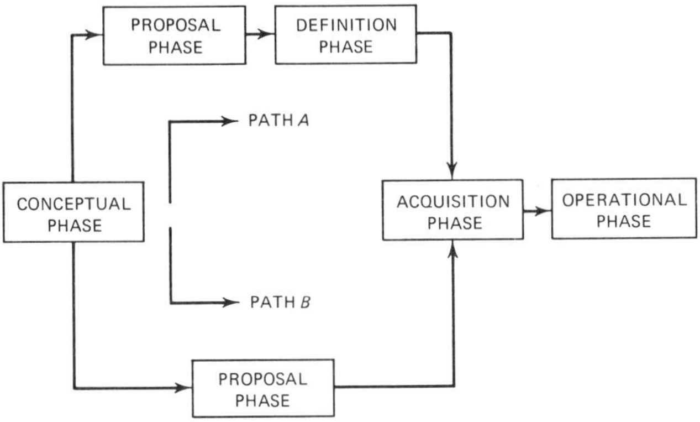

FIG. 3-1 Program Phases.

are defined, and these will be discussed in the following paragraphs. It is emphasized, however, that differences do exist between programs, and the allocation of tasks discussed in this chapter should not be considered valid for all situations. As an example, a system may be defined clearly during the conceptual phase of a program, thereby eliminating the need of a program-definition phase. In this event, the maintainability cover-

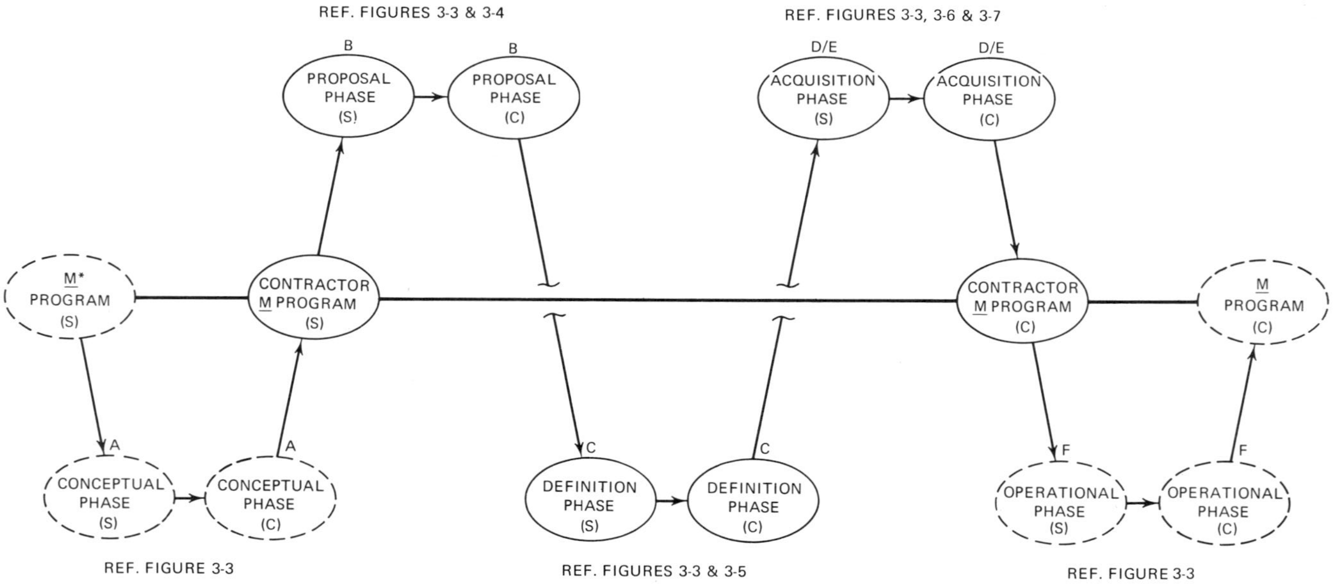

*The abbreviation for maintainability is M.

FIG. 3-2 Maintainability-program Effort.

age of the proposal phase will likely be increased by supplementing the effort with some tasks normally performed during the definition phase. Therefore, a reallocation of tasks will be necessary to adjust to this specific condition. One abbreviated sequence of program phases is illustrated by path *B* in Fig. 3-1. As a general rule, however, the maintainability effort is initiated during the conceptual phase of a program and is continued through to the end of the operational phase. This coverage is illustrated in Fig. 3-2, and the major maintainability tasks are listed in Fig. 3-3.

Conceptual Phase

The tasks performed during this phase provide the conceptual and technological foundation required to further define, acquire, and operate a total system or subsystem. This program base is formed by:

1. Recognizing and defining requirements for future systems.
2. Defining basic mission requirements (including system quantitative factors).
3. Performing system research and feasibility studies to conceive systems which will satisfy mission requirements.
4. Selecting, through analysis and trade-off studies, those system conceptual designs which will *best* satisfy mission requirements. This includes the optimization process (employing concepts of systems/cost effectiveness) that result in a proposed system level technical approach. (The basic support concept is developed from the proposed approach.)
5. Developing plans and generating documentation to define requirements necessary for definition and acquisition of the proposed system. Such documentation generally results in the generation of a request for proposal (RFP).

When all possible solutions are defined, the intent is to assure (through analysis and related studies) that at least one system configuration will satisfy the requirement, can be acquired (within a reasonable period of time), and is acceptable relative to cost (based on life-cycle considerations). In the event that two or more system approaches are deemed feasible, the trade-off optimization process is employed to define the best approach.

Maintainability-related effort in the conceptual phase primarily constitutes identification of maintenance quantitative requirements supplementing overall mission requirements, definition of the basic maintenance concept, provision of input data for trade-off study accomplishment, and the generation of system-design criteria (both qualitative and quantitative) as input to RFP documentation.

As a general rule, all conceptual-phase effort is accomplished by the customer. However, on specific occasions, industry may be contracted to assist the customer in performing feasibility and system trade-off

CONCEPTUAL PHASE	PROPOSAL PHASE	DEFINITION PHASE
PROGRAM MANAGEMENT A1 – DEFINE GENERAL PROGRAM REQMTS A2 – DEFINE MISSION REQMTS (System Quant. Factors) A3 – PREPARE REQMTS DOCUMENTATION (RFP) MAINTAINABILITY DESIGN A4 – ACCOMPLISH SYSTEM FEASIBILITY STUDIES A5 – ACCOMPLISH SYSTEM TRADE-OFF STUDIES A6 – DEFINE BASIC MAINTENANCE CONCEPT	PROGRAM MANAGEMENT B1 – DEFINE DETAILED M REQMTS. (Program & Technical). B2 – PREPARE M PROGRAM PLAN (Prelim. Policies & Procedures) B3 – DEFINE SUPPLIER REQMTS. MAINTAINABILITY DESIGN B4 – DEFINE MAINTENANCE CONCEPT B5 – ACCOMPLISH M ANALYSIS SYSTEMS ANALYSIS M ALLOCATION/APPORTIONMENT B6 – PARTICIPATE IN SYSTEM TRADE-OFF STUDIES B7 – DEFINE DESIGN CHARACTERISTICS (CRITERIA) B8 – ACCOMPLISH DESIGN LIAISON (SERVICES) SUPPORT PLANNING & DOCUMENTATION B9 – PREPARE PRELIM. LOGISTICS SUPPORT PLAN	PROGRAM MANAGEMENT C1 – IMPLEMENT M PROGRAM PLAN (Update Policies & Procedures) C2 – IMPLEMENT SUPPLIER ACTIVITY (Monitor & Control) C3 – ACCOMPLISH TRAINING & INDOCTRINATION C4 – PREPARE M REPORT (TDP) MAINTAINABILITY DESIGN C5 – UPDATE M ANALYSIS SYSTEMS ANALYSIS M ALLOCATION/APPORTIONMENT C6 – ACCOMPLISH SUBSYSTEM & M TRADE-OFF STUDIES C7 – DEFINE DESIGN CHARACTERISTICS (CRITERIA) C8 – ACCOMPLISH DESIGN LIAISON (SERVICES) C9 – ACCOMPLISH M PREDICTION C10– DEFINE MAINTENANCE POLICY C11– PARTICIPATE IN FORMAL DESIGN REVIEWS SUPPORT PLANNING & DOCUMENTATION C12– PREPARE LOGISTICS SUPPORT PLAN (Support Equipment Plan)

FIG. 3-3 Maintainability-program Tasks.

studies. In such instances, the scope of effort will vary from program to program. In view of these factors, additional coverage of maintainability tasks during the conceptual phase is not included in this text.

Proposal Phase

The proposal phase provides the mechanism for translating requirements defined during the conceptual phase into a planned approach for system definition and acquisition. The planned approach is reflected in the contractor's proposal to the customer.

Maintainability effort during the proposal phase results in two basic product outputs (each a part of the contractor's proposal). These products are (1) the maintainability-program plan and (2) proposal technical support. Individual program tasks supporting these products are summarized in Fig. 3-3, and their relationships are illustrated in Fig. 3-4.

Maintainability-program Plan. In all proposals, the contractor must describe how he plans to develop and conduct a maintainability program. This, the management aspect of a program, formulates the maintainability-program plan. The plan should cover the following basic areas.

ACQUISITION PHASE		OPERATIONAL PHASE
DESIGN & FABRICATION	INTEGRATION & TEST	
PROGRAM MANAGEMENT D1 – IMPLEMENT M PROGRAM (Monitor & Control) D2 – IMPLEMENT SUPPLIER ACTIVITY (Monitor & Control) D3 – ACCOMPLISH TRAINING & INDOCTRINATION MAINTAINABILITY DESIGN D4 – ACCOMPLISH DETAILED M TRADE-OFF STUDIES D5 – ACCOMPLISH DESIGN LIAISON (SERVICES) D6 – UPDATE M PREDICTION D7 – UPDATE MAINTENANCE POLICY D8 – PARTICIPATE IN FORMAL DESIGN REVIEWS MAINTENANCE ANALYSIS D9 – ACCOMPLISH MAINTENANCE ANALYSIS (Support Reqmts. Definition) SUPPORT PLANNING & DOCUMENTATION D10 – PREPARE SUPPORTING DOCUMENTATION MAINTAINABILITY DEMONSTRATION D11 – PREPARE M DEMONSTRATION PLAN	PROGRAM MANAGEMENT E1 – IMPLEMENT M PROGRAM (Monitor & Control) E2 – IMPLEMENT SUPPLIER ACTIVITY (Monitor & Control) E3 – PREPARE M FINAL REPORT MAINTAINABILITY DESIGN E4 – UPDATE M PREDICTION MAINTENANCE ANALYSIS E5 – UPDATE MAINTENANCE ANALYSIS (Support Reqmts. Definition) SUPPORT PLANNING & DOCUMENTATION E6 – UPDATE SUPPORTING DOCUMENTATION E7 – PARTICIPATE IN SUPPORT PLANNING & PROVISIONING ACTIVITIES MAINTAINABILITY DEMONSTRATION E8 – ACCOMPLISH M DEMONSTRATION E9 – ACCOMPLISH DATA COLLECTION & ANALYSIS E10 – INITIATE CORRECTIVE ACTION (As Required) E11 – PREPARE M DEMONSTRATION REPORT	PROGRAM MANAGEMENT F1 – MONITOR SYSTEM/ EQUIPMENT MODIFICATION (Approved Changes) MAINTAINABILITY DEMONSTRATION F2 – ACCOMPLISH DATA COLLECTION & ANALYSIS (Field Data as Available) F3 – INITIATE CORRECTIVE ACTION (As Required)

FIG. 3-3 (Continued)

1. *Maintainability assurance*—Describe briefly how maintainability characteristics in equipment design will be assured (management approach).

2. *Maintainability-program tasks*—Describe in detail each program task. Discuss task input/output requirements and time-phasing relationships (schedule).

3. *Organization*—Present an organizational chart showing company structure. Include a detailed organizational breakdown of the maintainability group. Indicate the personnel assigned and list detailed task assignments.

4. *Organizational interfaces*—Show relationships and lines of communication with related (supporting) organizations such as reliability, human factors, logistics (technical publications, spares provisioning, training and field service), systems engineering, design engineering, value engineering, data management, etc. Define proposed relationships with customer organizational elements.

5. *Policies and procedures*—Reference or incorporate management-policy directives which implement maintainability provisions on a company-wide basis. Reference maintainability detailed operating procedures. Describe specific techniques covering maintainability allocation,

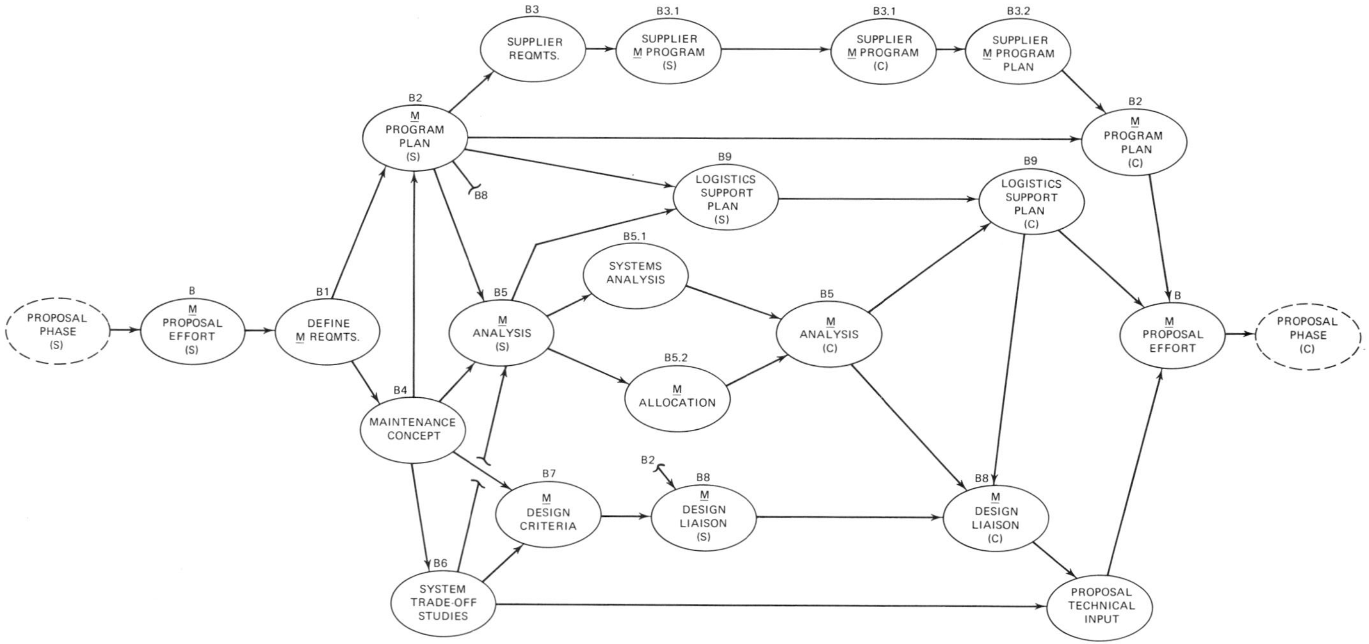

FIG. 3-4 Maintainability Tasks—Proposal Phase.

prediction, maintenance analysis, and demonstration methods. Discuss design liaison functions to cover participation in trade-off studies, design review and evaluation, utilization of design aids, and data collection, analysis, and corrective action.

6. *Control and evaluation*—Indicate methods of cost and schedule control, and means of status reporting. Indicate joint customer-contractor program-milestone review points.

7. *Supplier activity*—Define company-supplier relationships and methods of control.

8. *Maintenance concept*—Describe and illustrate the basic maintenance concept. Elaborate in sufficient detail to cover operational and support concepts, qualitative and quantitative maintainability/maintenance goals, personnel factors, facility/resource data, support-equipment criteria, cost constraints, etc. This area of coverage not only is an important aspect to the maintainability-program plan, but is a basis for the preparation of an integrated logistics support plan.

The above areas, along with references to applicable reports and documentation, are basic ingredients to a maintainability-program plan. These areas are further supplemented through the proposal technical support area discussed below. Specific depth of coverage, as well as organization of the program plan outline, are based on RFP requirements and may vary from program to program.

Proposal Technical Support. The maintainability technical input to a proposal consists of maintenance-policy statements, design criteria (packaging concepts, diagnostic test provisions, standardization requirements, etc.), quantitative factors (derived through a maintainability analysis), trade-off-study results, and information relative to support requirements. A review of the system or equipment technical characteristics will establish the limits or depth of coverage.

The development of the technical data evolves through accomplishment of the day-to-day maintainability-design tasks reflected in Fig. 3-3 (*B*4–*B*8). The resultant output is generally included in the design technical section of the proposal document in order to present a composite definition of the system or equipment. In addition, summary statements of this data are included in the maintainability-program plan.

Definition Phase

The definition phase is a logical step in accomplishing an orderly transition from the conceptual phase to the acquisition phase. The definition phase, generally applied as part of the development cycle for large-scale systems, provides for the definition of detailed cost, schedule, and technical-performance requirements as early as possible in the overall system life-cycle process. This is required to ensure that full-

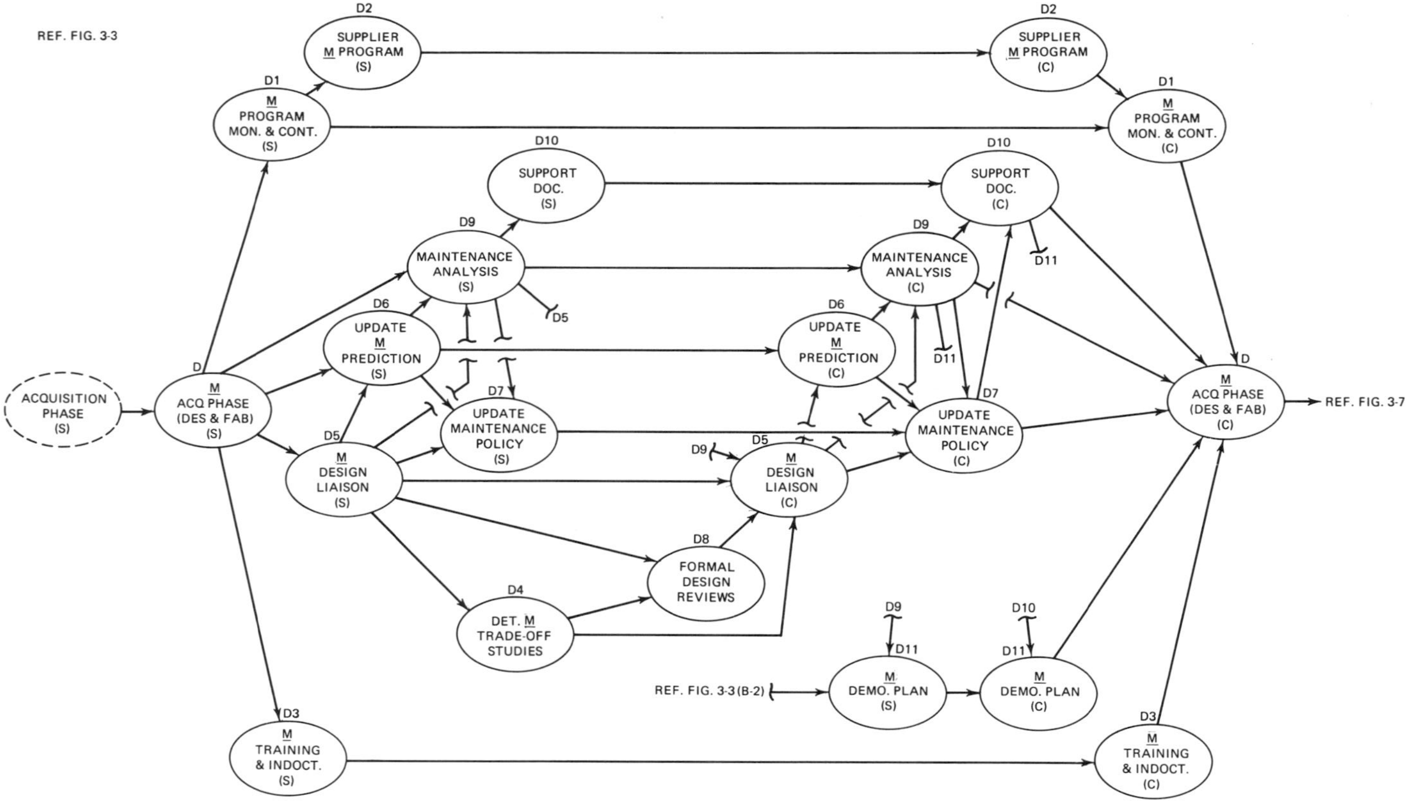

FIG. 3-5 **Maintainability Tasks—Definition Phase.**

scale design and development is not started until all program objectives (performance, cost, and schedule) have been thoroughly established and carefully evaluated. Specific phase objectives are to:

1. Definitize system performance/design requirements through functional analysis and system/subsystem trade-off studies. Validate overall system design and support concepts.

2. Develop firm equipment end-item specifications covering design, development, test, and qualification of each end item. Such specifications will contain both qualitative and quantitative design criteria.

3. Develop detailed plans, costs, and schedules against each end item. Coverage shall include requirements for production engineering, fabrication, and test, facilities acquisition, logistics support planning, and acquisition.

4. Identify customer-contractor areas of high risk.

The resultant output products of the definition phase are a technical-development plan (TDP) supplemented by a series of detailed equipment end-item specifications. These documents fully define both the technical and management approach to all subsequent program activities.

The maintainability tasks associated with the definition phase are summarized in Fig. 3-3, and their relationships are illustrated in Fig. 3-5. Such tasks, as defined in the maintainability-program plan, cover both management and technical activity, results of which are incorporated in a maintainability report (part of the TDP) and in individual equipment end-item specifications.

Acquisition Phase

The acquisition phase produces the total system as defined by system and equipment specifications developed during the definition and/or proposal phases. The acquisition phase may be divided into subphases to include design and fabrication and integration and test (refer to Fig. 3-3).

The design and fabrication activity primarily constitutes:

1. Review, updating, amplification, and verification of plans and specifications generated during the definition and/or proposal phase.

2. Design, development, and fabrication of system/equipment end items.

3. Development of integrated test plans and procedures.

4. Development of logistics-support requirements. This includes a plan to effect an orderly transition of the system from contractor to customer support.

The integration and test activity constitutes:

1. Implementation of integrated test plans and procedures

2. Accomplishment of category I and II test requirements (qualifica-

REF. FIG. 3-3

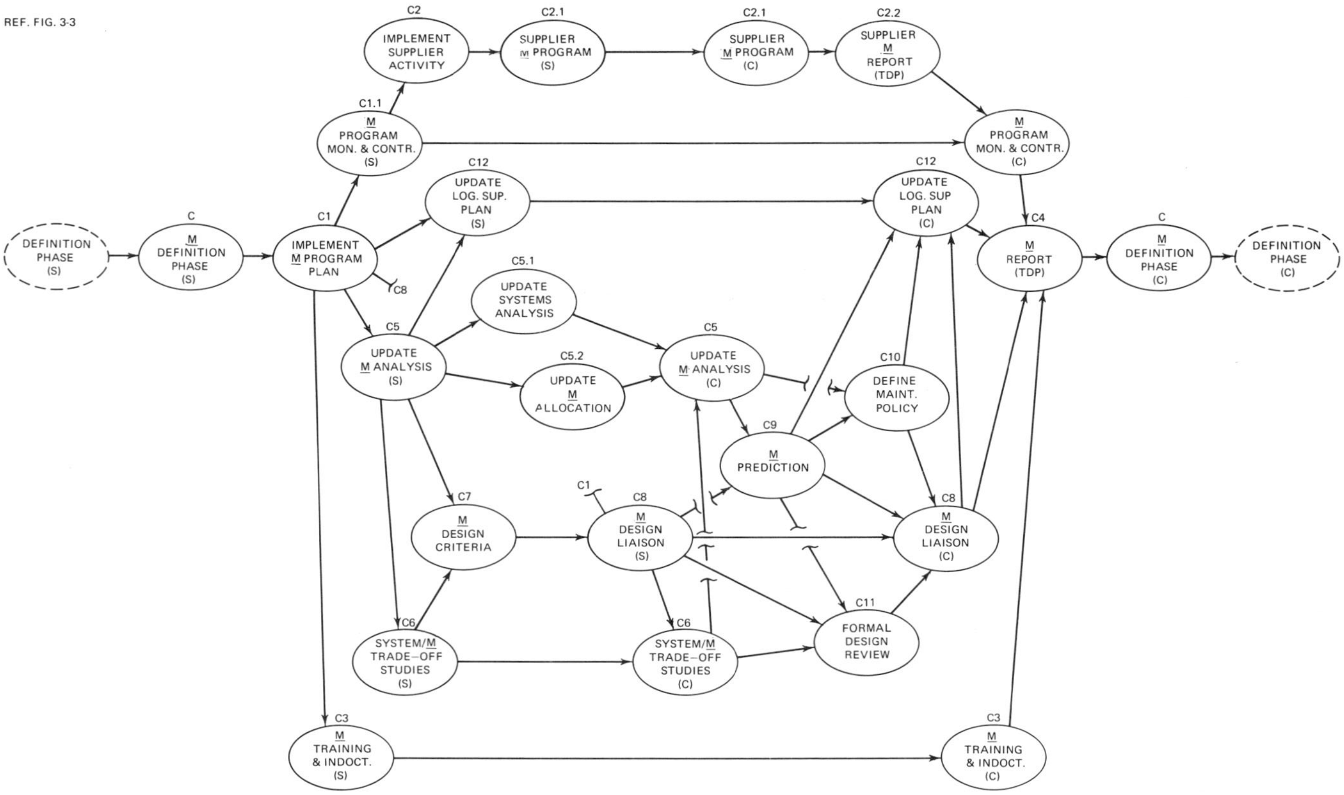

FIG. 3-6 Maintainability Tasks—Acquisition Phase (Design and Fabrication Stages).

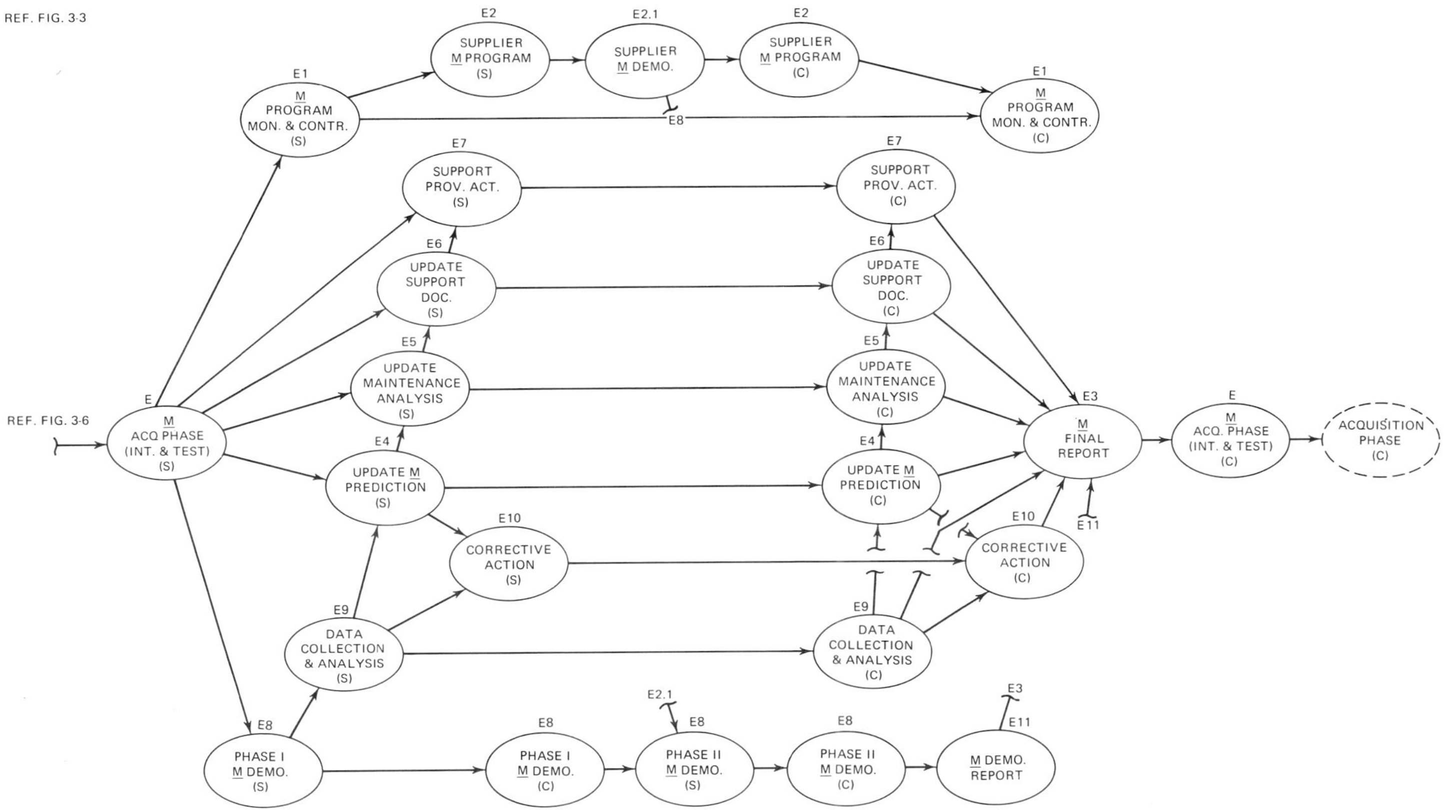

FIG. 3-7 Maintainability Tasks—Acquisition Phase (Integration and Test Stages).

tion tests, acceptance tests, reliability/maintainability demonstrations, support-equipment compatibility tests)

3. Collection, evaluation, and analysis of test data, and the preparation of final program/test reports

4. Delivery of equipment to the customer

The acquisition phase continues until the last system is produced and accepted by the customer.

Maintainability effort during the acquisition phase is summarized in Fig. 3-3, and the relationships of individual tasks are illustrated in Fig. 3-6 (design and fabrication) and Fig. 3-7 (integration and test). Such effort culminates in a final maintainability-program report (Fig. 3-7, *E*3) submitted at the end of the acquisition phase.

Operational Phase

The operational phase commences with the acceptance and deployment of the first operational system, and extends through phase-out of the last system in inventory. The operational and the acquisition phases overlap somewhat since the acquisition phase extends through delivery and acceptance of the last system into the operational inventory.

Contractor activity during this phase is somewhat limited to field-service support, the establishment (on an "as required" basis) of an interim depot-level maintenance capability, and the processing of change modifications. Maintainability effort includes the review and analysis

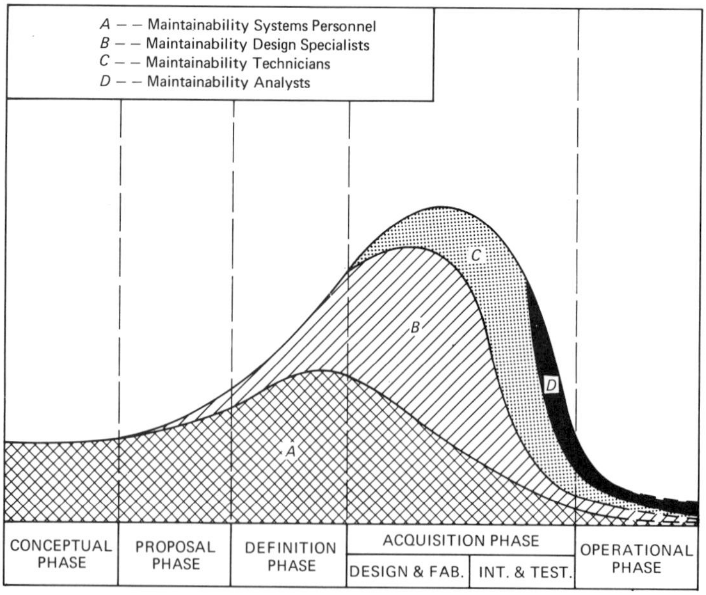

FIG. 3-8 Manloading Curve (Typical) of Maintainability Personnel.

of field-maintenance data for possible areas of corrective action, and the monitoring of design changes plus the development of modification kits.

PROGRAM MANLOADING

The manloading requirements associated with the implementation of a maintainability program are, of course, dependent on the type of equipment involved (full-scale system or small subsystem, newly designed or off-the-shelf equipment), specification requirements, and program schedules. In any event, the level of maintainability effort is generally proportional to the level of effort applied by other engineering disciplines.

Assuming a rather comprehensive program, maintainability-systems personnel of high capability and broad background (versed in a wide spectrum of engineering disciplines—operations analysis, systems analysis, design engineering, reliability, human factors, logistics, etc.) are needed most in the beginning during the proposal phase and the early part of the definition phase. The foundation for the total maintainability program is developed through their efforts and understanding. At this point in time, the quality of personnel must be high although the quantity is low.

As the program progresses through the latter part of the definition phase and the early acquisition phase (design and fabrication), there is an increased requirement relative to both quantity and type of personnel. Manloading at this time constitutes three broad categories to include maintainability-systems personnel (cognizant of systems- and support-engineering functions), design specialists knowledgeable of detailed design practices and specific types of equipment, and skilled technicians who possess the knowledge of equipment operation and maintenance procedures through actual field experience.

During the latter part of the acquisition phase (integration and test), the quantities of maintainability-systems personnel and design specialists decrease while the quantity of skilled technicians remains high for maintenance-analysis completion and demonstration-test participation. In addition, there is a need for skilled analysts to cover the requirements for data analysis and corrective action.

A typical curve relating projected manloading requirements against program phases is illustrated in Fig. 3-8.

COST ALLOCATION

The initial step in the determination of total maintainability-program costs is to assign estimated costs against each of the applicable tasks identified in Fig. 3-3. The individual tasks with associated costs are then grouped into elements of work, the sum being equivalent to total program costs.

As a second step, the total cost package is reviewed from the standpoint of organizational efficiency (elimination of redundancy, optimum procedures for task accomplishment, proper task assignment relative to personnel background and skills). The company with the most cost-effective methods of doing business is obviously in the most favorable position. Based on these considerations, the costs are adjusted to reflect the company's best interests.

When the total maintainability-program cost package has been established and is ready to be related to the customer, these costs are reallocated against the applicable tasks and elements of work. The resulting factors are subsequently applied to the overall program-cost reporting network (PERT/cost or equivalent). Cost control is maintained over each area of work through conventional reporting procedures.

PROGRAM MONITORING AND CONTROL

Once the initial maintainability requirements have been specified and a formal program has been established, management's prime role is relegated to a monitoring and control function. The purpose is to ensure to the maximum extent possible that all maintainability requirements are being met and to identify as early as possible any deviations from those requirements. Although the preferred approach to fulfilling this purpose is through an effective day-to-day design liaison effort, this is often precluded due to the fact that there are insufficient maintainability personnel available to adequately cover all design and support areas of activity. In addition, design-engineering personnel, in their haste to define equipment parameters and complete design in a minimum period of time, often concentrate on performance parameters only and tend to overlook or disregard maintainability requirements. Hence, the above purpose is best facilitated through proper initial planning, periodic checks and balances throughout the program, adequate measurement functions, and effective status reporting.

Basic program planning, reflected in the initial maintainability-program plan, establishes the methods for program monitoring and control. Program checkpoints, methods of measurement, and areas of documentation are specified.

Program checks and balances are established to allow a certain amount of control of end-item design through periodic review and approval functions accomplished by maintainability personnel (customer as well as contractor personnel). Such functions include:

1. System/equipment design-specification review and approval through sign-off
2. System/equipment design-drawing and change-data review and approval through sign-off
3. Supplier-data review and approval
4. Periodic formal design reviews
5. Maintainability-demonstration planning and testing

In each area above, maintainability provides a design-check function.

Measurement-type data is intended to provide management with an up-to-date comparison of predicted quantitative factors reflecting the equipment-design configuration and specified quantitative requirements. Immediate action is required in areas where predicted values exceed the initial allocated requirement. Measurement-type data includes maintainability-prediction reports, maintenance-analysis reports, maintainability-assessment data, supporting documentation (e.g., calibration-requirement data), and the final maintainability-demonstration/program report.

Status reporting provides management a continuous review of program-task status (cost and schedule relationships), accomplishments, and major problem areas. Documentation in this area includes periodic (monthly or quarterly) status or progress reports plus special maintainability problem-identification reports.

Relative to specific status-monitoring techniques, such systems as PERT/time and PERT/cost, as employed throughout industry, are quite appropriate from the standpoint of cost and schedule monitoring and control. On programs employing these systems, maintainability-task input data are incorporated into PERT charts along with reliability, human factors, design, and other related milestone data. Subsequently, the PERT system enables management to closely monitor program-task accomplishment (schedule and cost) and to identify major problem areas. The relationships of individual maintainability-program tasks for possible PERT application are illustrated in Figs. 3-4 through 3-7.

As conveyed above, there are a number of imposed requirements which facilitate the overall monitoring and control process. These requirements (e.g., planning documentation, review functions, measurement reports, periodic status reporting) are presented relative to typical program phasing in Fig. 3-9, and are briefly discussed below.

Program-planning Documentation

Maintainability-program Plan. The maintainability-program plan, prepared as part of the proposal effort, specifies the methodologies and techniques that the contractor intends to employ to ensure the accom-

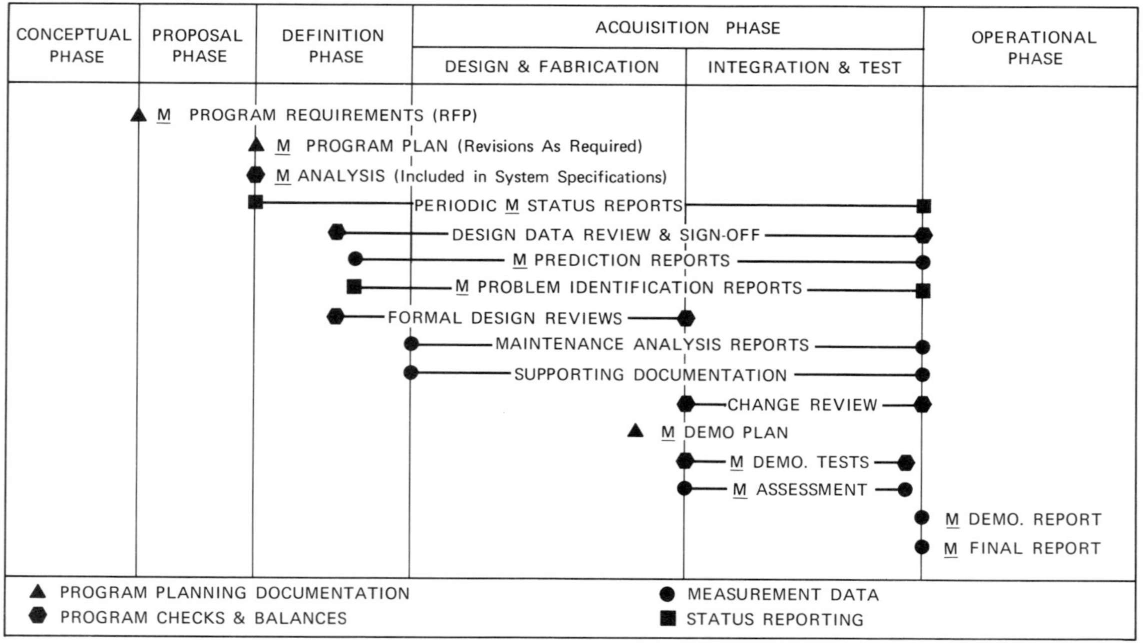

FIG. 3-9 Maintainability-program-control Documentation.

plishment of customer maintainability requirements specified in the RFP. The program plan is the focal point for all management planning and covers such areas as maintainability-program tasks, task input/output requirements, cost and schedule data, organizational responsibilities and interfaces, and policies and procedures (included or referenced). The program plan requires internal contractor management as well as customer approval. Once approved, the customer plus contractor management personnel utilize the plan to monitor maintainability progress (task completion) throughout the program. The plan is initially prepared as part of the contractor's proposal and is updated throughout the program on an "as required" basis.

Maintainability-demonstration Plan. The maintainability-demonstration plan describes the methods, techniques, and procedures that the contractor intends to employ relative to the accomplishment of maintainability-demonstration testing. Demonstration tests are accomplished to verify that maintainability requirements have been met. The demonstration plan is prepared and submitted to the customer approximately 60 days prior to the start of formal demonstration testing. Once again, the customer maintains a "check" function through required approval of the plan prior to the start of demonstration.

Program Checks and Balances

Maintainability Analysis. The maintainability analysis is the mechanism for translating system qualitative and quantitative requirements specified in the RFP into specific design criteria. The analysis includes the generation of maintenance functional flows and the allocation/apportionment of quantitative factors. Results of the analysis are inputs to system design and are generally included in system- and equipment-design specifications. Maintainability provides a review and approval (through sign-off) function for design specifications. Once incorporated into applicable specifications, maintainability-analysis data serves as a constraint on design.

Design-data Review and Sign-off. As design layouts and detailed drawings are completed, maintainability often provides a review and sign-off function prior to release to production. In this capacity, each drawing is checked by the maintainability engineer for the incorporation of maintainability characteristics in the applicable equipment item being reviewed. Concurrence with the design configuration reflected results in approval and sign-off. Rejection results in the generation of recommendations for product improvement processed through the corrective-action system. Next to an effective day-to-day design liaison effort, the drawing review and sign-off policy is one of the most effective methods of providing a design-check function.

Formal Design Reviews. At critical points throughout the overall design process, formal design reviews are conducted to ensure that all facets of design (to include maintainability) are being incorporated or adequately considered. The intent is to provide a systematic review of basic design and support data, to assure that the design presented can be economically manufactured, and to discuss outstanding problem areas. Relative to maintainability, prediction and analysis data are compared against initially allocated requirements. In addition, outstanding (unincorporated) maintainability design-improvement recommendations are reviewed and reevaluated. In areas where maintainability has been traded off for other features, applicable trade-off–study reports are made available for review.

The design-review function is applicable to different program phases and adaptable to different levels of coverage. During the definition phase and the early part of the acquisition phase, design reviews are oriented to the system level. As equipment design becomes more definitive, design reviews are oriented to a more detailed level. In any event, the design review includes representation from all affected factors within the contractor's organization (e.g., systems engineering, design engineering, maintainability, reliability, manufacturing, and quality control) as well as interested customer personnel and provides a formal check on design.

Maintainability-demonstration Tests. The maintainability-demonstration tests, conducted during the latter part of the acquisition phase, are accomplished to verify that the contractor has complied with all initially specified maintainability quantitative requirements. Demonstration tests constitute the simulation of random-type failures followed by appropriate repair actions which are timed against a specified *MTTR* requirement. Such tests are accomplished as part of a contractor's qualification test program on a preproduction system or equipment item. Once again, the design engineer is obligated to prove his design relative to the incorporation of maintainability features. Failure of the equipment to meet specified requirements may result in system/equipment rejection, a loss of money for the contractor, or a loss of future potential business.

Measurement Data

Maintainability-prediction Reports. Maintainability prediction is a tool, when accomplished on an iterative basis, to provide the maintainability engineer as well as management with an up-to-date estimate of system/equipment *MTTR*, *MDT*, *MMH/OH*, etc. Resultant predicted values are compared against system requirements.

Maintenance Analysis and Supporting Documentation. Maintenance-analysis data provides an estimate of all projected system-support requirements to include maintenance tasks, task times and frequencies, personnel requirements (type, quantity, and skill level), support equipment, spare/repair parts, facilities, and resources. Analysis data reflects the degree of overall system maintainability incorporated in design. Areas of poor design for maintainability readily become evident through the analysis effort.

Maintainability-assessment Data. Maintainability-assessment data is provided through a failure-data collection and analysis system. As equipment becomes available within the contractor's facility (through fabrication and test), all experienced failures and resultant repair actions are monitored. Those events characteristic of customer-type operations (in lieu of being peculiar only to factory-type operations) are analyzed, and resultant repair-time data is compared against the initially allocated system requirements.

Maintainability-demonstration Report. The final maintainability-demonstration report, generally submitted to the customer approximately 30 days after completion of demonstration testing, presents a summary of test accomplishment as well as test-data analysis and results. This document specifies the final degree of contractor compliance with the maintainability quantitative requirements initially specified in the RFP. Customer acceptance and approval of this document terminates the contractor's formal maintainability-program effort. This document is often combined with the formal maintainability final report.

Status Reporting

Periodic Status Reports. On a periodic basis throughout the program, maintainability status reports are prepared and submitted to the customer. These status reports often constitute a section of a larger overall program report generally prepared on a monthly or quarterly basis. Such reports present an up-to-date status (e.g., identification of active tasks, percent completion, cost and schedule reporting, summary of significant events) relative to maintainability-program effort as defined in the program plan. Status reporting constitutes one of the mechanisms providing management with the necessary information for accomplishment of the monitoring and control function. Management, through an analysis of task-completion data versus costs expended, is readily able to detect program problem areas.

Maintainability Problem-identification Reports. Problem-identification reports are prepared on an "as required" basis covering specific problem areas. Through maintainability prediction, maintenance analysis, and demonstration data, specific design configurations not in compliance with specified maintainability requirements (e.g., predicted *MTTR* exceeds allocated *MTTR*) are readily identified. Recommendations for product improvement are submitted to design engineering through the corrective-action system. In addition, management is notified of the possibility of not meeting contractual requirements.

Maintainability Final Report. A final maintainability-program report is prepared as a summary document after the completion of all formal program effort. The document, often combined with the maintainability-demonstration report, presents a summary of program tasks, major accomplishments, documentation submittals, and recommendations (as applicable) for future action. The report indicates the degree of compliance with the maintainability-program requirements initially specified in the RFP.

QUESTIONS AND PROBLEMS

1. What is *program implementation*?
2. Maintainability-task requirements are consistent throughout a program. True or false?
3. The method of program implementation directly influences the operating efficiency of the maintainability organization. True or false?
4. List the five program phases.
5. Describe each program phase in terms of the system/equipment stages of evolution.
6. What program phases include maintainability tasks?
7. The five program phases apply in all cases. True or false?
8. Trade-off studies are not normally performed during the conceptual phase. True or false?
9. A company maintainability organization is normally involved in the conceptual phase of a program. True or false?
10. During the proposal phase of a program, the company maintainability organization develops two products that are integrated into the company proposal. What are these two products?
11. Outline the content of a typical maintainability-program plan.
12. What does TDP mean?
13. The operational phase of a program begins at the time the first system/equipment is deployed by the customer. True or false?
14. Experienced and highly skilled maintainability engineers are required during the proposal and definition phases of a program. True or false? Why?
15. The manloading requirements for a maintainability organization will be greatest during the definition phase of a program. True or false?
16. List the four types of maintainability personnel required during the acquisition and operational phases of a program.
17. Describe in brief form how you would develop the total cost projection for the conduct of a maintainability program.
18. How do program-planning documents relate to the monitoring and control of maintainability-program tasks?
19. What are the four forms of checks and balances that are normally applied to a maintainability program?
20. List the four types of maintainability-measurement data.

FOUR

Quantitative Factors

Maintainability requirements must be specified in such a manner that compliance with the requirements can be readily confirmed. If such confirmation is not possible, the specified requirements are meaningless. Although general workmanship requirements can be determined qualitatively through a simple examination of the equipment item, other more definite requirements such as system availability, and operational readiness must be specified quantitatively in order to have meaning.

Maintainability requirements can be quantitatively specified, measured, and demonstrated. The methods of specifying maintainability may vary depending on the customer, the type of equipment, and the specification involved. A number of the factors currently employed throughout industry are *MTBM, MTTR, ERT*, $\overline{M}_{ct}$, $\overline{M}_{pt}$, etc. The purpose of this chapter is to present a summary of the more significant measures of maintainability. A brief review of statistical principles is also presented.

REVIEW OF STATISTICAL PRINCIPLES

Based on experience through the analysis of field data, it has been found that maintenance-task times associated with various types of equipment will generally distribute themselves statistically into one of two general patterns. The patterns, or distribution curves as they are sometimes called, are known as the *normal distribution curve* and the *log-normal distribution curve.* As an introduction to maintainability quantitative processes, a review of each distribution is presented below.

The Normal Distribution Curve

Table 4-1 presents data representing a sampling of 50 corrective-maintenance repair actions (in minutes) on a typical electronic equipment item.

Based on the set of raw data presented in Table 4-1, which represents a random sampling, a frequency table may be set up as follows:

1. Range of observations, 98 — 33 = 65.

2. Subdivide the range into at least seven class intervals. (The Sturgess rule[1] was used to determine optimum choice of class-interval width as follows:

$$\text{Class interval width} = \frac{\text{range of observations}}{K} \tag{1}$$

where $K = 1 + 3.3 \log_{10} N$ and N = number of data samples.)

A class-interval width of 10, beginning with 30 min, will just provide seven class intervals (e.g., 40, 50, . . . , 100).

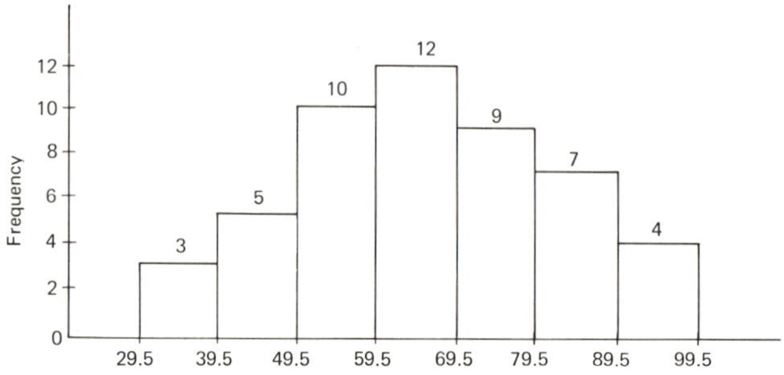

FIG. 4-1 Corrective-maintenance Time—Minutes.

3. Set up these class intervals as shown in Table 4-2, and include the 50 repair-time values.

Based on the frequency distribution indicated in Table 4-2, the data can be plotted into a histogram as shown in Fig. 4-1.

Referring to Fig. 4-1, the quantitative values on the abscissa were selected as midpoints between the class intervals. By determining the

[1] Richard H. Myers, Kam L. Wong, and Harold M. Gordy, "Reliability Engineering for Electronic Systems," 1st ed., p. 10, John Wiley & Sons, Inc., New York, 1964.

midpoints of the class intervals indicated in Fig. 4-1, a frequency polygon can be plotted as illustrated in Fig. 4-2.

If more data had been collected on the equipment being used, and additional data points had been plotted, the curve in Fig. 4-2 could be expected to take the shape of the normal distribution curve as shown in Fig. 4-3.

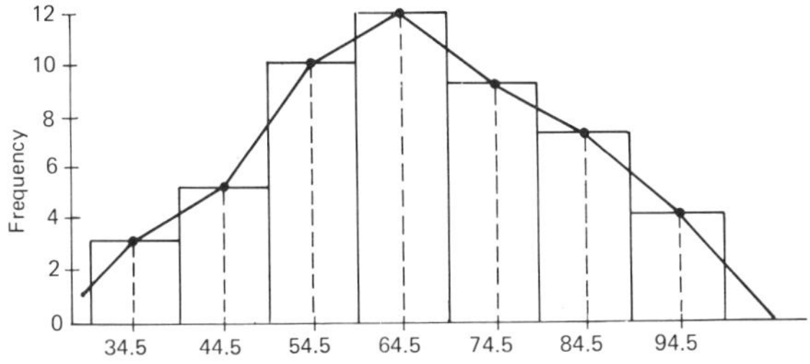

FIG. 4-2 Corrective-Maintenance Time—Minutes.

The two major parameters which describe a normal distribution are the arithmetic mean $\bar{X}$ and the standard deviation σ.

Arithmetic Mean $\bar{X}$ or Mean Repair Time $\bar{M}_{ct}$. An average of a distribution is more or less a typical value used to characterize the location of the distribution as a whole. An ordinate through the arithmetic mean passes through the centroid of the area of distribution (refer to Fig. 4-3).

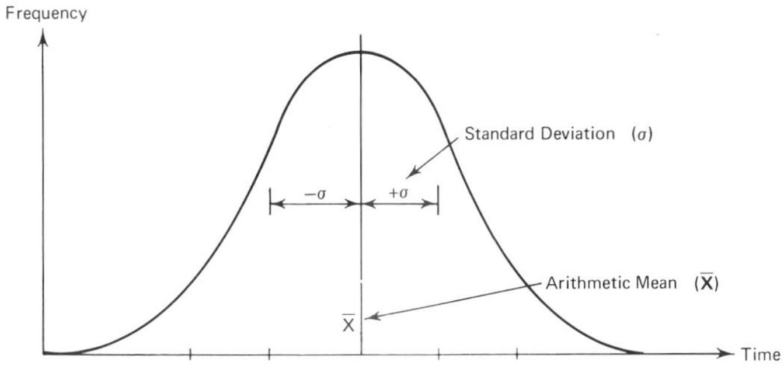

FIG. 4-3 Normal Distribution Curve.

Referring to the data presented in Table 4-1, the arithmetic mean of the repair times can be calculated as follows:

$$\bar{X} = \bar{M}_{ct} = \frac{\sum_{1}^{N} M_{ct_i}}{N} = \frac{\sum_{1}^{50} M_{ct_i}}{50} = \frac{3295}{50} = 65.9 \text{ (assume 66)} \qquad (2)$$

The above equation tells us that the average value for the specified set of 50 corrective-maintenance actions is equal to 66 min.

To obtain information regarding the spread of repair-time values on either side of the mean value, it is necessary to calculate the standard deviation σ.

TABLE 4-1 Corrective-maintenance Repair Times

60	71	80	86	94	57	74	59	67	65
33	81	41	78	66	95	47	53	39	61
85	56	61	55	98	57	68	45	76	89
52	50	91	48	66	74	54	77	60	88
65	35	55	69	73	77	64	73	85	42

Standard Deviation σ. The standard deviation is a measure of the dispersion of values within a distribution. The dispersion is indicated by the extent to which observed values of the variate tend to spread over an interval rather than cluster closely around a central value.

The traditional symbol for the standard deviation is the Greek letter sigma (σ).

The standard deviation becomes important in the quantitative analysis of maintainability because it is used in estimating upper and lower limits of the mean repair time with a specified degree of confidence. The standard deviation is also useful in determining the percentage of the total population sampled that falls within a specified band of values.

TABLE 4-2 Frequency Distribution of Repair Times

Class Interval	Frequency	Cumulative Frequency
30 - 39	3	3
40 - 49	5	8
50 - 59	10	18
60 - 69	12	30
70 - 79	9	39
80 - 89	7	46
90 - 99	4	50

Before the standard deviation can be calculated, a table should be generated (in a manner similar to the data presented in Table 4-3) giving the deviation of each task time from the mean. For simplicity, Table 4-3 is presented as showing only six individual task times although all 50 tasks should be considered; however, the total value (i.e., 13,073) does cover all 50 tasks.

The standard deviation σ of the sample normal distribution curve can now be calculated as follows:

$$\sigma = \sqrt{\frac{\sum_{1}^{N} (M_{ct_i} - \bar{M}_{ct})^2}{N - 1}}$$

$$\sigma = \sqrt{\frac{\sum_{1}^{50} (M_{ct_i} - \bar{M}_{ct})^2}{49}} \tag{3}$$

$$\sigma = \sqrt{\frac{13073}{49}}$$

$$\sigma = 16.3 \text{ min (assume 16)}$$

The standard deviation can also be calculated using the following formula:

$$\sigma = \sqrt{\frac{N \sum_{1}^{N} M_{ct}^2 - \left(\sum_{1}^{N} M_{ct}\right)^2}{N(N - 1)}} \tag{4}$$

Both $\bar{M}_{ct}$ and σ may be computed from sample data organized into a frequency table, such as Table 4-2, rather than from the raw, ungrouped data. Procedures for computing these statistics from frequency table data are explained and illustrated in most statistic textbooks.

TABLE 4-3 Variance Data

M_{ct_i}	$M_{ct_i} - \bar{M}_{ct_i}$	$(M_{ct_i} - \bar{M}_{ct_i})^2$
60	- 6	36
71	+ 5	25
80	+ 14	196
41	- 25	625
73	+ 7	49
85	+ 19	361
etc.	etc.	etc.
TOTAL		13073

Confidence Limits. Since the 50 repair times represent only a sample of all possible repair actions on the equipment item being considered, it is possible that another sample of 50 repair actions on the same equipment could have a mean value either greater or less than 66 min. The original set of 50 repair times were selected at random, however, and are statistically representative of the entire population. Using the standard deviation, an upper and a lower limit can be placed on the population mean value depending on the degree of risk the customer is willing to accept. For instance, if the customer will accept the chance of being wrong 10 percent of the time (90 percent confidence limit), then

$$\text{Upper limit (U.L.)} = \bar{M}_{ct} + Z\sigma_{\bar{M}_{ct}} \tag{5}$$

where $\bar{M}_{ct}$ = mean repair time

Z = confidence limit factor

$\sigma_{\bar{M}_{ct}} = \dfrac{\sigma}{\sqrt{N}}$ = standard deviation of the mean (standard error)

σ = standard deviation of sample data

N = sample size

The Z value is a constant obtained from a table of cumulative normal-distribution functions available in most textbooks on statistics (see Table 4-5 for an example). Table 4-4 presents several typical values for Z for locating an upper confidence limit.

For the purpose of discussion, assume that the equipment specification includes a mean maintenance-time value $\bar{M}_{ct}$ that shall not exceed 70 min.

In this instance, the upper confidence limit is of concern since the customer must be assured that the true mean does not exceed the specified value of 70 min. The upper confidence limit marks a probable extreme range of the true mean, and can be calculated using Eq. (5).

TABLE 4-4 Confidence Limits/Risks

Z	Confidence	Risk
1.65	95%	5%
1.30	90%	10%
1.05	85%	15%
0.85	80%	20%

Using the values from Eq. (2) and (3), the upper confidence limit is as follows

$$\text{Upper limit (U.L.)} = \bar{M}_{ct} + Z\sigma_{\bar{M}_{ct}}$$

where

$$\sigma_{\bar{M}_{ct}} = \frac{\sigma}{\sqrt{N}} = \frac{16}{\sqrt{50}} = 2.30$$

Therefore

$$\text{U.L.} = 66 + (1.30)(2.30) = 69.0 \text{ min}$$

TABLE 4-5 Cumulative Normal Distribution

z	X	Area	z	X	Area
1.05	$\mu + 1.05\sigma$	.8531	−4.265	$\mu - 4.265\sigma$	.00001
1.10	$\mu + 1.10\sigma$	.8643	−3.719	$\mu - 3.719\sigma$	.0001
1.15	$\mu + 1.15\sigma$	.8740	−3.000	$\mu - 3.000\sigma$	.001
1.20	$\mu + 1.20\sigma$	.8849	−2.576	$\mu - 2.576\sigma$	.005
1.25	$\mu + 1.25\sigma$	.8044	−2.326	$\mu - 2.326\sigma$	.01
1.30	$\mu + 1.30\sigma$	.9032	−2.054	$\mu - 2.054\sigma$	.02
1.35	$\mu + 1.35\sigma$	.9115	−1.960	$\mu - 1.960\sigma$	.025
1.40	$\mu + 1.40\sigma$	.9192	−1.881	$\mu - 1.881\sigma$	.03
1.45	$\mu + 1.45\sigma$	.9265	−1.751	$\mu - 1.751\sigma$	.04
1.50	$\mu + 1.50\sigma$	.9332	−1.645	$\mu - 1.645\sigma$	.05
1.55	$\mu + 1.55\sigma$	.9394	−1.555	$\mu - 1.555\sigma$	.06
1.60	$\mu + 1.60\sigma$	.9452	−1.476	$\mu - 1.476\sigma$	.07
1.65	$\mu + 1.65\sigma$	.9505	−1.405	$\mu - 1.405\sigma$	.08
1.70	$\mu + 1.70\sigma$	.9554	−1.341	$\mu - 1.341\sigma$	.09
1.75	$\mu + 1.75\sigma$	.9599	−1.282	$\mu - 1.282\sigma$	.10
1.80	$\mu + 1.80\sigma$	.9641	−1.036	$\mu - 1.036\sigma$	.15
1.85	$\mu + 1.85\sigma$	.9678	− .842	$\mu - .842\sigma$	.20
1.90	$\mu + 1.90\sigma$	.9713	− .674	$\mu - .674\sigma$	.25
1.95	$\mu + 1.95\sigma$	.9744	− .524	$\mu - .524\sigma$	.30
2.00	$\mu + 2.00\sigma$	.9772	− .385	$\mu - .385\sigma$	.35
2.05	$\mu + 2.05\sigma$	.0798	− .253	$\mu - .253\sigma$	.40
2.10	$\mu + 2.10\sigma$	.9821	− .126	$\mu - .126\sigma$	.45
2.15	$\mu + 2.15\sigma$	.9842	0	μ	.50
2.20	$\mu + 2.20\sigma$	.9861	.126	$\mu + .126\sigma$	.55
2.25	$\mu + 2.25\sigma$	.9878	.253	$\mu + .253\sigma$	.60
2.30	$\mu + 2.30\sigma$	.9803	.385	$\mu + .385\sigma$	.65
2.35	$\mu + 2.35\sigma$	.9906	.524	$\mu + .524\sigma$	.70
2.40	$\mu + 2.40\sigma$	.9918	.674	$\mu + .674\sigma$	.75
2.45	$\mu + 2.45\sigma$	.9929	.842	$\mu + .842\sigma$	.80
2.50	$\mu + 2.50\sigma$	.9938	1.036	$\mu + 1.036\sigma$	.85
2.55	$\mu + 2.55\sigma$	.9946	1.282	$\mu + 1.282\sigma$	.90
2.60	$\mu + 2.60\sigma$	.9953	1.341	$\mu + 1.341\sigma$	.91
2.65	$\mu + 2.65\sigma$	.9960	1.405	$\mu + 1.405\sigma$	.92
2.70	$\mu + 2.70\sigma$	.9965	1.476	$\mu + 1.476\sigma$	.93
2.75	$\mu + 2.75\sigma$	.9970	1.555	$\mu + 1.555\sigma$	.94
2.80	$\mu + 2.80\sigma$	.9974	1.645	$\mu + 1.645\sigma$	.95
2.85	$\mu + 2.85\sigma$	.9978	1.751	$\mu + 1.751\sigma$	.96
2.90	$\mu + 2.90\sigma$	.9981	1.881	$\mu + 1.881\sigma$	.97
2.95	$\mu + 2.95\sigma$	.9984	1.960	$\mu + 1.960\sigma$	.975
3.00	$\mu + 3.00\sigma$	.9987	2.054	$\mu + 2.054\sigma$	.98
3.05	$\mu + 3.05\sigma$	.9980	2.326	$\mu + 2.326\sigma$	.99
3.10	$\mu + 3.10\sigma$	.9990	2.576	$\mu + 2.576\sigma$	.995
3.15	$\mu + 3.15\sigma$	.9992	3.090	$\mu + 3.090\sigma$	.999
3.20	$\mu + 3.20\sigma$	.9993	3.719	$\mu + 3.719\sigma$	.9999
3.25	$\mu + 3.25\sigma$	.9994	4.265	$\mu + 4.265\sigma$	.99999

SOURCE: (*Wilfrid J. Dixon and Frank J. Massey, Jr., "Introduction to Statistical Analysis," 2d ed., McGraw-Hill Book Company, New York, 1957, Table A-4, pp. 382–383.*)

TABLE 4-5 (Continued)

z	X	Area	z	X	Area
−3.25	$\mu - 3.25\sigma$	.0006	−1.00	$\mu - 1.00\sigma$	.1587
−3.20	$\mu - 3.20\sigma$	.0007	− .95	$\mu - .95\sigma$	.1711
−3.15	$\mu - 3.15\sigma$	.0008	− .90	$\mu - .90\sigma$	.1841
−3.10	$\mu - 3.10\sigma$	.0010	− .85	$\mu - .85\sigma$	.1977
−3.05	$\mu - 3.05\sigma$	.0011	− .80	$\mu - .80\sigma$	.2119
−3.00	$\mu - 3.00\sigma$	.0013	− .75	$\mu - .75\sigma$	.2266
−2.95	$\mu - 2.95\sigma$	.0016	− .70	$\mu - .70\sigma$	.2420
−2.90	$\mu - 2.90\sigma$	.0019	− .65	$\mu - .65\sigma$	.2578
−2.85	$\mu - 2.85\sigma$	.0022	− .60	$\mu - .60\sigma$	.2743
−2.80	$\mu - 2.80\sigma$	.0026	− .55	$\mu - .55\sigma$	.2912
−2.75	$\mu - 2.75\sigma$	.0030	− .50	$\mu - .50\sigma$	.3085
−2.70	$\mu - 2.70\sigma$	.0035	− .45	$\mu - .45\sigma$	.3264
−2.65	$\mu - 2.65\sigma$	.0040	− .40	$\mu - .40\sigma$	.3446
−2.60	$\mu - 2.60\sigma$	.0047	− .35	$\mu - .35\sigma$	.3632
−2.55	$\mu - 2.55\sigma$	.0054	− .30	$\mu - .30\sigma$	.3821
−2.50	$\mu - 2.50\sigma$	.0062	− .25	$\mu - .25\sigma$	.4013
−2.45	$\mu - 2.45\sigma$	.0071	− .20	$\mu - .20\sigma$	.4207
−2.40	$\mu - 2.40\sigma$	.0082	− .15	$\mu - .15\sigma$	.4404
−2.35	$\mu - 2.35\sigma$	.0004	− .10	$\mu - .10\sigma$	.4602
−2.30	$\mu - 2.30\sigma$	.0107	− .05	$\mu - .05\sigma$	.4801
−2.25	$\mu - 2.25\sigma$	.0122			
−2.20	$\mu - 2.20\sigma$	.0130			
−2.15	$\mu - 2.15\sigma$	.0158	.00	μ	.5000
−2.10	$\mu - 2.10\sigma$	.0179			
−2.05	$\mu - 2.05\sigma$	.0202			
−2.00	$\mu - 2.00\sigma$	.0228	.05	$\mu + .05\sigma$	.5199
−1.95	$\mu - 1.95\sigma$	.0256	.10	$\mu + .10\sigma$	.5398
−1.90	$\mu - 1.90\sigma$	.0287	.15	$\mu + .15\sigma$	.5596
−1.85	$\mu - 1.85\sigma$	.0322	.20	$\mu + .20\sigma$	.5793
−1.80	$\mu - 1.80\sigma$	.0359	.25	$\mu + .25\sigma$	.5987
−1.75	$\mu - 1.75\sigma$	.0401	.30	$\mu + .30\sigma$	.6179
−1.70	$\mu - 1.70\sigma$	.0446	.35	$\mu + .35\sigma$	.6368
−1.65	$\mu - 1.65\sigma$	.0495	.40	$\mu + .40\sigma$	.6554
−1.60	$\mu - 1.60\sigma$	.0548	.45	$\mu + .45\sigma$	.6736
−1.55	$\mu - 1.55\sigma$	.0006	.50	$\sigma + .50\sigma$	.6915
−1.50	$\mu - 1.50\sigma$	.0668	.55	$\mu + .55\sigma$	.7088
−1.45	$\mu - 1.45\sigma$	.0735	.60	$\mu + .60\sigma$	.7257
−1.40	$\mu - 1.40\sigma$	.0808	.65	$\mu + .65\sigma$	.7422
−1.35	$\mu - 1.35\sigma$	.0885	.70	$\mu + .70\sigma$	.7580
−1.30	$\mu - 1.30\sigma$	.0968	.75	$\mu + .75\sigma$	.7734
−1.25	$\mu - 1.25\sigma$	.1056	.80	$\mu + .80\sigma$	.7881
−1.20	$\mu - 1.20\sigma$	.1151	.85	$\mu + .85\sigma$	.8023
−1.15	$\mu - 1.15\sigma$	.1251	.90	$\mu + .90\sigma$	.8159
−1.10	$\mu - 1.10\sigma$	.1357	.95	$\mu + .95\sigma$	.8280
−1.05	$\mu - 1.05\sigma$	.1469	1.00	$\mu + 1.00\sigma$	.8413

This means that the upper limit is 69.0 min at a 90 percent confidence level. And the true average population repair time is not more than 69 min, a statement that is either 100 percent correct or 100 percent wrong, but the odds favoring its correctness are 9 to 1. Since the specification limit for repair time is 70 min, it can be stated that a 90 percent confidence limit exists relative to meeting the specification requirement.

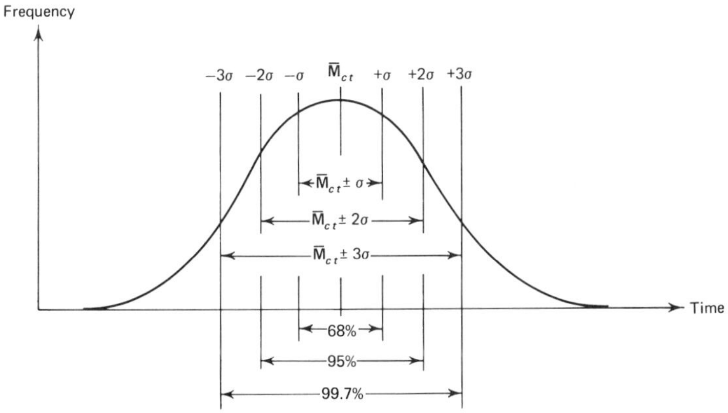

FIG. 4-4 Characteristics of Normal Distribution.

Additional Application of Standard Deviation. The normal distribution curve will always exhibit the relationships illustrated in Fig. 4-4.

In the event that it is desirable to know what percentage of the total repair actions fall within the range of 50 to 82 min, for example, it can be calculated by converting to multiples of standard deviation σ. In this instance, the interval of 50 to 82 equals $\bar{M}_{ct} \pm \sigma$ or 66 min plus or minus the original standard deviation of 16 min. Based on Fig. 4-4, it can be stated that 68 percent of the total population sampled will fall within the range of 50 to 82 min. Also in Fig. 4-4, it can be seen that 99.7 percent of the sample population will lie within the range of $\bar{M}_{ct} \pm 3\sigma$ or 66 ± 48 min. For simplicity, only whole multiples of standard deviation σ have been employed herein; however, through reference to a statistics handbook, the percentage of any segment of the normal distribution can be determined.

Considering a typical application, one might wish to determine the percent of total population of repair times that lies between 30 and 40 min. Graphically, this is represented in Fig. 4-5.

The shaded area under the curve represented by A_3 can be determined mathematically as follows:

$$A_1 = \bar{M}_{ct} - 30 \text{ min} = 36 \text{ min} \tag{6}$$

$$A_2 = \bar{M}_{ct} - 40 \text{ min} = 26 \text{ min} \tag{7}$$

See Eq. (2) for calculation of $\bar{M}_{ct}$.

Using the value 16 for standard deviation σ, areas A_1 and A_2 can be stated in terms of standard deviations.

$$A_1 = \bar{M}_{ct} - \frac{36}{16} = \bar{M}_{ct} - 2.25\sigma \tag{8}$$

$$A_2 = \bar{M}_{ct} - \frac{26}{16} = \bar{M}_{ct} - 1.62\sigma \tag{9}$$

See Eq. (3) for calculation of standard deviation.

With the areas expressed in this form, it is convenient to make use

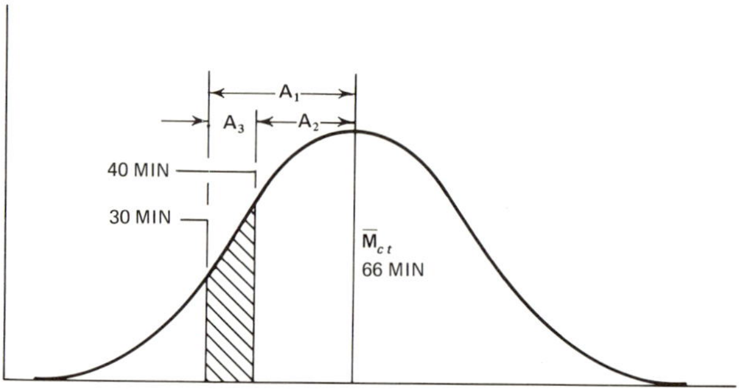

FIG. 4-5 Normal Distribution Sample.

of a statistical table for normal distribution. See Table 4-5. Table 4-5 values have been normalized to unity, which means that the entire area under the normal distribution curve is represented by $A = 1.000$ or 100 percent. Using Table 4-5 data

$$\text{Area } A_1 = \bar{M}_{ct} - 2.25\sigma = (\mu \pm 2.25\sigma) - 0.5000 = 0.9878 - 0.5000 = 0.4878 \tag{10}$$

$$\text{Area } A_2 = \bar{M}_{ct} - 1.62\sigma = (\mu \pm 1.62\sigma) - 0.5000 = 0.9474 - 0.5000 = 0.4474 \tag{11}$$

$$\text{Area } A_3 = \text{area } A_1 - \text{area } A_2 = 0.0404 \text{ (or } 4.04\%) \tag{12}$$

This means that 4.04 percent of the population of the maintenance times are estimated to lie between 30 and 40 min.

The Log-normal Distribution Curve

Table 4-6 presents data representing a sampling of 27 corrective-maintenance repair times (in minutes) on a typical electronic equipment item.

Following the same procedure as given in the preceding paragraphs, a frequency table is generated. This is presented in Table 4-7.

Based on the information presented in Table 4-7, a histogram is plotted. See Fig. 4-6 for the developed histogram.

TABLE 4-6 Corrective-maintenance Repair Times

61	71	85	41	105	57	80	67	65
33	98	41	78	66	112	47	53	39
68	102	35	55	120	57	45	29	57

TABLE 4-7 Frequency Distribution of Repair Times

Class Interval	Frequency	Cumulative Frequency
29 - 44	6	6
45 - 60	8	14
61 - 76	5	19
77 - 92	3	22
93 - 108	3	25
109 - 125	2	27

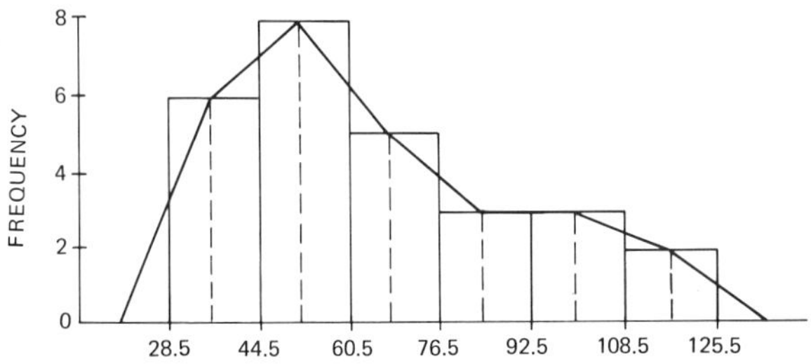

FIG. 4-6 Repair-time Histogram.

In this instance, if more data were added to the sample size, it is likely that the distribution curve would approach the log-normal shape as illustrated in Fig. 4-7.

Utilizing the data presented in Table 4-6, the median will give the best average location of the sample data. In a log-normal distribution, the median is also the geometric mean. The geometric mean can be computed through the data presented in Table 4-8.

$$\tilde{M}_{ct} = \text{geometric mean} = \text{antilog} \frac{\sum_{1}^{27} \log M_{ct_i}}{27}$$

$$= \text{antilog} \frac{48.216}{27} = \text{antilog } 1.786 = 61 \text{ min} \qquad (13)$$

When specifically defining the median of a corrective-maintenance log-normal distribution, the symbol used is $\widetilde{M}_{ct}$. For preventive maintenance, the symbol $\widetilde{M}_{pt}$ is used. Since only corrective-maintenance is being discussed in this example, the median will be represented by $\widetilde{M}_{ct}$.

On a log-normal distribution, the 95th percentile point is specified as

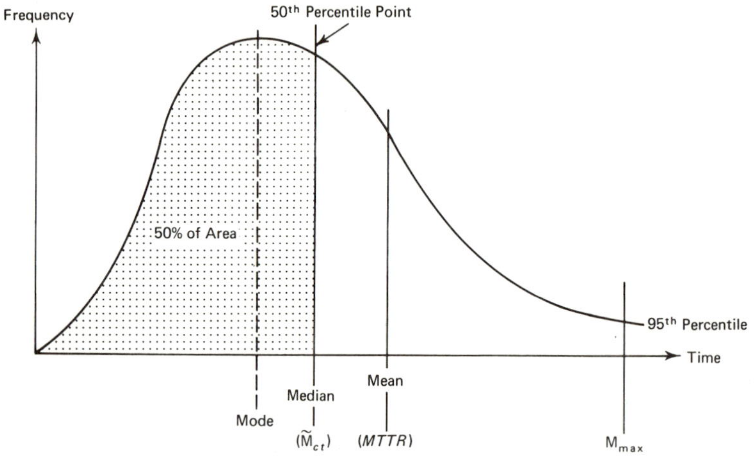

FIG. 4-7 Log-Normal Distribution.

TABLE 4-8 Geometric-mean Computations

M_{ct_i}	Log M_{ct_i}	(Log M_{ct_i})^2	M_{ct_i}	Log M_{ct_i}	(Log M_{ct_i})^2
61	1.785	3.186	112	2.049	4.198
71	1.851	3.426	47	1.672	2.795
85	1.929	3.721	53	1.724	2.972
41	1.622	2.631	39	1.591	2.531
105	2.021	4.084	68	1.832	3.356
57	1.756	3.083	102	2.009	4.036
80	1.903	3.621	35	1.544	2.384
67	1.826	3.334	55	1.740	3.027
65	1.813	3.287	120	2.079	4.322
33	1.519	2.307	57	1.756	3.083
98	1.991	3.964	45	1.653	2.732
41	1.622	2.631	29	1.462	2.137
78	1.892	3.579	57	1.756	3.083
66	1.819	3.308			
GRAND TOTAL: $\sum_{1}^{27} \log M_{ct_i} = 48.216$; $\sum_{1}^{27} (\log M_{ct_i})^2 = 86.818$					

M_{max}. Since maintenance times have been observed to be distributed log-normally, M_{max} (95th percentile) cannot be derived directly by using the observed maintenance times. However, by taking the logarithm of each of the repair values, the resultant distribution becomes normal, permitting utilization of the data in a normal manner. The formula for computing M_{max} is

$$M_{max} = \text{antilog}\,[\log \tilde{M}_{ct} + (1.645)\sigma_{\log M_{ct_i}}] \tag{14}$$

where $\log \tilde{M}_{ct}$ = the log of geometric mean (median), and $\sigma_{\log}\ M_{cti}$ = the standard deviation of the logarithms of all M_{ct} values. (Note that $\log \tilde{M}_{ct}$ is also equal to $\overline{\log M_{ct}}$.)

Before M_{max} can be computed using the above formula, it is necessary to determine $\sigma_{\log M_{ct_i}}$ as follows:

$$\begin{aligned} \sigma_{\log M_{ct_i}}{}^2 &= \frac{\sum_1^N (\log M_{cti})^2 - \left(\sum_1^N \log M_{cti}\right)^2/N}{N-1} \\ \sigma_{\log M_{ct_i}} &= \sqrt{\frac{86.818 - 48.216^2/27}{26}} \\ \sigma_{\log M_{ct_i}} &= 0.1658 \end{aligned} \tag{15}$$

Substituting the standard deviation value and $\tilde{M}_{ct}$ into Eq. (14)

$$\begin{aligned} M_{max} &= \text{antilog}\,[\log \tilde{M}_{ct} + (1.645)(0.1658)] \\ M_{max} &= \text{antilog}\,(1.786 + 0.272) \\ M_{max} &= \text{antilog}\,(2.058) \\ M_{max} &= 115 \text{ min} \end{aligned}$$

DERIVATION OF MAINTAINABILITY QUANTITATIVE FACTORS

The maintainability quantitative requirements of a system or equipment item may be specified in a number of ways. They could be set forth in the system-criteria statements in the form of (1) availability factors or (2) maintenance indices (e.g., *MTBM, MMH/OH*) or a combination of these forms of measurement. In any event, the maintainability engineer will likely be called upon to (1) evaluate the specific requirements, (2) determine the associated or dependent factors, (3) assess the impact of the quantitative requirement on the total program, and (4) define the risk in fulfilling the requirements. To accomplish this, the maintainability engineer must be familiar with the elements that compose each quantitative factor.

The maintainability quantitative forms of measurement are defined in Chap. 1. They will be discussed further in the following paragraphs to illustrate the relationship of the elements that compose each form of measurement.

Availability Factors

Inherent Availability (A_i). Inherent availability is a function of the reliability and the mean active corrective-maintenance-time characteristics of the system. It can be expressed as

$$A_i = \frac{MTBF}{MTBF + MTTR} \quad \text{or} \quad A_i = \frac{MTBF}{MTBF + \bar{M}_{ct}} \tag{16}$$

when

$$MTBF = \frac{1}{\text{failure rate}} \quad \text{or} \quad MTBF = \frac{1}{\lambda} \tag{17}$$

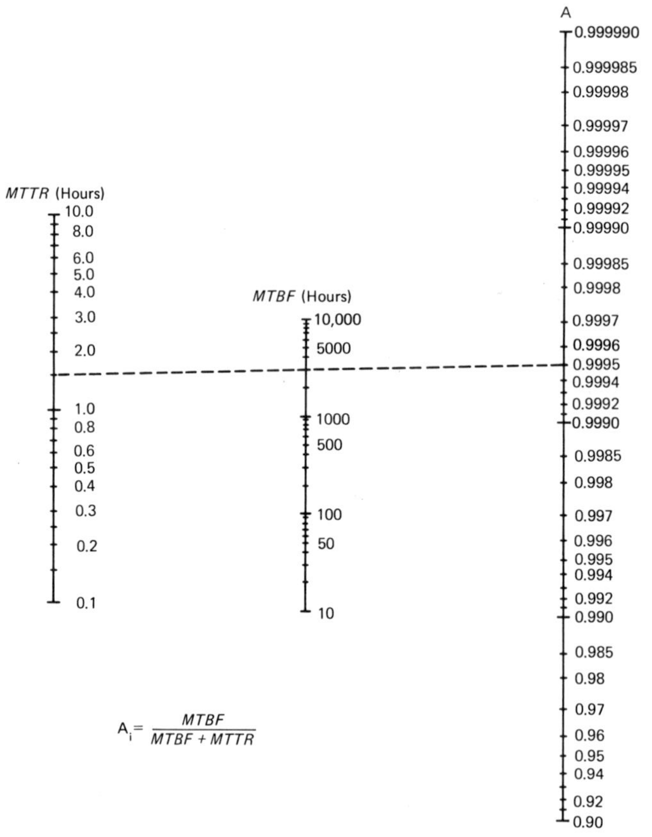

FIG. 4-8 Availability Nomograph.

Failure rate λ is the number of failures of the system per hour of operation. And

$$\bar{M}_{ct} = \frac{\sum_{1}^{N} M_{ct_i}}{N} \tag{18}$$

Achieved Availability (A_a). Achieved availability is a function of the frequency of maintenance and the mean maintenance time. It can be expressed as

$$A_a = \frac{MTBM}{MTBM + \bar{M}} \tag{19}$$

when

$$MTBM = \frac{1}{1/MTBM_u + 1/MTBM_s} \tag{20}$$

and

$$\bar{M} = \frac{\bar{M}_{ct} \times 1/MTBM_u + \bar{M}_{pt} \times 1/MTBM_s}{1/MTBM_u + 1/MTBM_s} \tag{21}$$

$MTBM_u$ is the mean interval of unscheduled maintenance and $MTBM_s$ is the mean interval of scheduled maintenance.

Operational Availability (A_o). Operational availability is a function of the frequency of maintenance, active maintenance time, waiting time,

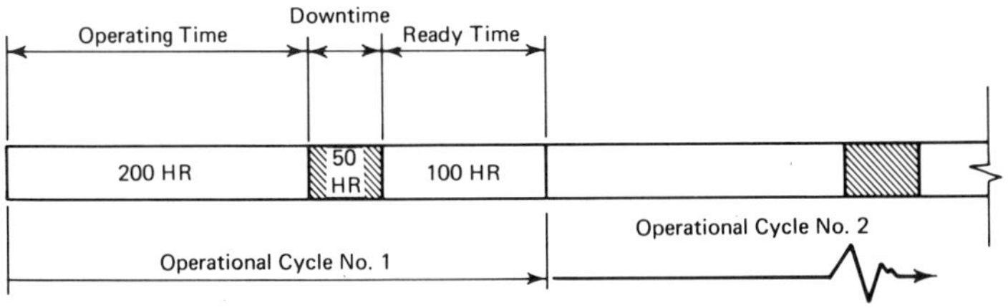

FIG. 4-9. Operational Cycle.

logistics time, administrative time, and the ready time of the system. It can be expressed as

$$A_o = \frac{MTBM + \text{ready time}}{(MTBM + \text{ready time}) + MDT} \tag{22}$$

when $MTBM$ is the mean interval of all maintenance requirements (see Eq. (20)), and $MDT = \bar{M}$ + mean waiting time + mean logistics time + mean administrative time.

and Ready time = operational cycle — ($MTBM + MDT$).

The waiting time, logistics time, and administrative time are equated with the frequency of maintenance.

The operational cycle is the total of all operating time, ready time, and downtime. This is illustrated in Fig. 4-9.

Therefore

$$A_o = 1 - \frac{MDT}{\text{operational cycle}} \tag{23}$$

Maintenance Indices

Maintenance Man-hours. The factor of maintenance man-hours per equipment or system operating hour is a function of the reliability and maintainability of the system. It can be expressed as

$$MMH/OH = \frac{\text{maintenance-labor hours per period}}{\text{operating time per period}} \tag{24}$$

Maintenance-labor hours is a relative term so it should be defined prior to specifying any element. It could mean some or all of that labor expended at the (1) organizational level of maintenance, (2) intermediate level of maintenance, or (3) depot level of maintenance.

Maintenance Man-minutes. This measurement equates to that of maintenance man-hours. The only difference is the unit of measure.

Cost Per Operating Hour. The factor of maintenance cost per equipment operating hour can be used to measure the effectiveness of an integrated logistics support program and to assess the design attributes of the system.

As with other maintenance indices, the maintenance-cost factor must be defined to have meaning.

The maintenance-cost factor can be expressed as

$$\text{Cost}/OH = \frac{\text{total support cost per period}}{\text{operating time per period}} \tag{25}$$

The support costs per period can include expenditures for the training of maintenance personnel, spares, utilities, depreciation of facilities and equipment, and labor.

Mean-time-between-maintenance. The *MTBM* is a significant maintenance index since this factor directly influences the measures of achieved availability and operational availability. In addition, the *MTBM* indirectly contributes to the measurements of the other maintenance indices discussed above. The expression of *MTBM* is conveyed in Eq. (20).

Total Maintenance Downtime. The *MDT* factor is a measure of the overall support program since it includes the elements of active-maintenance downtime, logistics time, and waiting or administrative time. It permits an assessment of one element in relationship to the total, and aids in establishing the problem areas of a support program. *MDT* is also significant from the standpoint that it directly contributes to the measurement of operational availability. Refer to Eq. (22).

Achieved Availability. In addition to being a prime measurement of a system, the achieved-availability factor is another significant maintenance index. This is due to the fact that its measurement is comprised of maintenance elements, namely, *MTBM* and $\bar{M}$. Refer to Eq. (19).

Maintenance Times

Item Repair Time R_P. The repair time is the total time required to perform the following maintenance actions in correcting an equipment malfunction due to the failure of one particular part:

1. Preparation (R_O)
2. Localization (R_L)
3. Isolation (R_I)
4. Disassembly (R_D)
5. Interchange (R_X)
6. Reassembly (R_R)
7. Alignment (R_A)
8. Verification checkout (R_V)

Item repair time may be expressed as

$$R_P = R_O + R_L + R_I + R_D + R_X + R_R + R_A + R_V \tag{26}$$

Refer to Chap. 1 for definitions.

Mean-time-to-repair MTTR. This is the arithmetic mean of the distribution of individual repair times R_P. *MTTR* is the summation of active repair times during a given period of time divided by the total number of malfunctions during the same time interval or

$$MTTR = \frac{\Sigma\lambda R_P}{\Sigma\lambda} \tag{27}$$

where λ is the failure rate of the item.

Active Maintenance Time per Corrective-maintenance Task M_{ct}. This expression is equivalent to repair time R_P.

Mean Active Corrective-maintenance Time $\bar{M}_{ct}$. This expression is equivalent to *MTTR* and is sometimes expressed as follows:

$$\bar{M}_{ct} = \frac{\Sigma\lambda M_{ct}}{\Sigma\lambda} \tag{28}$$

Active Maintenance Time per Preventive-maintenance Task M_{pt}. The active time consumed in the accomplishment of a single preventive-maintenance action.

Mean Active Preventive-maintenance Time $\bar{M}_{pt}$. The arithmetic mean of the distribution of individual preventive-maintenance tasks M_{pt}.

$$\bar{M}_{pt} = \frac{M_{pt_1} + M_{pt_2} + M_{pt_3} + \cdots + M_{pt_N}}{N} \tag{29}$$

where N is the number of preventive-maintenance actions.

Mean Active Corrective- and Preventive-maintenance Time $\bar{M}$. This index is established to represent all system active-maintenance downtime to include both corrective- and preventive-maintenance actions. See Eq. (21).

Equipment Repair Time ERT. Equipment repair time is the median of individual repair times. When repair times follow a normal distribution, the median is equal to the mean, and the $MTTR$ value will be a satisfactory measure of the true ERT. When repair times follow the log-normal distribution, the median and mean values no longer coincide. In this instance, ERT can be expressed as follows:

$$ERT = \frac{MTTR}{\text{antilog } (1.15\sigma^2)} \tag{30}$$

where σ is the standard deviation of the logarithms, to the base 10, of repair times. If the average value of σ is 0.55 (based on experience data), then

$$ERT = 0.45MTTR \tag{31}$$

Although the above relationship is used in many instances, it is better to calculate the ERT using the actual standard deviation.

Geometric Mean-time-to-repair $MTTR_G$. When repair times follow a log-normal distribution, the geometric mean-time-to-repair $MTTR_G$ occurs at the median; therefore, it is a true measure of ERT, and can be expressed as follows:

$$MTTR_G = \text{antilog} \left[\frac{\Sigma(\lambda \log R_P)}{\Sigma\lambda} \right] \tag{32}$$

Maximum Maintenance Time M_{max}. Maximum maintenance time M_{max} refers to the repair time represented at the 95th-percentile point on a log-normal distribution, and can be expressed as follows:

$$M_{max} = \text{antilog } [\log \tilde{M}_{ct} + (1.645)\sigma_{\log M_{ct_i}}]$$

where $\sigma_{\log M_{ct_i}}$ is the standard deviation of the logs of all M_{ct} values. See Eq. (15). For the purposes of comparison, the following relationship is generally acceptable:

$$\log M_{max} \approx \log \tilde{M}_{ct} + 0.5 \tag{33}$$

The Air Force specification MIL-M-26512 relates the derivation of this equation.

Maximum Equipment Repair Time ERT_{max}. This factor is defined as the maximum allowable value of median equipment repair time. Assuming the log-normal distribution, ERT_{max} can be related to M_{max} by the following expression:

$$ERT_{max} = 0.45\ M_{max} \tag{34}$$

This factor considers a standard deviation of 0.55. The above relationship is not precise; however, small errors resulting from the indicated approximation will not normally be significant.

Further information on the derivation of ERT_{max} is contained in the Navy document, NAVSHIPS 94324.

QUESTIONS AND PROBLEMS

1. What characteristics of a system determine its inherent availability?
2. What is a failure rate?
3. If the inherent availability is specified, what trade-off can be exercised to determine the associated quantitative factor?
4. The factor of maintenance man-hours per operating hour is based on the *MTBM* of the system and the labor hours expended on all maintenance tasks. True or false? Why?
5. What is the difference between maintenance man-minutes per operating hour and *MMH/OH*?
6. The *MTBM* factor is not a significant maintenance index. True or false?
7. The following corrective-maintenance task times were observed:

Task time (min)	*Frequency*
10	1
12	2
14	3
16	4
18	6
20	10
22	12
24	13
38	1
36	2
34	3
32	6
30	8
28	10
26	12

a. What is the range of observations?
b. What is the class interval?
c. Plot the data and construct a curve. What type of distribution is indicated by the curve?

d. What is the mean repair time?
e. What is the standard deviation of the sample data?
f. The specified mean repair time is 26 min, and the stated confidence level is 95 percent. Does the data reveal that the specification requirements will be met? Why?

8. The following corrective-maintenance task times were observed:

Task time (min)	*Frequency*
12	1
14	4
16	8
18	12
20	13
22	14
24	13
28	10
26	11
44	2
42	2
40	3
38	4
36	5
34	6
32	7
30	8

a. What is the range of observations?
b. What is the class interval?
c. Plot the data and construct a curve. What type of distribution is indicated by the curve?
d. What is the geometric mean of the repair times?
e. What is the standard deviation of the sample data?
f. What is the M_{max} value?
g. What is the *MTTR* value?

9. Given an *MTTR* of 30 min and an *MTBF* of 1,000 hr, what is the factor of inherent availability? (Use the availability nomograph, Fig. 4-8.)
10. With a specified inherent availability of 0.9990 and a calculated *MTBF* of 400 hr, what is the *MTTR* requirement?
11. Given the following data, calculate the factor of achieved availability:
a. $\bar{M}_{ct} = 0.5$ hr
b. $MTBM_u = 250$ hr
c. $\bar{M}_{pt} = 2.0$ hr
d. $MTBM_s = 1{,}000$ hr
12. Illustrate a typical operational cycle.
13. If an operational cycle consists of 1,000 hr, and the total downtime is 100 hr, what is the factor of operational availability?

FIVE

Maintenance Concept

The maintenance concept defines criteria covering repair philosophies, maintenance-support levels, personnel factors, maintenance-time constraints, etc. The concept serves two purposes:

1. It provides the basis for the establishment of maintainability requirements in system/equipment design. For example, if the repair philosophy dictates that no external support equipment is allowed at the organizational level of maintenance, then the equipment design must incorporate some provision for built-in test.

2. It provides the basis for the establishment of requirements for total support. The basic maintenance concept, supplemented by the maintenance analysis, leads to the identification of support requirements to include maintenance tasks, task frequencies and times, personnel quantities and skill levels, spare/repair parts, support equipment and tools, facilities, and other resources.

Complete fulfillment of both purposes in an effective and economical manner requires that the maintenance concept be developed at the inception of a program (conceptual phase) prior to the start of actual equipment design. The relative timing of concept definition in the overall system/equipment development cycle is illustrated in Fig. 5-1 (block

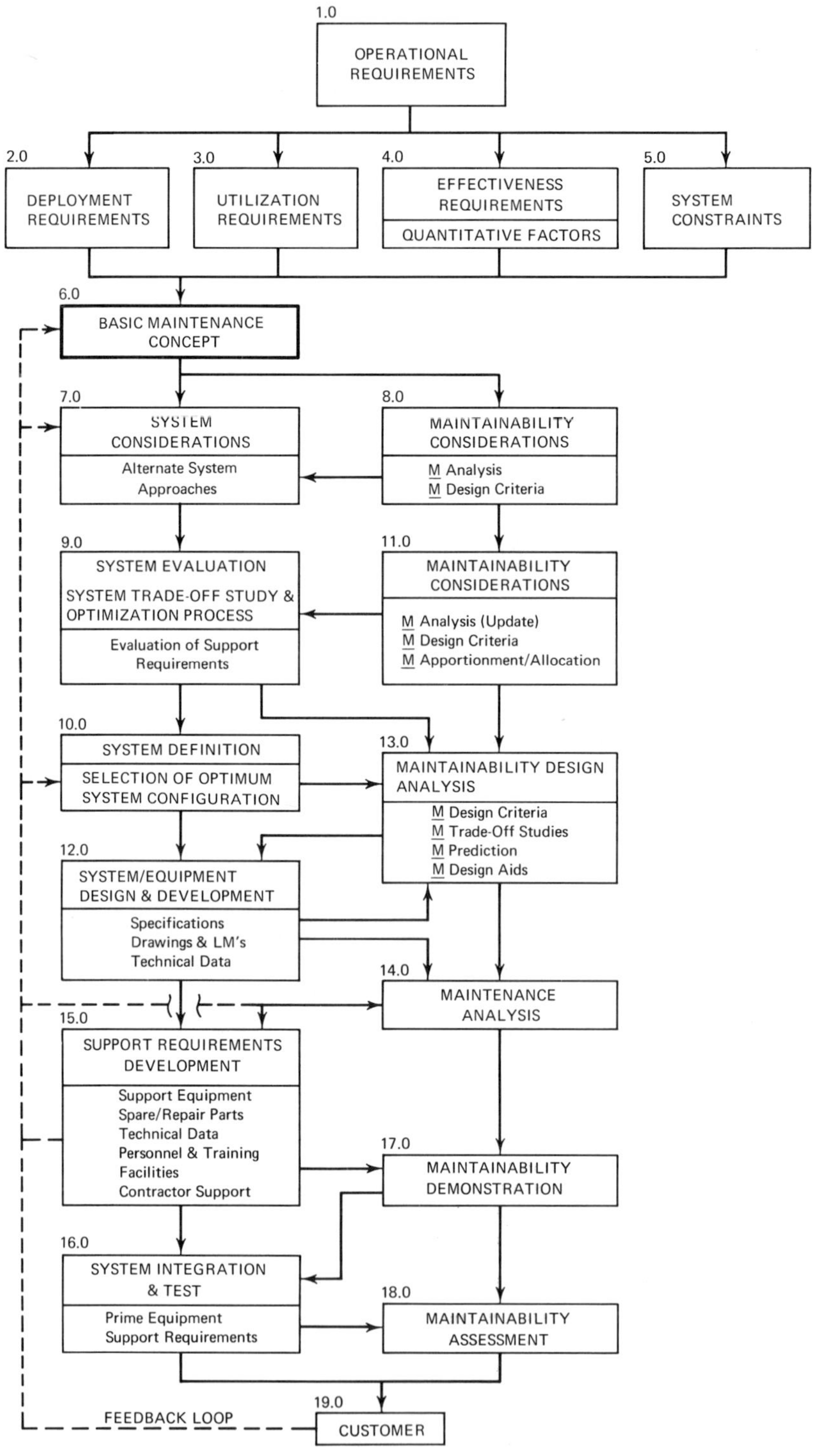

FIG. 5-1 System Development Cycle.

6.0). Initiation of concept development at this stage will tend to ensure that all functions of design and support are integrated and oriented to the *same* concept. If development of the maintenance concept is not considered at this stage in the system development cycle, individual subsystem-design personnel may reflect different approaches in their design (lack of standardization), and the various functions of support (technical data, spare/repair parts, support equipment, etc.) may not be oriented to an integrated approach. The results of such a situation are often quite costly.

MAINTENANCE CONCEPT DEVELOPMENT

The maintenance concept serves a twofold purpose: (1) it is a significant factor in formulating the maintainability requirements for a system to be developed; and (2) it provides the basis for supporting that system after it has been developed. Hence, the maintenance concept must be both realistically and sufficiently definitive to satisfactorily meet the needs of the system-design engineers and the requirements of the logistics-support planners.

Formulation of the maintenance concept for a new system must be systematically undertaken during the early conceptual or proposal phases of a program. It cannot be considered as a routine matter or as an action distinct from the establishment of system-design philosophies. The concept is basic to design work, and when the concept has been translated into maintainability requirements for the system, it may demonstrate the need for changes in an initially conceived design approach.

The first step in formulating the desired maintenance concept is to define a basic repair policy that will best meet the needs of a proposed new system. The repair policy will consider the support to be provided at various echelons or levels of maintenance (organizational, intermediate or field, and depot) as applicable. An understanding of each level is helpful to the maintenance-concept definition process.

Levels of Maintenance

Organizational Maintenance. Organizational maintenance encompasses tasks performed by the using organization on its own equipment. This maintenance consists of functions and repairs within the capabilities of authorized personnel, skills, tools, and test equipment. Organizational-level personnel are generally occupied with the operation and use of the equipment, and have minimum time available for detailed maintenance or diagnostic check-out. Usually, the least-skilled maintenance men are associated most closely with the operation of the equipment.

	ORGANIZATIONAL MAINTENANCE	INTERMEDIATE MAINTENANCE		DEPOT MAINTENANCE
		Mobile Units	Fixed Units	
Done Where?	Wherever the Equipment is Located	Mobile Unit Facility	Fixed Field Shop	Depot Facility
Done By Whom?	Equipment Operator	Squadron, Division, Ship Personnel		Theater or Zone of Interior Personnel
On Whose Equipment?	Own Equipment	Other Organization's Equipment		
Basis	Repair and Retain It	Repair and Return to User		Repair and Return to Stock Inventory
Type of Work Accomplished	Inspection Servicing Adjustment Minor Repairs and Modifications	Detailed Inspection Major Servicing Complicated Adjustments Major Repairs and Modifications Major Replacements Calibration Employing "Working Standards" Overload from Organizational Maintenance Level		Complicated Factory Type Adjustments Inspection, Overhaul, and Rebuild Complex Repairs and Modifications Calibration Employing "Primary and "Secondary" Standards Overload from Intermediate Maintenance Level

TABLE 5-1 Major Echelons of Maintenance

Maintenance at the organizational level is normally restricted to periodic checks of equipment performance, cleaning of equipment, front panel adjustments, and the removal and replacement of some components. Personnel at this level usually do not repair the removed components, but forward them to the intermediate level. Mobility requirements generally limit the amount of tools, test equipment, and supplies available at the organizational level. The design engineer can, therefore, expect to find limited personnel skills and related support at this level and should plan equipment maintenance and servicing requirements accordingly.

Intermediate Maintenance. Intermediate- or field-maintenance tasks are performed by mobile, semimobile, or fixed specialized organizations and installations.

Mobile or semimobile units are often designated to provide close support to combat troops and facilitate tactical operations. Maintenance is limited to the repair of end items or unserviceable assemblies in support of using organizations on a return-to-user basis.

Although the mobility requirement limits the maintenance tasks that can be undertaken, the available maintenance personnel are generally more skilled and better equipped than those at the organizational level and are charged with performing more-detailed maintenance. At this level, end items may be repaired by replacement of major modules or assemblies. These units are authorized larger quantities of spare parts and maintenance equipment than the using organization, and are often deployed to support more than one organizational entity.

Fixed installations (permanent shops) are generally established to support lower levels of maintenance (i.e., organizational) within specified geographical areas. These units perform maintenance tasks that cannot, due to limited personnel skills and test equipment, be accomplished by mobile units. High personnel skills, an additional test-equipment capability, and better facilities are provided at this level. Equipment repair is generally accomplished to the small module or piece part level. Rapid maintenance turnaround-time requirements are not as imperative here as at the lower levels of maintenance.

Depot Maintenance. The depot constitutes the highest level of maintenance, and provides support for maintenance tasks above and beyond the capabilities provided at the intermediate level. Depot maintenance-level organizations are fixed and mobility is not a problem. Equipment of extreme bulk and complexity can be employed if required. The high volume possible in depot shops lends itself to effective use of assembly-line techniques. This, in turn, permits the use of relatively unskilled labor for the greater part of the workload, with a concentration of highly skilled specialists in key positions.

The depot level of maintenance provides facilities for completely overhauling and rebuilding equipment as well as the performance of highly complex maintenance actions. In addition, the depot provides the necessary standards for equipment-calibration purposes. This level of maintenance also serves as a major supply base for intermediate-level maintenance facilities. Depot maintenance facilities are generally located remotely from individual theaters of operation and provide services for several such theaters.

The three levels of maintenance, as discussed above, are summarized in Table 5-1.

System Operational Requirements

In the selection of a repair policy to be adopted in formulating the maintenance concept for a new system, the basic frame of reference is systems operational requirements. System operational requirements, generally defined through a feasibility study, include:

1. A general description of the system relative to basic performance parameters. The system description is generally supported by operational functional-flow diagrams.

2. A definition of system-deployment requirements (Fig. 5-1, block

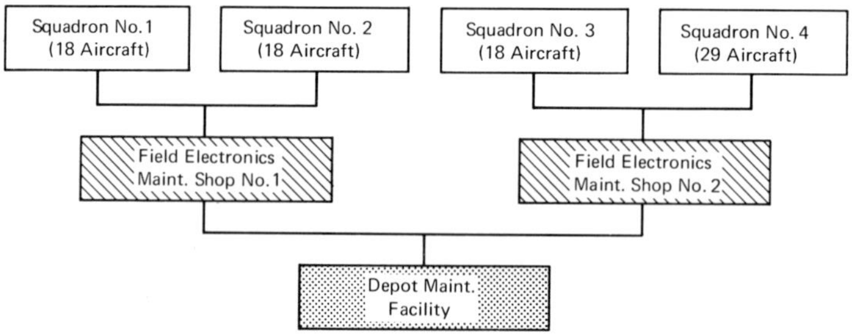

Operational Factors

No. of *XYZ* Systems per Aircraft -- 1
No. of Squadrons -- 4
No. of Field Maintenance Shops -- 2
No. of Training Facilities -- 1 (Location -Squadron No. 4)
Average Distance
Squadron - Field Maintenance Shop - - 100 miles
Field Maintenance Shop - Depot -- 1,000 miles
Operational Time Period -- 10 years
System Operational Availability -- 95%
Flight Hours/Aircraft
Peacetime - 30 Hours per A/C per month (Squadrons 1 and 3)
Wartime: 60 Hours per A/C per month (Squadrons 2 and 4)
XYZ System Utilization - - 1.3 *XYZ* Operational hours per A/C Operational Hour
Average Aircraft Mission Duration -- 2 hours
Crew to Aircraft Ratio -- 1.5:1

FIG. 5-2 Operational Requirements.

2.0). Such data cover quantity of operational installations, number of systems per installation, location of operational installations and distances from support facilities, operational environments (are systems installed outside or in a sheltered environment?), etc.

3. A definition of system-utilization requirements (Fig. 5-1, block 3.0) to include the system operational cycle (on-time versus off-time), operational time per cycle for peacetime requirements, and operational time per cycle for combat conditions. The system operational cycle may be expressed in terms of months, days, or operating hours.

4. An identification of effectiveness requirements or system quantitative factors (Fig. 5-1, block 4.0). Such requirements may be expressed in terms of single factors or a combination of availability, operational readiness, dependability, *MTBF*, *MTTR*, maintenance man-hours per operating hour, etc.

5. An identification of system constraints (Fig. 5-1, block 5.0) covering specific operational limits. For instance, it may be specified that organizational maintenance is to be accomplished by the operator and that system operation is possible with a single operator within a particular career field of a specified skill level.

Figure 5-2 illustrates a typical operational-deployment posture.

Definition of system operational requirements is one of the first steps in the system development cycle and it is accomplished during the conceptual phase of a program. As a general rule, all conceptual-phase effort is completed by the customer. However, in certain instances, industry may be contracted to assist the customer in performing operational analyses, feasibility studies, etc. In any event, system operational requirements, as covered herein, must be defined from either known or specified data in order to effectively proceed in the development of the basic maintenance concept.

Repair Policies

There are a number of possible repair policies applicable to a system design. A system may be designed to be nonmaintainable, partially repairable, or fully repairable.

1. A nonmaintainable item, usually modular in construction, is a unit (when it malfunctions) not repaired but replaced by a like spare. If this policy is selected, design criteria promote a positive self-test capability plus the hermetic sealing or encapsulation of all item components. The positive self-test capability ensures that good units are not discarded. The sealing protects against humidity and corrosion, and tends to improve item reliability. In this case, there is no requirement for internal accessibility, plug-in subassemblies, test points, or additional maintainability-design criteria. In addition, there is no requirement for

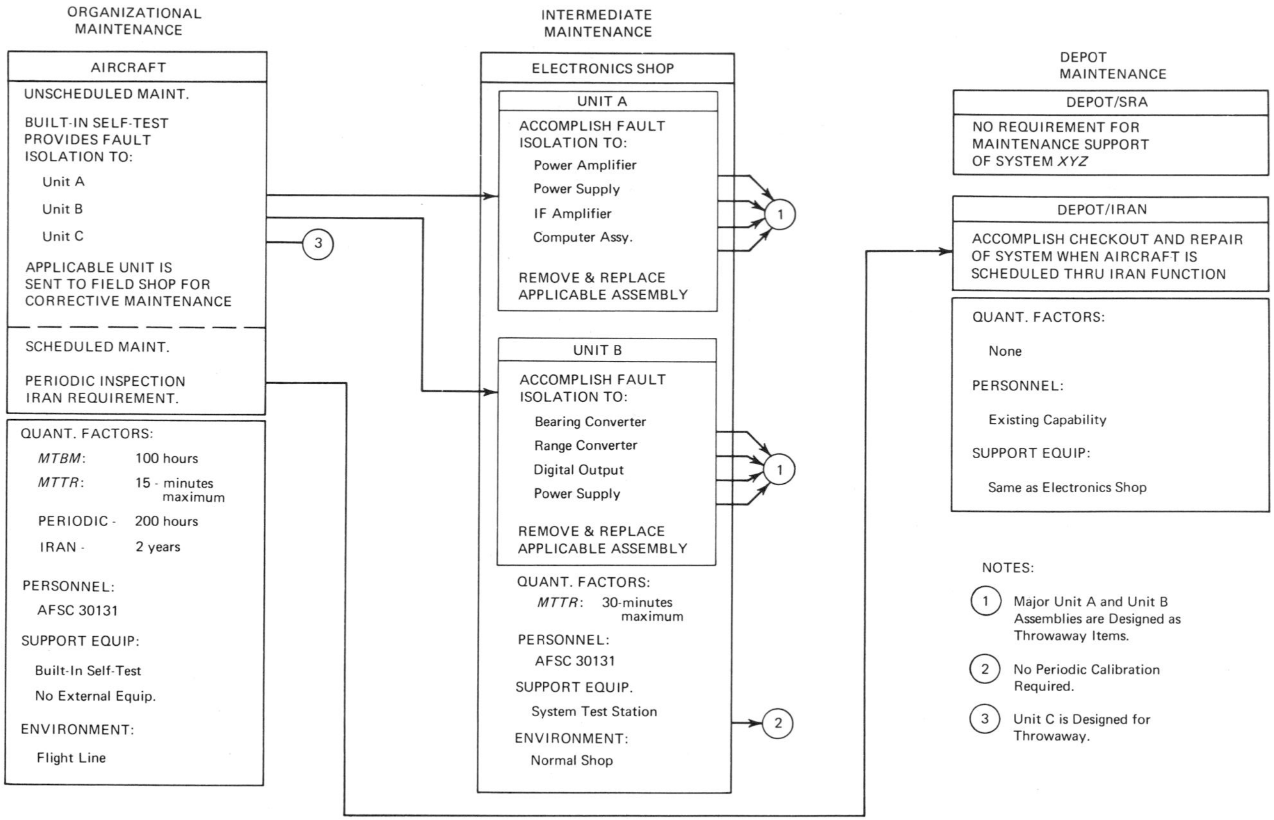

FIG. 5-3 *XYZ* **System Maintenance-concept Flow. Alternate 1.**

ancillary test equipment, specific personnel skills, or elaborate maintenance-support facilities and resources.

2. A partially repairable system may be considered in varying degrees. For the purpose of discussion, two alternates are covered.

Alternate 1, as illustrated in Fig. 5-3, considers the two major system units (i.e., unit *A* and unit *B*) of the basic system as being repairable through removal and replacement of applicable assemblies such as the power amplifier, the power supply, or the bearing converter. The assemblies are designed as throwaway items.

Alternate 2, as illustrated in Fig. 5-4, considers a more detailed level of repair since the major assemblies (e.g., power amplifier, power supply, bearing converter are considered repairable through removal and replacement of subassemblies or piece parts. Subassemblies such as circuit boards are considered nonrepairable.

As noted earlier, selection of the repair policy stems from the system operational requirements. Referring to Figs. 5-3 and 5-4, both policies reflect the same operational requirements. For example, the system operational availability (i.e., 95 percent, as specified in Fig. 5-2) may dictate a mean downtime requirement of such short duration that it can be met only by providing for maximum repair capability at the organizational level. Since the personnel skills and equipment available at the user's organizational level are limited, a need for designing the system for easy and positive fault identification is established. In addition, a requirement has been identified for rapid removal and replacement of the applicable item once the fault has been identified. Thus, the specific operational requirements at the organizational level promote the incorporation of built-in self-test plus system design for modularization. At the intermediate level of maintenance, a different requirement exists and is designed primarily to support the organizational maintenance requirements necessary to meet system operational needs. Combining both sets of requirements results in the identification of specific maintainability-design criteria as listed below:

a. The system shall incorporate a positive built-in self-test capability providing fault isolation to unit *A*, unit *B*, or unit *C*.

b. The three units (units *A*, *B*, and *C*) shall be monitored and easily removed and replaced within 10 min without the use of ancillary test equipment or special tools.

c. The three units shall contain test points to enable fault isolation to the assembly level within a given period of time (in order to meet the shop level 30-min *MTTR* requirement).

d. Major assemblies (e.g., power amplifier, power supply, bearing converter) shall be modularized, plug-in, and directly accessible. Like functional modules will be interchangeable without the

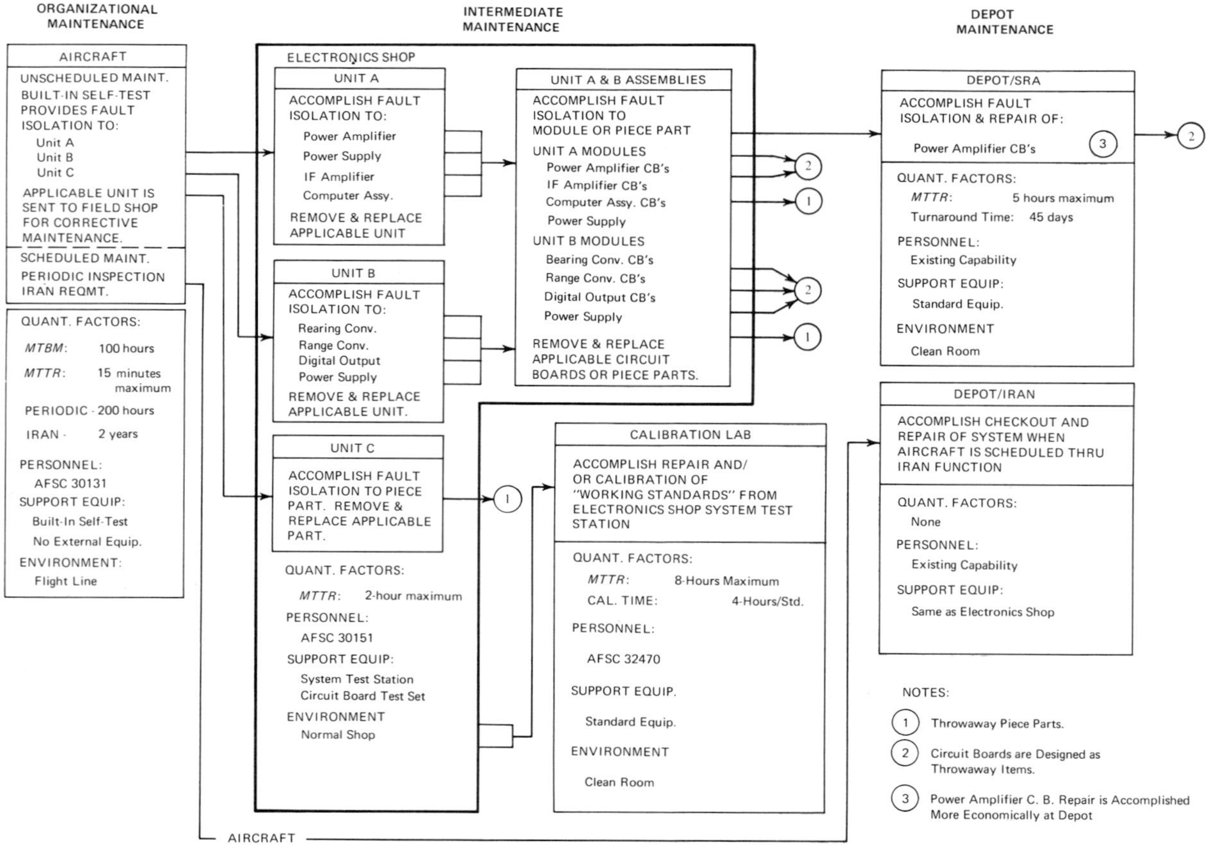

FIG. 5-4 *XYZ* **System Maintenance-concept Flow. Alternate 2.**

requirement of maintenance alignments or adjustments subsequent to the interchange action.

In regard to alternate 1 (Fig. 5-3), specific design criteria need not be specified to a higher level. The major subassemblies of units *A* and *B* are considered throwaway and may be hermetically sealed or encapsulated (nonmaintainable item). For alternate 2 (Fig. 5-4), the same type of maintainability-design criteria listed above should be applied to the assemblies of unit *A* and unit *B* since these items are considered repairable. In the area of support, alternate 2 requires more test equipment, more personnel with higher skill levels, more data, and more facilities and resources than alternate 1.

3. A fully repairable system promotes the level of maintenance one step beyond the repair policy illustrated for alternate 2 in Fig. 5-4. In other words, individual subassemblies or circuit boards within unit *A* and unit *B* are deemed repairable. In this instance, maintainability-design criteria, similar to the areas covered above, are applicable to the subassembly level. This policy reflects the greatest amount of support requirements in terms of personnel and training, tools and test equipment, spare/repair parts, data, and facilities.

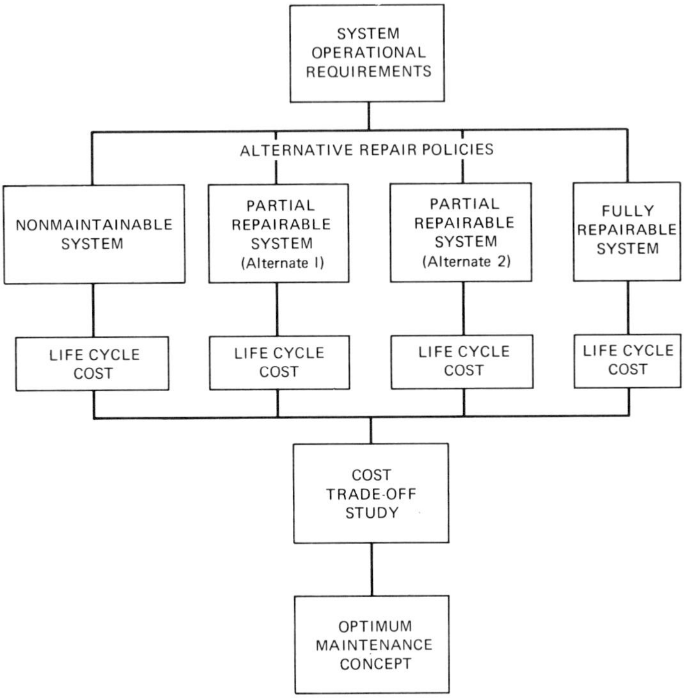

FIG. 5-5 Maintenance-concept Optimization.

As an aid in establishing the various repair policies, the following questions may be employed as a guide to ensure complete coverage:

1. What are the quantitative maintenance/maintainability requirements at the various levels of maintenance in terms of *MDT, MTBM, MTTR,* maintenance man-hours per operating hour, preventive-maintenance frequency, turnaround times, etc.?
2. What are the anticipated scheduled and unscheduled maintenance functions for each level of maintenance?
3. What are the estimated personnel types and skill levels for each level of maintenance?
4. What are the basic constraints associated with support-equipment requirements definition?
5. What are the facility requirements for each level of maintenance?
6. What are the maintenance environments (e.g., transportation, storage, working conditions) associated with each level of maintenance?

The customer may elect to closely control system design and development by providing detailed criteria to ensure compatibility with an existing operational requirement. On the other hand, the customer may allow the contractor to recommend a specific repair policy. Such a recommendation evolves through the maintenance-concept optimization process.

Maintenance-concept Optimization

In the development of a maintenance concept for a given new system, it is possible (through the analysis of all relevant information) that two or more repair policies will satisfactorily meet all system operational requirements. When this occurs, the optimum approach is determined only by making a trade-off study evaluating the various alternative approaches in terms of cost. The basic trade-off approach in terms of the possible alternative repair policies discussed is illustrated in Fig. 5-5.

The accomplishment of the actual trade-off study, in terms of methodologies and techniques, is covered in Chap. 7 of this text. When completed, the trade-off study will identify the concept reflecting the lowest total life-cycle cost.

SUMMARY

Development of the maintenance concept is one of the most important program functions. It is from this concept that maintainability-design requirements are determined and, consequently, from which all support criteria and requirements evolve. Concept development must be relatively complete (at least to the organizational and intermediate levels of maintenance) during the conceptual and/or proposal phases of a program. Concept development is facilitated and supplemented through such efforts as maintainability analyses and system trade-off studies.

QUESTIONS AND PROBLEMS

1. The maintenance concept is the basis for establishing maintainability-design requirements and total support requirements. True or false?
2. The maintenance concept should be developed immediately after design definition. True or false?
3. The maintenance concept defines the functions that will be performed at each level of maintenance. True or false?
4. What would you consider in the allocation of maintenance tasks?
5. What are the three main levels of maintenance? Briefly describe each.
6. Personnel skill requirements are higher at the organizational level than at the other levels of maintenance. True or false?
7. The equipment operator always performs the maintenance tasks assigned to the organizational level of maintenance. True or false?
8. The repair policy for a specific item remains the same regardless of what system it may be in. True or false?
9. More than one maintenance concept may complement the operational needs of a system. True or false?
10. Illustrate two possible maintenance concepts for each of the following items:

    a. Military vehicle (truck or jeep)
    b. Military field telephone
    c. Military field radio (vehicle-mounted)
    d. Circuit-board assembly
    e. Power-supply assembly

11. Outline the factors that will probably determine the life-cycle cost of any particular item.
12. The development of a maintenance concept is only pertinent to military systems. True or false?
13. Develop a maintenance concept for a major home appliance.
14. Why does the *MTTR* factor influence the maintenance-concept definition?
15. Why does the *MTBM* factor influence the maintenance-concept definition?

SIX

Maintainability Analysis

The maintainability-analysis function constitutes the process of translating system-level operational and maintenance requirements (as specified in the customer's RFP) into specific qualitative and quantitative equipment-design requirements. Basically, the analysis provides:

1. Comprehensive documentation of specific maintainability requirements to be met in designing the system
2. A list of checkpoints (design criteria) to ensure that the appropriate features necessary to meet the above maintainability requirements are incorporated
3. A summary of basic maintenance functions and support requirements specified for the applicable system

The maintainability analysis, based on the definition of operational and maintenance functions and on the maintenance concept, is accomplished during the proposal and/or definition phases of a program prior to the start of actual equipment design. The relative timing of maintainability-analysis development in the overall system/equipment development cycle is illustrated in Fig. 6-1 (blocks 8.0 and 11.0). As such, the analysis serves as the starting point for maintainability system design.

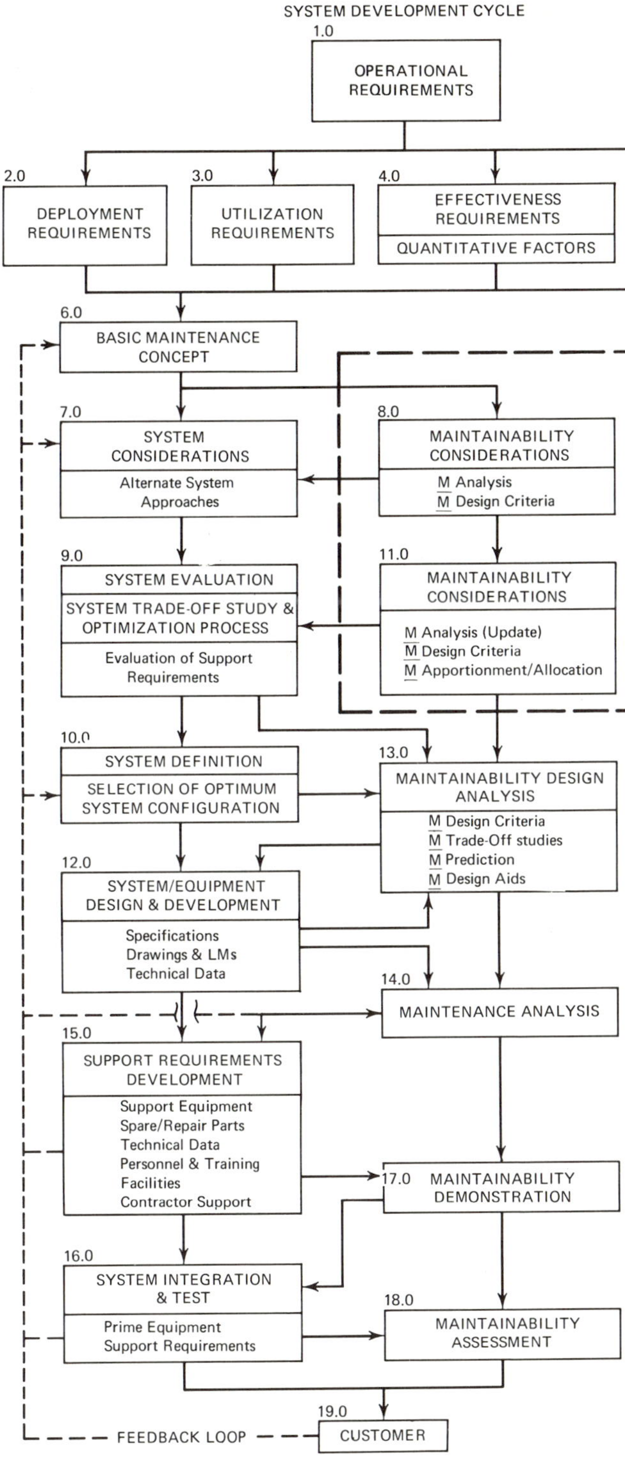

FIG. 6-1 System Development Cycle.

The purpose of this chapter is to review the maintainability-analysis process commencing with the development of operational functional-flow diagrams and extending through the allocation of specific maintainability quantitative factors to equipment assemblies and the identification of basic design criteria.

FUNCTIONAL ANALYSIS

An essential element of the overall systems-engineering process is the employment of a functional approach as a frame of reference for the identification of initial requirements at each design stage of the system. The functional approach, which defines basic operational and maintenance functions, assures:

1. That all elements of the system are fully recognized and defined
2. A means of relating equipment-packaging concepts and support requirements to given functions
3. A common reference point for all subsequent engineering system design and development effort

Operational Functions

The initial step in the system-engineering process consists of formulating a functional description of the system. The functional description represents a gross level portrayal of the functions which must be met to satisfy total system requirements. This is the genesis of detailed requirements. Once basic system functions have been defined, they are translated into functional-flow diagrams. A typical example of a top-level system operational functional-flow diagram is illustrated in Fig. 6-2. For the purpose of discussion, the system covered will henceforth be noted as the *XYZ* system.

Referring to Fig. 6-2, the major operational functions of the *XYZ* system are specified, and the respective blocks are numerically identified for reference purposes. Each major function, in turn, is then analyzed and supported through the identification of subfunctions as indicated in Fig. 6-3.

Maintenance Functions

Once system operational functions are specified, the systems-engineering process leads into the development of gross maintenance functions. For instance, for each operational function identified there are specified performance requirements (signal level, tolerances and accuracies, etc.). A check of the applicable function will indicate either a *GO* or *NO-GO* decision. A *GO* decision leads to the check of the next function. A

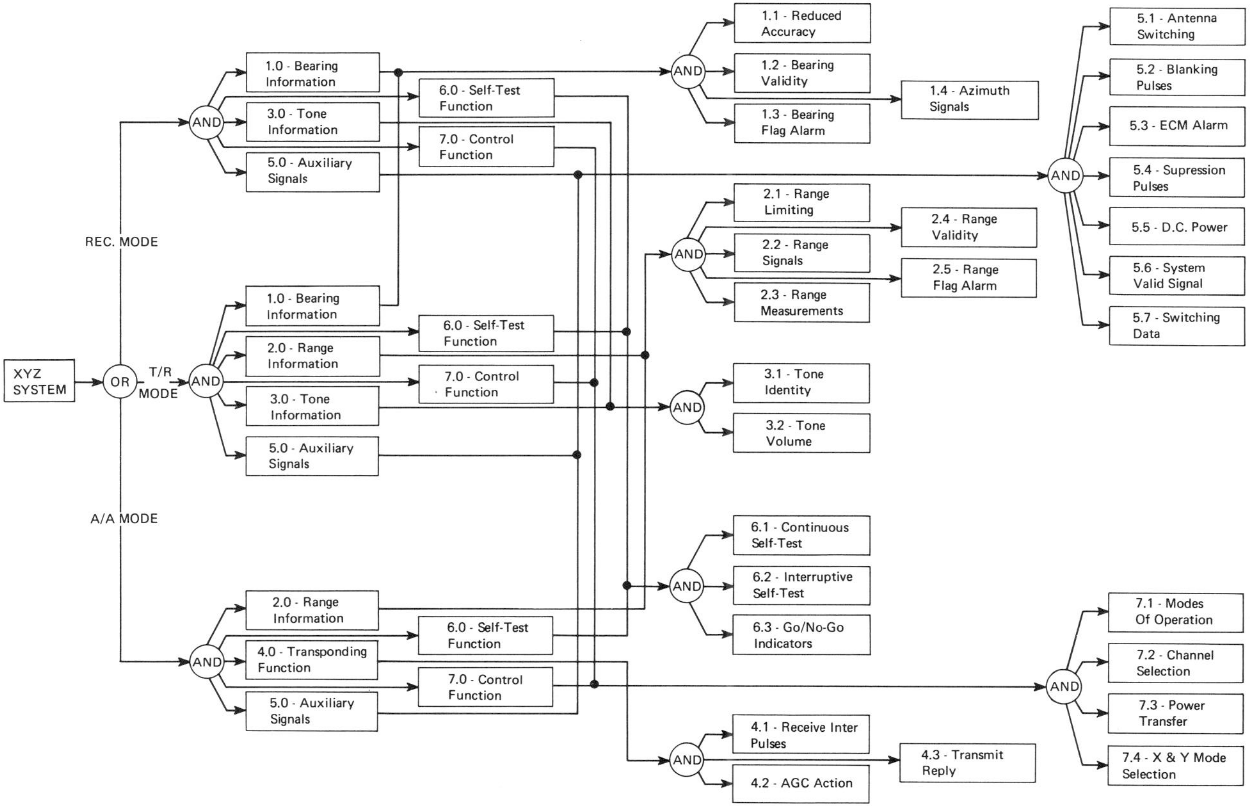

FIG. 6-2 *XYZ* **System Operational Functions (Stage 1).**

NO-GO indication (i.e., symptom of malfunction) provides a starting point for the development of detailed maintenance functional flows and logic troubleshooting diagrams. Maintenance functions identified at this point will reflect the consideration of customer-specified system readiness and downtime requirements as well as available maintenance resources. In many cases, only an estimate of first-level maintenance functions

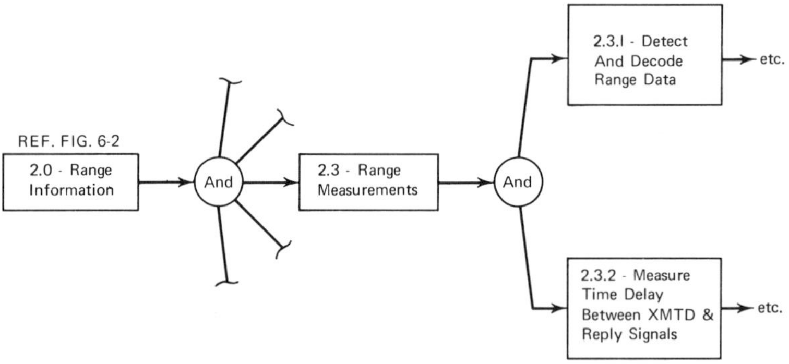

FIG. 6-3 Range-information Functions.

can be made, since the availability of preliminary engineering data is limited. As the systems-engineering process continues, the basic design configuration is established through trade-off studies of various alternative packaging concepts, unit sizes and weights, etc. Gross system-level reliability and maintainability requirements are considered in arriving at a basic system-packaging scheme.

Considering the desirability of packaging by function in conjunction with system size and weight constraints (as specified by the customer), the packaging scheme for the *XYZ* system results in three basic units (units *A*, *B*, and *C*). An illustration of the operational functions incorporated in each of the three units is presented in Fig. 6-4. The functions presented are the same major system functions as identified in Fig. 6-3. Developing the design concept further, an analysis of the function contained within each unit will lead to the identification of major assemblies. For the purpose of illustration, it is assumed that unit *A* contains a power amplifier, a power supply, an i-f amplifier, and a computer assembly. The details (e.g., assembly size, weight, quantity of parts) relative to each assembly are not known at this time. Having progressed thus far, the basic maintenance concept (discussed in Chap. 5) is updated to reflect the above system-packaging approach in terms of customer-specified maintainability qualitative and quantitative requirements. Figures 5-3 and 5-4 reflect two alternate concept approaches based on system-design information. The gross level maintenance functional flows can be expanded to the degree illustrated in Fig. 6-5.

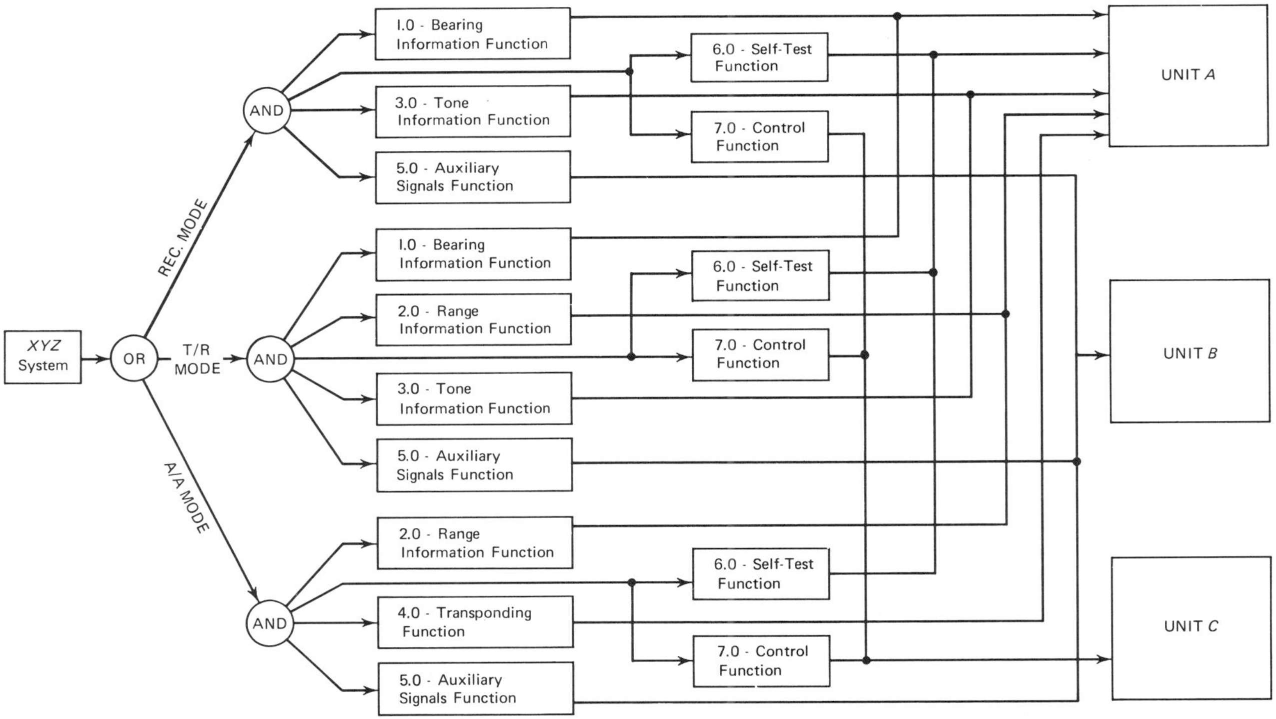

FIG. 6-4 *XYZ* **System Operational Functions (Stage 2).**

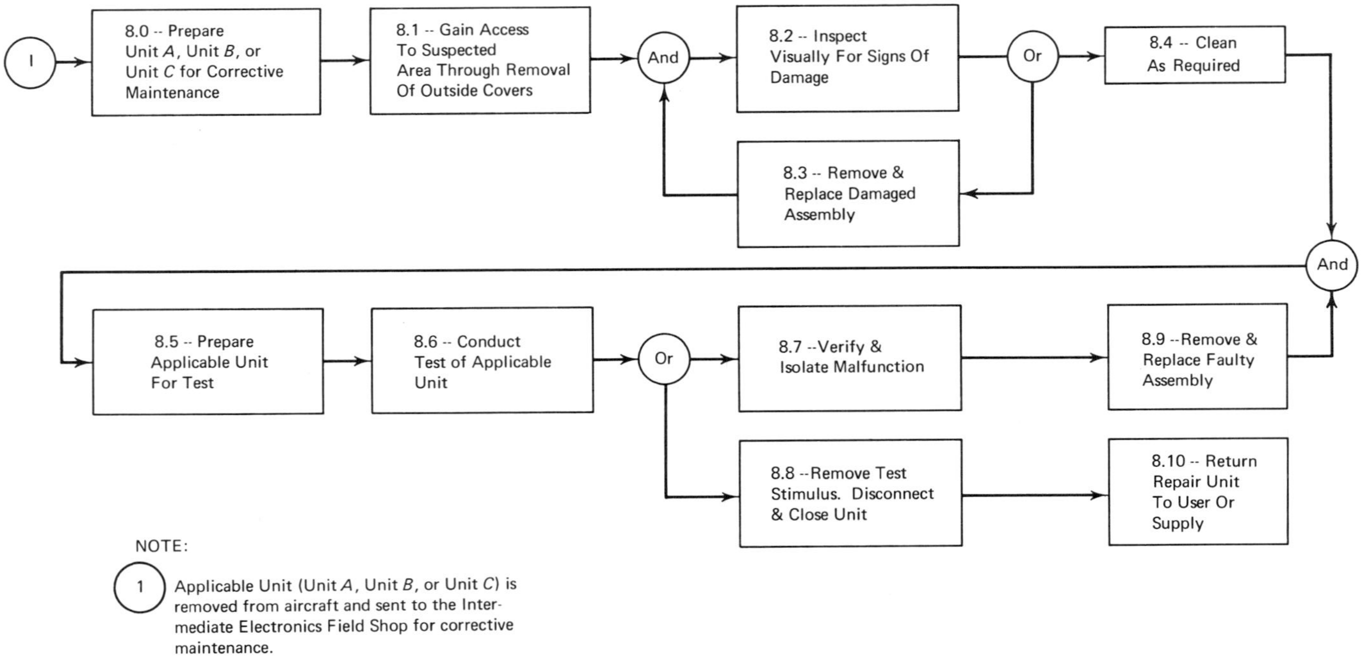

FIG. 6-5 Unit-assembly Maintenance Functions.

The maintenance-concept/functional-flow development process is continued throughout the definition phase.

MAINTAINABILITY ALLOCATION

As stated earlier, the intent of the maintainability analysis is to translate system functional-analysis and maintenance-concept data into maintainability controls in the form of qualitative and quantitative requirements. If quantitative requirements (e.g., availability factor) are not specified in the customer's RFP, they must now be determined.

When the maintainability-program plan has been prepared and functional-analysis and maintenance-concept data have been updated, design and performance specifications for the system are prepared. These specifications become the primary means whereby maintainability requirements operate as constraints on the system designer. Accordingly, the maintainability engineer must ensure that relevant maintainability requirements are incorporated in every specification. In order to accurately determine the quantitative maintainability requirements for each system or equipment item, the maintainability engineer must budget the *MTTR* of the various equipment items such that the statistical mean is equal to or less than the overall system *MTTR* requirement. Budgeting is accomplished by means of maintainability allocation.

Maintainability-allocation Procedure

An allocation consists of determining the contribution of active downtime of each equipment item, and the evaluation of the contributions for all such equipment items against the established *MTTR* for the system. A typical problem exercise is presented below.

For the purpose of illustration, it is assumed that the *XYZ* system is required to meet an inherent availability requirement of 99.89 percent and a reliability *MTBF* requirement of 450 hr. The *MTTR* equation is

$$MTTR = \frac{MTBF(1 - A_i)}{A_i} \tag{1}$$

and by substituting the specified values

$$MTTR = \frac{450(1 - 0.9989)}{0.9989} = 0.5$$

Thus, the system's *MTTR* requirement is 0.5 hr.

The next step is to prepare a functional system diagram as illustrated in Fig. 6-6 which shows the three proposed major units of the *XYZ* system and the four major assemblies of unit *A*. The objective is to allocate an *MTTR* quantitative value against each indentured level of the equipment.

The allocation process is facilitated through the use of a form similar to that presented in Table 6-1. Referring to Table 6-1, each type of unit of the *XYZ* system is listed and the quantity Q of units per system is indicated. The reliability engineer provides an allocated failure rate λ against each unit and the applicable data is entered in the third column. The extent to which the failure rate of each unit contributes to the overall failure rate (represented by C_f) is the product of the failure rate and the quantity of each item type. The ratio of unit failure-rate contribution to the total-system failure rate is equal to C_f divided by the sum of all unit failure rates ΣC_f. The resultant value is entered in the fifth column of Table 6-1. The mean corrective-maintenance time $\bar{M}_{ct}$ allocated for each unit is entered in the sixth column. These times are apportioned initially by anticipated order of equipment complexity (indicated by failure rate). Repair times are the result of qualitative maintainability features incorporated into the design. Due to the fact that such design features are not known at this time, the failure-rate information must be the criteria for estimating $\bar{M}_{ct}$. Ideally,

TABLE 6-1 *XYZ* **System Allocations**

1 Item	2 Quantity Per System Q	3 Failure Rate λ (1,000 HRS)	4 Contribution Of Total Failures $C_f = (Q)(\lambda)$	5 Percentage Contribution Of Total Failures $C_t = C_f/\Sigma C_f$	6 Average Corrective Maint. Time $\bar{M}_{ct}$ (HR)	7 Contribution Of Total Corrective Maint. Time $C_m = (C_f)(\bar{M}_{ct})$
1. UNIT A	1	1.71	1.71	0.76	0.5	0.855
2. UNIT B	1	0.48	0.48	0.21	0.4	0.192
3. UNIT C	1	0.06	0.06	0.03	0.8	0.048
TOTAL			2.25	1.00		1.095

$$MTTR = \frac{C_m}{C_f} = \frac{1.095}{2.250} = 0.487 \text{ Hours (Requirement: 0.5 Hours)}$$

units or equipment items which contribute the highest percentage to the total failures will require a low $\bar{M}_{ct}$, and those with low contributions will require a higher $\bar{M}_{ct}$. On certain occasions, however, it may be feasible to alter this approach somewhat owing to costs as long as the end result (*MTTR* at the system level) falls within the specified quantitative requirement.

The next step in the allocation process involves the calculation of C_m, the contribution of each unit to the total system corrective-maintenance time. The formula for C_m is

$$C_m = C_f \bar{M}_{ct} \tag{2}$$

when C_f represents contribution to overall unit failure rate, and $\bar{M}_{ct}$ is the mean corrective-maintenance time (assigned).

The value of C_m for each unit is entered in the seventh column. The sum of these contributions for all units can then be entered into the equation for determining the overall system's $MTTR$ as follows:

$$MTTR = \frac{\Sigma C_f \bar{M}_{ct}}{\Sigma C_f} = \frac{\Sigma C_m}{\Sigma C_f} = \frac{1.095}{2.250} = 0.487 \tag{3}$$

In the example presented in Table 6-1, the computed $MTTR$ for the system is within the initial system requirement of 0.5 hr. The $\bar{M}_{ct}$ values for units A, B, and C provide a downtime goal for each unit design, and the values are included in equipment-design specifications.

Once allocation is accomplished at the unit level, the resultant $\bar{M}_{ct}$ values can be allocated to the next lower equipment indenture level. For instance, the 0.5-hr $MTTR$ value for unit A can be allocated to the power amplifier, power supply, i-f amplifier, and computer assembly. The method of allocation is accomplished in accordance with the same procedure as employed above. An example of allocated values for the assemblies of unit A is presented in Table 6-2.

TABLE 6-2 Unit *A* Allocations

UNIT *A*						
1. Power Amplifier	1	0.55	0.55	0.33	0.4	0.224
2. Power Supply	1	0.25	0.25	0.14	0.4	0.096
3. IF Amplifier	1	0.05	0.05	0.03	0.3	0.015
4. Computer Assembly	1	0.86	0.86	0.50	0.6	0.516
TOTAL			1.71	1.00		0.851

$MTTR = \frac{C_m}{C_f} = \frac{0.851}{1.71} = 0.498$ Hours (Requirement: 0.5 Hours)

Following the completion of quantitative allocations for each indenture level of equipment, all values are entered in the functional system diagram illustrated in Fig. 6-6. The functional system diagram provides an overall view of total system requirements.

Revision of Maintainability Allocations

In the event that the $MTTR$ allocation (indicated in Table 6-1) for the system exceeds the system's $MTTR$ goal of 0.5 hr, the maintainability engineer must revise his allocations. Three possible approaches may be considered. They are a decrease in reliability failure rates through reliability improvement (selection of a different design), a decrease in corrective-maintenance times through tighter requirements on diagnostics, accessibility, modularization, etc., or a decrease in both through the process of trade-offs. Any one of the above approaches is feasible. When this occurs, the allocation process often results in a series of proposed alternatives, the best of which is selected through cost trade-offs.

MAINTAINABILITY CRITERIA

With the identification of the maintenance concept and maintenance functions, and the accomplishment of a maintainability allocation, it is possible to generate detailed design criteria. Such criteria constitute specific maintainability goals in the areas of standardization, interchangeability, degree of self-test features, quantity and placement of test points, throwaway versus repair levels, modularization, accessibility, cables and connectors, safety features, labeling, etc. These criteria may be stated qualitatively or quantitatively, and employed as guidelines for the design

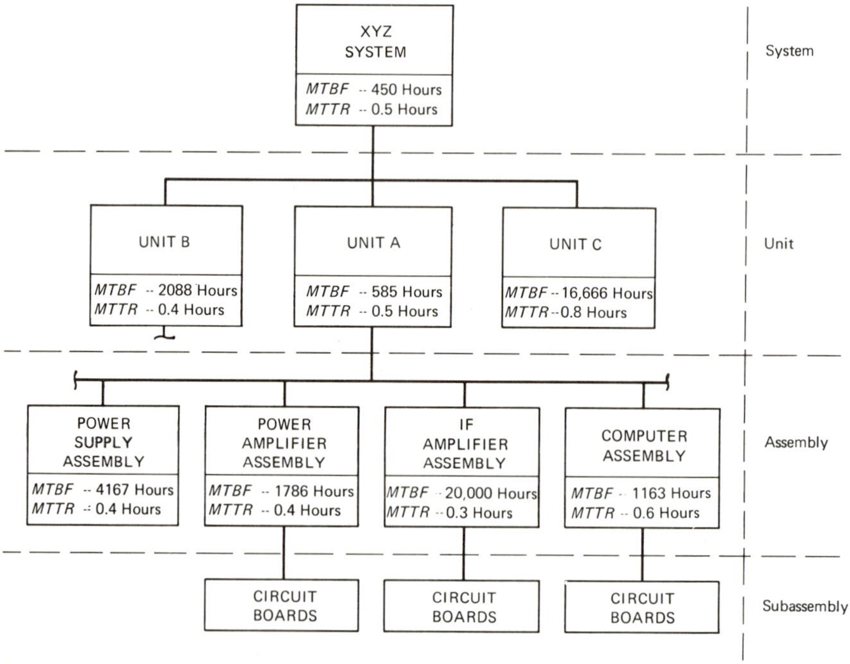

FIG. 6-6 Functional System Diagram.

engineer. Qualitative criteria must support the quantitative goals derived through the maintainability allocation.

In regard to the development of design criteria, the example of unit *A* of the *XYZ* system (refer to Fig. 6-6) can be cited. The allocated *MTTR* of unit *A* is 0.5 hr. This means that in the event of malfunction, the maintenance technician must be able to accomplish the complete maintenance cycle on the unit (localization, isolation, disassembly, remove and replace, reassembly, alignment and/or adjustment, and verification checkout) within 30 min. Based on experience, corrective-maintenance time for the average maintenance cycle on electronic equipment is spent in the proportions indicated in Fig. 6-7. The localization and isolation functions generally account for about 60 percent of the total expended corrective-maintenance time. Assuming that the percentage is valid for the *XYZ* system, an estimate of the allowable localization and isolation time for unit *A* would be 18 min. Meeting the 18-min

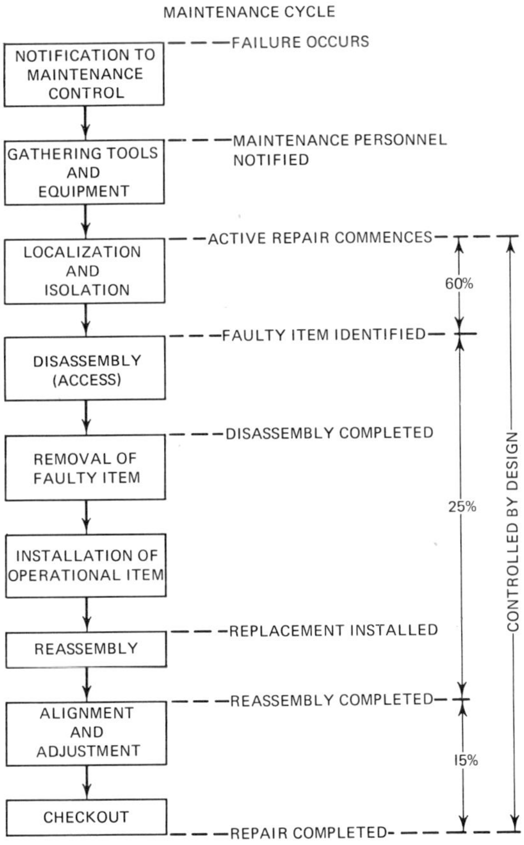

FIG. 6-7 Typical Maintenance Cycle.

goal would require the availability of a minimal quantity of readily accessible test points to allow positive fault isolation to any one of the four assemblies of unit *A* (power amplifier, power supply assembly, i-f amplifier assembly, computer assembly). The exact quantity of test points depends on the degree of functional packaging (number of functions packaged in each of the four assemblies).

Troubleshooting and test-point analysis stem from the maintenance functional-flow diagrams. Evaluation of the applicable functional flows against the time goal will lead to the identification of quantity and placement of test points.

Assuming that the time percentages in Fig. 6-7 are characteristic, the disassembly, removal and replacement, and reassembly functions must be accomplished within 0.125 hr or 7½ min. This infers that major assemblies (e.g., power amplifier) must be directly accessible and must not require the removal of another assembly to gain access. In addition, the assembly should be plug-in and/or should be held to the base plate by not more than four quick-release-type fasteners. (Connectors should be of the quick-disconnect variety.)

Relative to the assembly itself, the question of whether the item should be designed for discard-at-failure or for repair is raised. Although detailed design information is not available at this point in time, gross-level reliability and maintainability estimates (through the allocation process) of each assembly have been made. In addition, by similarity with like items in other already-existing systems, the maintainability engineer can estimate anticipated support requirements. With this information at hand, a trade-off study can be accomplished to determine the best approach (refer to Chap. 7 for an example of a discard-at-failure versus repair trade-off). If the decision is discard, each assembly should be a module (sealed). Accessibility, internal test points, etc., are not necessary. If the decision is repair, then specific maintainability-design criteria should be established.

Employing an approach similar to that discussed above, the maintainability engineer, by analyzing functional-analysis and quantitative data, is in a position to derive initial criteria for all elements of the system to be developed. Criteria should comply with the basic principles of maintainability within the quantitative goals stated to assist the design engineer in producing a system which will ultimately meet the requirements of the customer. As system design and development progresses and more-detailed data is made available, the maintainability engineer must reevaluate the design criteria. Maintainability-design criteria may be modified to reflect additional needs.

Once design criteria have been defined, the maintainability engineer assumes the role of monitoring design progress against the specified design criteria. This monitoring process is accomplished through day-to-day design-liaison activity, maintainability-prediction tasks, and periodic design reviews.

QUESTIONS AND PROBLEMS

1. The maintainability analysis is performed during the acquisition phase of a program. True or false?
2. Describe the relationship of the maintainability-analysis function and the maintainability system-design function.
3. A functional analysis of a system permits definition of all operational and maintenance functions. True or false?
4. What is the purpose of an operational functional-flow diagram?
5. What is the purpose of a maintenance functional-flow diagram?
6. What is the relationship of the operational and maintenance functional-flow diagrams?
7. What is a *maintainability allocation*? How is it accomplished?
8. As a general rule, a downtime allocation is greatest for an item with a high failure rate. True or false?
9. In your own words, define *failure rate*.
10. How do the failure rates of the system items influence the *MTTR* of the system?
11. Given the following information, calculate the *MTTR* of the system comprised of items *A* thru *E*.

Item	Failure rate	$\bar{M}_{ct}$ (min)
A	40	15
B	60	20
C	100	10
D	2	90
E	4	60

12. In reference to the system items listed in problem 11, above, what item will require the most attention as regards maintainability design? Why?
13. The maintainability allocation for a specific item should be entered into the design specification. True or false?
14. What is the purpose of maintainability-design criteria?
15. The specified design criteria should complement the quantitative requirements. True or false?

SEVEN

Trade-off Techniques

The trade-off is a technique of decision making. When considering the increasing complexities and resultant high costs of our operational systems and the importance of making correct decisions, one can no longer depend strictly on intuitive processes. A poor decision can be extremely detrimental in terms of performance, cost, time and material. Therefore, the need for formal trade-offs has evolved to provide evidence that all related factors have been systematically and adequately considered prior to making a decision.

TRADE-OFF APPLICATIONS

The trade-off technique can be applied to a variety of situations. In the business world, the top-management executive can employ the methodology in making a decision for a large industrial firm on whether to expand by purchasing a new plant facility or by modifying an existing facility. To the average individual, the technique can be applied to decide on the purchase of an automobile (e.g., Volkswagen versus Cadillac), etc. As regards maintainability, the trade-off methodology is commonly applied to a number of areas to include:

1. *Man versus machine*—Should the operational/maintenance function be accomplished by man or machine?

2. *Built-in versus portable test equipment*—Should the test capability be an inherent characteristic of the prime equipment or should it be external and portable?

3. *Semiautomatic versus manual*—Should the system be designed for semiautomatic operation or should it be manually operated?

4. *Repair versus throwaway*—Should the subassembly or module be designed such that repair can be readily accomplished or should it be designed for discard upon failure?

5. *Reliability versus maintainability*—Given an availability requirement, should the equipment be enhanced through improvement of the reliability characteristics in design or through the maintainability characteristics in design?

6. *Intermediate-level maintenance versus depot-level maintenance*—Is it more economical to support the system or equipment item at the intermediate level or depot level of maintenance?

7. *Operational spares versus support equipment*—During system deployment is it more economical to carry unit spares and no support equipment, or spares and support equipment, or a mixture of unit spares, component spares, and support equipment?

8. *Modular versus nonmodular packaging*—Should the equipment contain plug-in modules or piece parts?

9. *Make or buy*—Is it feasible to make the item, or should the item be purchased from an outside source?

The above applications are typical. It can be seen that the problems or alternatives stated are quite varied. In each case a decision must be made, and in each case the trade-off–study methodology is applicable. However, the specific criteria, parameters, weighting values, etc., relative to trade-off implementation are different and must be tailored to the problem at hand.

The application of trade-off studies with respect to specific program phases is covered in Chaps. 2 and 3. Basically, trade-off–study evaluations at the system level are accomplished during the conceptual and proposal phases of a program. Trade-offs at the subsystem level are normally conducted during the proposal and definition phases, while trade-offs at the detailed level will in all probability be accomplished late in the definitions phase or early in the acquisition phase. In any event, the accomplishment of trade-off studies must be early in the program cycle to support design decisions in a timely and effective manner.

TRADE-OFF APPROACH

Definition of Problem

Unless the problem is clearly and precisely defined, meaningful evaluation parameters and quantitative figures cannot be established. In many instances, although the nature of the problem appears to be obvious, precise definition of the problem may be the most difficult part of the entire process. In any event, the problem must be stated as precisely as possible before proceeding with the trade-off.

Proposed Alternative Solutions

The second step in the process is to identify possible alternative solutions to the problem. Initially, it may be that some of the possible alternatives are by no means clear at the outset and may be overshadowed because of personal preferences. To ensure against inadvertent omissions, list *all* possible solutions.

On the other hand, the more alternatives that are considered, the more complex the trade-off–study analysis becomes. Therefore, having listed all of the possible alternatives, the engineer or analyst may proceed to eliminate those that are clearly unattractive, leaving only a few for evaluation. The outcome may result in the evaluation of anywhere from two (the simplest situation) to five alternative courses of action.

Establishment of Trade-off Parameters

There are a number of parameters involved in accomplishing a trade-off study. Such parameters will vary depending on the level and complexity of the trade-off. For example, at the top system level, parameters of primary importance include cost effectiveness, system effectiveness, and operational availability. At the detailed level, the order of parameters will be different. This is illustrated in Fig. 7-1. Once all of the applicable parameters are established, the analyst should review each parameter from the standpoint of relevance or degree of importance. The parameters may be weighted to emphasize degree of importance (heaviest weighting for most important items). The application of weighting factors will depend on the type of trade-off.

Evaluation Process

Data Generation and Application. The identification of trade-off parameters leads to the definition of requirements for data generation and application. For instance, when comparing two major systems, it is necessary to consider operational deployments (quantity and locations of systems in the field), maintenance concepts, utilization rates, anticipated frequency and extent of system-maintenance requirements, etc. In such cases, a preliminary operations analysis, supported with prediction data, may be required. At the lower level of trade-off, input-data requirements may constitute a comparison of performance characteristics, detailed equipment predictions, and determination of support requirements. When cost-effectiveness evaluations are required, the identification of cost categories and quantitative cost factors is necessary. In any event, data requirements are numerous and varied. Major information sources required for data generation include customer RFPs, specifications, special study reports, engineering drawings, and lists of material.

Analytical Processes. Fulfillment of the evaluation process can often be accomplished through computerized-model applications, particularly where trade-offs are relatively complex in nature (involving a number of variables). Where manual procedures are time consuming, the use of computerized models, generated to simulate operational conditions, results in considerable time savings as well as added flexibility when reruns are required due to data changes.

In developing a model, the first step is to generate a comprehensive list of system/subsystem parameters that will describe the situation being simulated. Second, it is necessary to generate a matrix portraying the

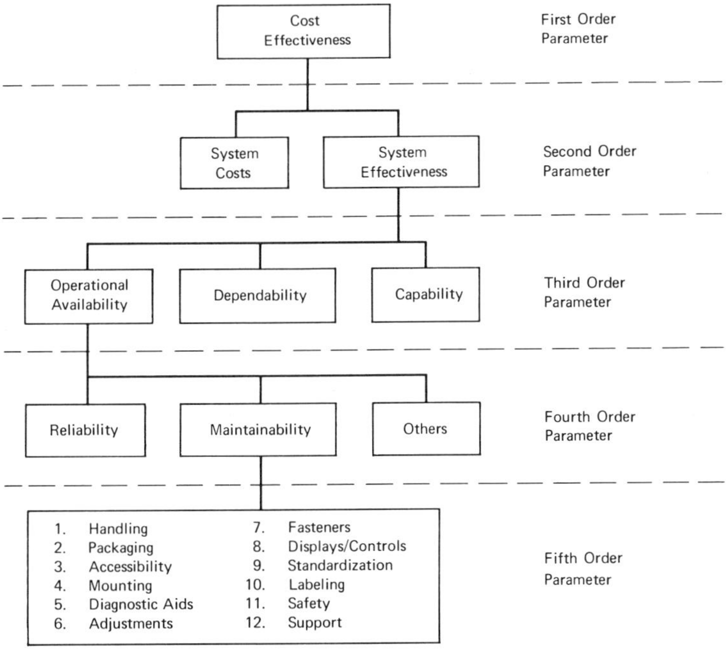

FIG. 7-1 Order of Trade-off Parameters.

direct relationships between the trade-off parameters. Each parameter should be analyzed with respect to every other parameter to determine the magnitude of relationship. In addition, each parameter should be analyzed to define the input and feedback-loop requirements as illustrated in Fig. 7-2. And finally, testing is required to validate the computer program. The development of a *standard model* for trade-off

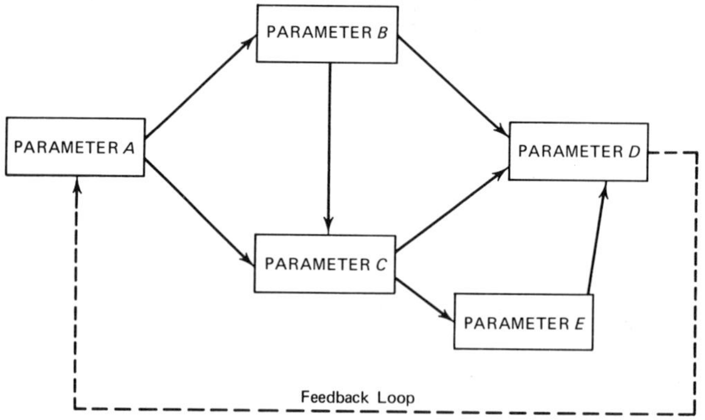

FIG. 7-2 Model Relationships.

accomplishment is impossible since the basic system-parameter relationships vary extensively from one situation to the next. A separate computer program is required for each trade-off application.

Decision

Having undertaken the above steps, the analyst has two choices: to seek additional information, or to arrive at a decision based on the information at hand. Without question, most decisions could be improved with additional data. However, the acquisition of data requires effort and is time consuming. Therefore, there comes a point in time when the analyst concludes that it is better to act than to defer a decision until more data has been collected.

The decision, with supporting assumptions, parameters, weighting values, and a description of the evaluation technique, should be documented in a trade-off–study report.

Validity of the Trade-off

In accomplishing trade-off studies, care must be taken to eliminate bias during any stage of the process. Trade-off studies involving a single design problem should be conducted independently by more than one person. The greater the number of evaluators, the more accurate and unbiased the trade-off study will be.

TYPICAL TRADE-OFF SUMMARIES

As a means of illustrating the trade-off process, six examples have been selected that represent typical problems that will confront a maintainability engineer. The topics of discussion are:

1. Life-cycle–cost study of two avionic subsystems of a new aircraft
2. System study of three different configurations of support equipment (support-system selection)
3. Reliability-maintainability trade-off study
4. Throwaway versus repair study
5. Evaluation of maintenance concepts
6. Evaluation of three computers

Life-cycle–cost Study. This example illustrates a cost-effectiveness evaluation where maintainability is one of many considerations. The example as presented emphasizes the total systems approach.

1. DEFINITION OF THE PROBLEM
   A new aircraft in the design and development phase requires the incorporation of a tactical air-navigational system (TACAN). A decision is required as to the type of system deemed most feasible from the standpoint of total life-cycle cost.
2. PROPOSED ALTERNATIVE SOLUTIONS
   a. A micro-TACAN system employing integrated circuits, microelectronics, and advanced packaging techniques.
   b. A conventional TACAN system employing electronic tubes and standard packaging techniques.
3. TRADE-OFF APPROACH
   a. The accomplishment of a cost-effectiveness evaluation requires the selection of cost categories. Cost categories not only cover initial investment costs but include operational (sustaining) costs. All applicable cost categories are listed in Table 7-1.
   b. Once major cost categories have been established, the next step is to determine the detailed costs associated with each category. In determining costs, certain specific factors are required. Table 7-2 lists some of these factors.
   c. The next step involves the generation of operational and maintenance data. Operational data covers quantity of aircraft, operational deployment, TACAN system-utilization rates, system-availability requirements, etc. Figure 7-3 illustrates the operational-deployment posture. Subsequent to defining basic operational factors, the maintenance concept and detailed maintenance policy must be defined (to the maximum degree

TABLE 7-1 Cost-category Checklist

INITIAL INVESTMENT COSTS	OPERATIONAL (SUSTAINING) COSTS
A1. Design & Development includes: contractor management & control, conceptual studies, prime equipment design and development, support equipment design and development, and technical (engineering) data.	B1. Replenishment (Spare/Repair Parts) a. Prime equipment b. Support equipment c. Training equipment
A2. Manufacturing a. Prime equipment b. Support equipment c. Prime equipment spare/repair parts d. Support equipment spare/repair parts	B2. Annual Facility Maintenance a. Field Maintenance b. Depot Maintenance
A3. Integrated Testing includes contractor and customer qualification-type testing.	B3. Technical Data includes technical manual revision and maintenance data.
A4. Facilities includes operational and maintenance facilities.	B4. Modification Kits
A5. Technical Data Includes: engineering drawings, technical manuals, and logistics data	B5. Support Equipment
A6. Personnel & Training a. Initial training b. Training equipment & spares	B6. Personnel & Training a. Direct personnel – prime equipment & support equipment b. Indirect personnel – prime equipment & support equipment c. Replenishment training
A7. Initial Support Services a. Initial inventory management b. Field service support c. Interim contractor depot maintenance capability	B7. Annual Support Services a. Sustaining inventory management b. Contractor services
A8. Transportation (first destination) a. Packaging b. Shipping	B8. Annual Transportation a. Prime equipment b. Support equipment c. Spare/Repair parts

TABLE 7-2 Maintenance-cost Factors

A. PERSONNEL & TRAINING – ORGANIZATIONAL & FIELD MAINTENANCE

1. Direct Labor Cost for Maintenance: $3.62/Hour. Includes Base Pay, Quarters Allotment, Clothing Allowance, Separate Ration, Retirement Annuity.

2. Burden Rate for Direct Supervision: $1.80/Direct Labor Hour.

3. Cost of Training (Including Indirect Costs): $1.60/Week/Person.

B. PERSONNEL & TRAINING – DEPOT MAINTENANCE

1. Direct Labor Cost for Maintenance: $2.75/Hour

2. Burden Rate for Direct Supervision, Non Technical Support Personnel, Buildings, and Test Equipment: $6.13/Hour.

C. SUPPORT SERVICES (SUPPLY & CONTRACTOR MAINTENANCE)

1. Cost of Introducing a Non-FSN Line Item Into the Supply System: 20% of Average Annual Inventory Value.

2. Cost of Maintaining a Line Item in the Supply System: 20% of Average Annual Inventory value.

3. Depot Handling Costs: $14/Repairable Item Handled By Depot.

4. Cost of Expendable Material (solder, wire, etc.): 0.5% of Total Labor Cost of Repair Action.

5. Cost of Contractor Field Service Support: $800/Man-Month.

D. SUPPORT EQUIPMENT

1. Burden Rate for Support Equipment Use: $0.60/Direct Labor Hour for Maintenance.

E. FACILITIES

1. Burden Rate for Facilities: $0.19/Direct Labor Hour for Maintenance.

F. TRANSPORTATION

1. Motor & Rail Express: $5.00/100 Pounds.
2. Air Express: $65/100 Pounds.
3. Packaging: $1.70/Pound

G. TECHNICAL DATA

1. Overall Cost of Technical Manuals: $200/Page.
2. Documentation Cost/Repair Action: $14.00.

possible) for each proposed alternative. Definition should include extent of maintenance accomplished at each echelon, estimated frequencies of maintenance and downtime requirements, turnaround time between squadron and field electronics shop, and between field electronics shop and depot. This information, supplemented by a maintenance analysis, leads to the identification of support equipment, spares, personnel, and facility requirements.

The operational and maintenance data are the basis for determining prime equipment requirements as well as the requirements for total support.

d. The quantitative data describing the operational and maintenance characteristics of each proposed system are applied to arrive at the sustaining cost of the systems. This is accomplished by utilizing the information in Fig. 7-3 and Table 7-2. These costs are added to the system acquisition costs to arrive at the total cost for the system. These factors are summarized in Table 7-3 which reveals the life-cycle cost for each of the proposed systems. This same data is applied to a time reference in Fig. 7-4, which graphically highlights the difference in the sustaining costs of the two systems.

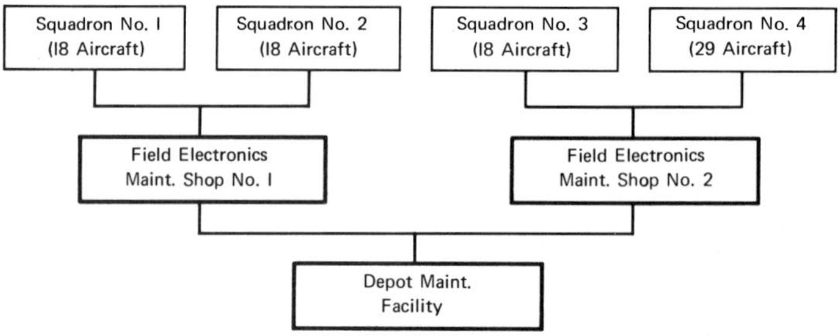

Operational Data

Number of Squadrons: 4
Number of Field Maintenance Shops: 2
Number of Depot Maintenance Facilities: 1
Number of Training Facilities: 1
System Operational Life: 10 years
Aircraft Availability Requirement: 85%
Flight Hours per Aircraft Requirement:
 Peacetime: 30 hours per month
 Wartime: 60 hours per month
TACAN System Utilization Factor:
 1.3 hours of TACAN operation for each hour of aircraft operation
Duration of Aircraft Mission: 2 hours (average)

FIG. 7-3 Operational Concept.

4. DECISION

Based on the total cost comparison of the two proposed solutions presented in Table 7-3, and the point in time when "break even" occurs (Fig. 7-4), the decision is to purchase the micro-TACAN system.

Support-system Selection. This example illustrates the selection of a support system for the depot level of maintenance based on the trade-off of performance, utilization, equipment characteristics, and other factors that define the measure of a system configuration. As with other cases of this nature, the trade-off involves the analysis of many elements to include technical, cost, and schedule considerations.

1. DEFINITION OF THE PROBLEM

An aircraft system requires support equipment for maintenance operations at the depot. The operations analysis has been accomplished and maintenance requirements (items assigned to depot, loading factors, etc.) have been specified. A specific configuration of support equipment must be selected to fulfill this requirement.

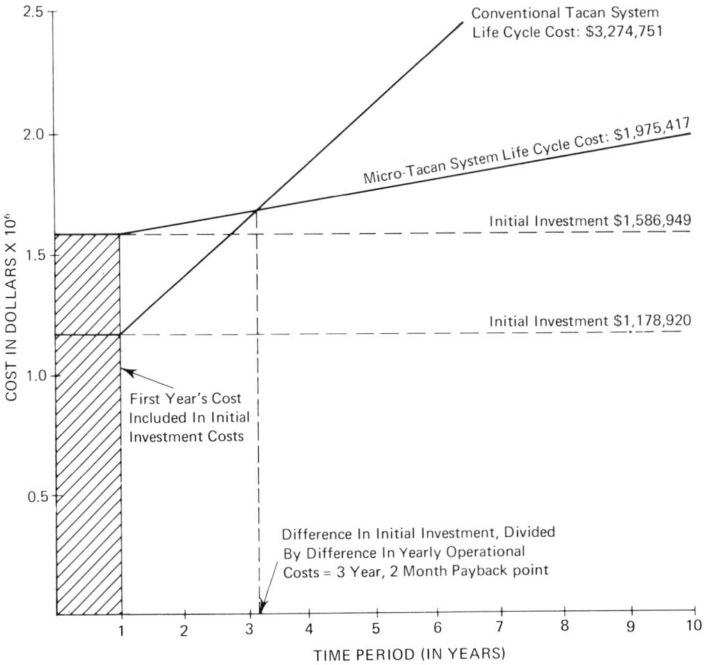

FIG. 7-4 Investment Payback.

2. PROPOSED ALTERNATIVE SOLUTIONS

Three alternative support-equipment configurations are being considered. Configuration *A* is completely automatic, configuration *B* is semiautomatic, and configuration *C* is manual. Each configuration is in existence and is supporting other systems. Through various degrees of redesign, each configuration could be made com-

TABLE 7-3 Cost Summary

INITIAL INVESTMENT COSTS

CATEGORY	FIG. 7-1 REFERENCE	MICRO-TACAN SYSTEM	CONVENTIONAL TACAN SYSTEM
Manufacturing			
Prime Equipment	$A2_{\underline{a}}$	$1,420,960	$ 988,530
Support Equipment	$A2_{\underline{b}}$	------	58,880
Prime Equipment Spares	$A2_{\underline{c}}$	30,670	103,935
Operational Spares		(23,542)	(95,353)
Maintenance Spares		(7,128)	(8,582)
Support Equipment Spares	$A2_{\underline{d}}$	------	589
Personnel & Training			
Initial Training	$A6_{\underline{a}}$	2,560	
Training Equipment	$A6_{\underline{b}}$	44,512	
Initial Support Services			
Initial Inventory Management	$A7_{\underline{a}}$	64,584	
Field Service Support	$A7_{\underline{b}}$	9,600	
Transportation			
Packing & Shipping	$A8_{\underline{a}}$ & $_{\underline{b}}$	14,063	26,980
TOTAL		$1,586,949*	$1,178,920*

OPERATIONAL COSTS

Replenishment Spare/Repair Parts			
Prime Equipment	$B1_{\underline{a}}$	$ 276,020	$ 935,413
Operational Spares		(211,873)	(858,177)
Maintenance Spares		(64,147)	(77,236)
Support Equipment	$B1_{\underline{b}}$	------	5,299
Facility Maintenance	B2	221	3,263
Technical Data	B3	11,256	85,456
Support Equipment	B5	699	10,306
Personnel & Training	$B6_{\underline{a}}$ & $_{\underline{b}}$	10,054	431,427
Support Services	B7	61,338	209,047
Transportation	B8	28,880	415,620
TOTAL		$ 388,468*	$2,095,831*

Total COSTS

Total (Initial Investment & Operational)	$1,975,417*	$3,274,751*

*The cost values presented are hypothetical

patible with the new requirements. Schedule and cost requirements preclude the design and development of a new support system.

3. TRADE-OFF APPROACH
   a. The first step is to list the major *evaluation parameters* (e.g., performance, utilization, physical-design requirements, supportability, etc.). These parameters are summarized in Table 7-4.
   b. Quantitative *weighting factors* (from 0 to 100) are assigned, according to degree of importance, to each evaluation parameter. The sum of all weighting factors is 100.
   c. Each of the three support-equipment configurations is then in-

TABLE 7-4 Trade-off Evaluation Summary

ITEM	EVALUATION ITEM	WEIGHTING FACTOR	CONF. *A*		CONF. *B*		CONF. *C*	
			Base Rate	Score	Base Rate	Score	Base Rate	Score
1	PERFORMANCE Maintenance Capability, Maintenance Load, Method/Mode of Testing, Versatility of Testing	13	8	104	10	130	4	52
2	UTILIZATION Application-Maintenance Level, Application-Systems	6	6	36	8	48	4	24
3	SUPPORT EQUIPMENT DESIGN Number of Test Stations, Number of Packs, Functional Grouping, Self-Contained, Transportability & Mobility, Interchangeability	7	10	70	8	56	6	42
4	TEST AIDS General Test Equipment, Maintenance Tapes, Technical Manuals	4	6	24	8	32	4	16
5	FACILITIES Environment, Space, Weight, Volume, Power	1	10	10	10	10	5	5
6	FLEXIBILITY/GROWTH POTENTIAL Station Accuracies, Performance Range, Space, Reconfiguration, Design Change Acceptability	5	8	40	10	50	5	25
7	SUPPORTABILITY Reliability, Maintainability, Human Engineering, Personnel, Training, Support Equipment, Spares, Facilities	3	8	24	5	15	3	9
8	TECHNICAL DATA Drawings, Specifications, Reliability/Maintainability Data, Logistics Data, Facilities Data, Test Data	6	5	30	8	48	3	18
9	SCHEDULE RDT&E, Production	30	7	210	8	240	4	120
10	COST Initial Costs, Operational Costs	25	10	250	8	200	5	125
SUBTOTAL		--	--	798	--	829	--	436
DERATING FACTOR (DEVELOPMENT RISK)		--	--	80 (10%)	--	41 (5%)	--	87 (20%)
GRAND TOTAL		100	--	718	--	788	--	349

dependently evaluated. Specific checklist criteria are used to support each evaluation. Based on the criteria employed, *base-rating* values (from 0 to 10) are applied according to the degree of compatibility with the new requirement.

d. The base-rating value of each configuration item is then multiplied by each weighting factor to obtain the *score*. The total score is then determined through the addition of all individual scores for each configuration. A special derating factor is applied against each configuration depending on the degree

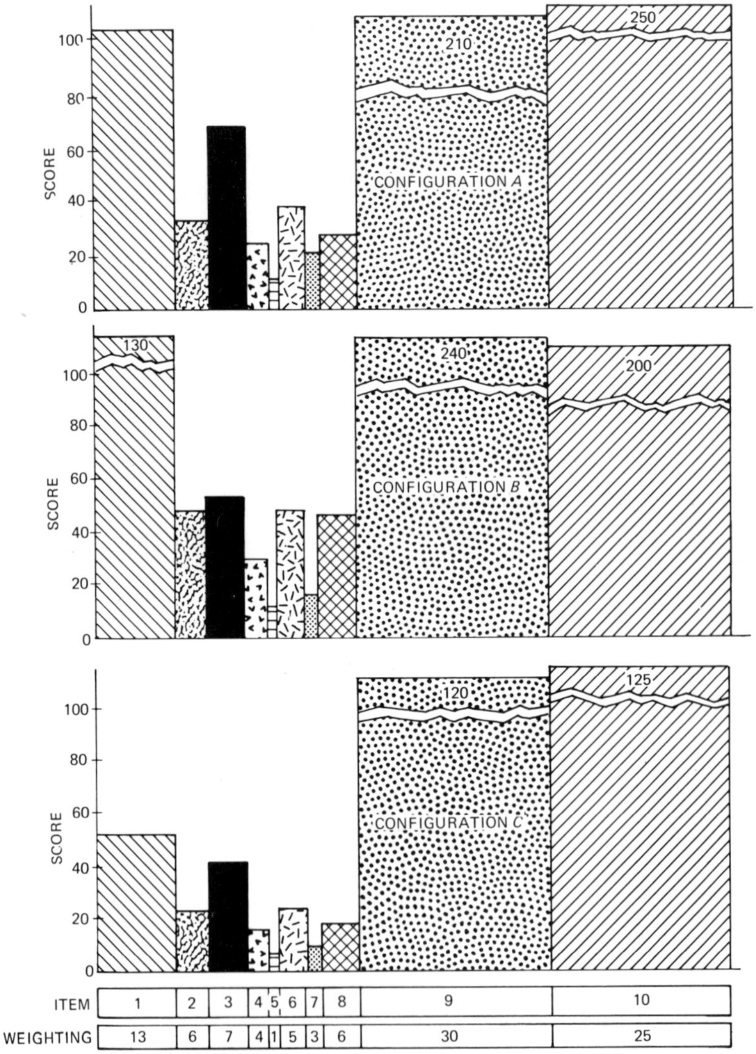

FIG. 7-5 Weighting-factor Graphs.

of redesign and the risk anticipated. The resultant quantitative values are summarized in Table 7-4. The illustrations in Fig. 7-5 emphasize the distribution of weighting factors for each configuration.

4. DECISION
   Referring to Table 7-4, configuration *B* represents the best solution based on the highest total score. This shop configuration is recommended.

Reliability-Maintainability Study. This example illustrates a trade-off between reliability and maintainability based on a specified inherent availability. The initial and operational (sustaining) costs are controlling considerations.

1. DEFINITION OF THE PROBLEM
   A requirement exists to design a radar receiver which will meet an inherent availability of 0.990, a minimum *MTBF* of 200 hr, and an *MTTR* not to exceed 4.0 hr. Existing design with the use of military standard parts meets an availability of 0.97, an *MTBF* of 150 hr, and an *MTTR* of 4.64 hr. An anticipated quantity of 300 radar receiver units are to be ordered.
2. PROPOSED ALTERNATIVE SOLUTIONS
   Three different alternative design configurations are being considered to satisfy availability requirements. Design configuration 1

TABLE 7-5 Data on Alternative Designs

CONFIGURATION AND CHARACTERISTICS	A_i	*MTBF*	*MTTR* (hrs)
RADAR RECEIVER NO. 1: Modular design with capability for automatic testing. Standard parts used in construction.	0.990	200	2.02
RADAR RECEIVER NO. 2: Some modular design. Some provisions for semiautomatic testing. Military approved parts used extensively.	0.990	300	3.03
RADAR RECEIVER NO. 3: Manual test capability only. Very limited modular design. Redundant functional circuits used throughout. Military approved parts used extensively. Parts are inspected and tested 100% before assembly of the receiver	0.990	350	3.54

emphasizes the maintainability aspects in design while design configuration 3 emphasizes reliability. Specific, detailed equipment-design characteristics have been established. The factors of reliability-maintainability trade-off are illustrated in Fig. 7-6.

3. TRADE-OFF APPROACH

a. Cost data are developed against each configuration. Such data include both initial investment and operational (sustaining) costs.

b. Apply cost data against a cost-category checklist similar to that presented in Table 7-1. Only those cost categories which appear to be significant should be included.

c. Initial costs are presented in Table 7-6. Table 7-7 presents the net incremental value (cost) of reliability and maintainability in equipment design as derived from the figures in Table 7-6. Operational (sustaining) costs are presented in Table 7-8.

d. Referring to Table 7-8, a net operational cost savings is realized for each proposed design configuration. This savings is derived through the improvement in equipment availability resulting in a reduction in required maintenance support. The cost savings for each design configuration when compared with the existing design are:

Configuration 1: $966,700
Configuration 2: $997,104
Configuration 3: $984,816

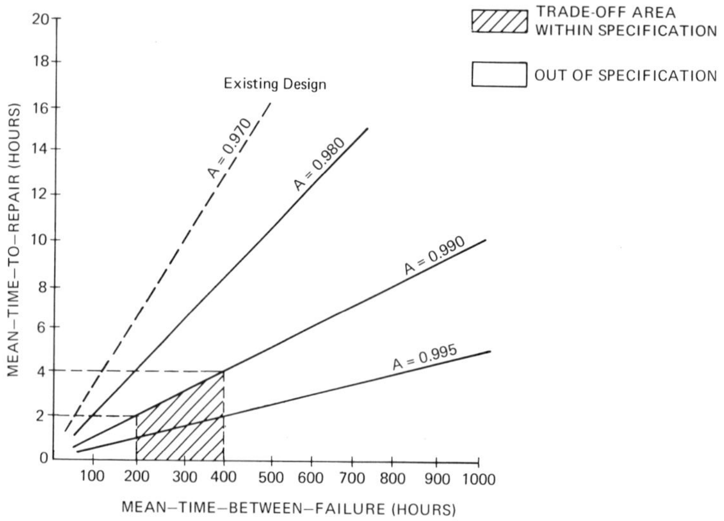

FIG. 7-6 Trade-off Factors.

e. The net effect is obtained by subtracting the added costs for reliability and maintainability from the overall sustaining-cost savings:

Configuration 1: \$966,700 — \$59,187 = \$907,513
Configuration 2: \$997,104 — \$43,825 = \$953,279
Configuration 3: \$984,816 — \$51,750 = \$933,066

4. DECISION

Configuration 2 satisfies the required equipment availability with maximum overall-cost savings. This configuration is recommended. The design goal, therefore, is to have a radar receiver with an *MTBF* of 300 hr and an *MTTR* of 3.03 hr.

Throwaway Versus Repair Study. This example illustrates some of the factors that should be considered in determining the feasibility of designing an item so that it can be repaired upon failure. From an

TABLE 7-6 Initial-investment Cost Factors

ITEM	EXISTING CONF.	CONF. NO. 1	CONF. NO. 2	CONF. NO. 3
Design & Development	\$ 300,000	\$ 324,937	\$ 319,125	\$ 321,500
Reliability	—	1,187	6,625	16,750
Maintainability	—	23,750	12,500	4,750
Manufacturing*	4,500,000	4,534,250	4,524,700	4,530,250
Reliability	—	1,750	9,200	22,500
Maintainability	—	32,500	15,500	7,750

*Manufacturing cost is based on 300 units.

TABLE 7-7 Incremental Cost Factors

ITEM	CONF. NO. 1	CONF. NO. 2	CONF. NO. 3
Reliability	\$ 2,937	\$ 15,825	\$ 39,250
Maintainability	56,250	28,000	12,500
TOTAL COST*	\$ 59,187	\$ 43,825	\$ 51,750

*The cost values presented are hypothetical.

economic standpoint, it may be practical to discard the item when it fails. A logical decision can only be reached after a full evaluation of the life-cycle–cost factors.

1. DEFINITION OF THE PROBLEM

A new tactical air navigational system (TACAN) in the design and development phase will incorporate a number of modules and module subassemblies (circuit boards). Through trade-off analysis, the repair of the TACAN system at the field level of maintenance has been deemed the most feasible approach. The problem is to determine the maintenance disposition of 33 circuit boards.

2. PROPOSED ALTERNATIVE SOLUTIONS

This trade-off study involves consideration of the following alternatives:

a. Discard the circuit boards at failure. No repair required.
b. Repair the circuit boards at field level.
c. Repair the circuit boards at depot level.

3. TRADE-OFF APPROACH

The decision to discard or repair the circuit boards is dependent upon the operational deployment of the system, circuit-board cost, circuit-board reliability, and the maintenance requirements.

a. The first step in the trade-off is the generation of operational data to include quantity of systems deployed, location (deployment), system-utilization rates, availability requirements, and system service life. Data presentation should follow the approach illustrated in Fig. 7-3.

TABLE 7-8 Operational Cost Factors

ITEM	EXISTING CONF.	CONF. NO. 1	CONF. NO. 2	CONF. NO. 3	REMARKS
Cost of Spares	$ 210,000 (14@15,000)	$ 150,970 (10@15,097)	$ 105,420 (7@15,060)	$ 90,402 (6@15,067)	Freq. of maint./10 yr: Existing Conf--139; Conf. 1--104; Conf. 2--69; Conf. 3--59; Assume 10% spares
Cost of Repair					
Administrative	385,065	125,370	124,773	124,176	Assume $1.99/direct labor hour
Active Repair	700,470	151,200	188,100	212,160	One tech./maint. action Existing Conf.--E6@3.62/hr. Conf. 1--E3@2.40/hr. Conf. 2--E4@3.00/hr. Conf. 3--E5@3.40/hr.
Test Equipment And Associated Spares	174,150	56,700	56,430	56,160	Assume 3 items--std. equip. (burden--$0.90/direct labor hr.).
Spares (piece parts)	1,390	1,040	690	590	Assume $10/repair action
Facilities	36,765	11,970	11,913	11,856	Use existing facilities (burden--$0.19/direct labor hour)
Cost of Training	140	80	100	120	Based on $20/day
Cost of Tech. Manuals	20,000	14,000	16,000	18,000	Assume $200/page
Cost of Provisioning					
Provisioning	–	30,000 (30 items)	15,000 (15 items)	5,000 (5 items)	Provisioning – assume cost of $1000/line item
Handling	475,000 (2500 pts)	494,950 (2605 pts)	487,450 (2565 pts)	499,700 (2630 pts)	Handling – assume $19/line item/year.
TOTAL COST	$2,002,980	$1,002,980	$1,005,876	$1,018,164	

DISCARD-AT-FAILURE VS. REPAIR TRADE-OFF EVALUATION — CIRCUIT BOARD 1A7

EVALUATION	FIELD		DEPOT	REMARKS
	REPAIR	DISCARD	REPAIR	
Unit Cost of Circuit Board	$ 145	$ 100	$ 145	Production Price. Circuit Board design is simplified in discard case.
Circuit Board Spares	435	5500	870	Based on 55 replacements (λ = 0.4617/1000 hours) per 10 yr operational period. Spares: Field Repair: 3; Discard: 55; Depot Repair: 6.
Cost of Repair				
Maintenance Labor Costs	1351	N/A	1954	Field Maint. labor rate: $6.14/hr (direct & indirect) Depot Maint. labor rate: $8.88/hr (direct & indirect) MTTR: 2 hr. Two maintenance technicians
Test Equipment	8787	N/A	2000	Special Test equipment required. Cost amortized over number of installations.
Test Equipment Repair	879	N/A	200	Assume 10% of test equipment cost.
Piece Parts for Repair	275	N/A	275	Assume $5/ repair action.
Depot Handling Cost	N/A	N/A	770	Assume $14/repair action.
Facilities	42	N/A	42	Use existing facilities for repair (assume burden rate of $0.19/direct labor hour).
Cost of Retrofit or Modification	80	1120	140	Assume one retrofit at $20 for prime plus spare circuit boards.
Cost of Packing & Transportation	N/A	N/A	72	110 one-way trips @ $65/100 pounds. Assume circuit board weighs one pound.
Cost of Technical Training	16	N/A	16	Delta training cost to cover maintenance of circuit board. Assume $4/hour/person ($160/week/person).
Cost of Technical Manuals	800	N/A	800	Assume 4 pages for troubleshooting diagrams & text covering repair of circuit board. Assume $200/page.
Inventory Management				
Initial Inventory management	2000	N/A	2000	Assume 2 parts are nonstandard. Cost of entering non-FSN item in inventory: $1,000/item.
Sustaining Inventory cost	142	1100	229	Assume 20% of total inventory value (C. B. spares + piece parts).
TOTAL ESTIMATED COST	$14,952	$7,820	$9,513	

FIG. 7-7 Trade-off Summary for Board 1A7.

b. Next, a detailed list of evaluation items (for both the discard and repair cases) is prepared. Figure 7-7 presents a summary list of these items for board 1A7.

c. Supporting maintenance factors and requirements are then assigned. Such data, based on operational needs, include circuit-board replacement per application during the operational service life of the system, circuit-board mean-time-to-repair (*MTTR*), supporting tools/test equipment to accomplish circuit-board repair, and personnel types (quantity and skill level). These data stem from a maintainability prediction and analysis of the system.

d. The next step involves the evaluation of the data on each of the 33 circuit boards. The total cost to implement each of the three concepts for a particular circuit board is determined through the addition of the various cost factors outlined in Fig. 7-7.

e. When the evaluation process is accomplished for each of the 33

TABLE 7-9 Summary of Costs

Circuit Board Designation	Module Designation	Field		Depot Repair
		Repair	Discard	
1A6	Range Computer	$14,516	$ 6,525	$ 9,099
1A7		14,952	7,820	9,513
1A8		14,524	6,705	9,119
1A9	Bearing Computer	18,825	24,960	16,642
1A10		14,828	9,680	9,922
1A11		19,558	37,785	18,283
1A12		14,496	7,155	9,172
1A13	Synthesizer	15,686	6,885	11,064
1A14		14,718	6,655	9,591
1A15		16,673	9,775	12,974
1A16		14,796	8,800	9,155
1A17		14,228	6,020	8,574
1A18	Decoder	21,584	25,500	21,533
1A19		20,590	27,090	19,679
1A20		14,819	8,200	9,356
1A21		14,188	6,975	8,751
1A22		20,610	34,850	19,986
1A23		14,514	7,560	9,216
1A24	Encoder	21,701	40,250	21,732
1A25		14,132	4,410	8,390
1A26		14,106	3,955	8,338
1A27	Self-Test	20,975	34,615	20,449
1A28		14,819	8,200	9,356
3A1	Range D/A	15,607	12,320	10,385
3A2		19,737	43,200	18,463
3A3		18,046	21,140	15,278
3A4	Digital Output	14,124	4,200	8,366
3A5		20,081	26,400	18,132
3A6	Bearing D/A	14,231	4,040	8,179
3A7		14,285	5,120	8,010
3A8		26,079	42,375	29,308
3A9		26,079	42,375	29,308
3A10		14,159	2,600	8,390
TOTAL ESTIMATED COST		$562,266	$544,140	$443,713

circuit boards, the individual costs are listed and summarized (see Table 7-9).

4. DECISION

The most economical approach is to repair the circuit boards at the depot level of maintenance. This is evident by reviewing the total-cost figures in Table 7-9. It should be noted, however, that this decision is based on an evaluation of the circuit boards as a class. For a discussion on the optimum policy for each circuit board, refer to the following paragraphs.

5. ALTERNATE COURSE OF ACTION

During the evaluation of the individual circuit boards the associated unit-cost factors were plotted in relationship to the failure rates (see Fig. 7-8). Based on cost criteria, the distribution can be segmented and identified with the various levels of maintenance. This provides a basis for making a design decision which could be any one of the following:

a. The reliability of the items designated for *field discard* should be increased, with no significant unit-cost increase, so that in-

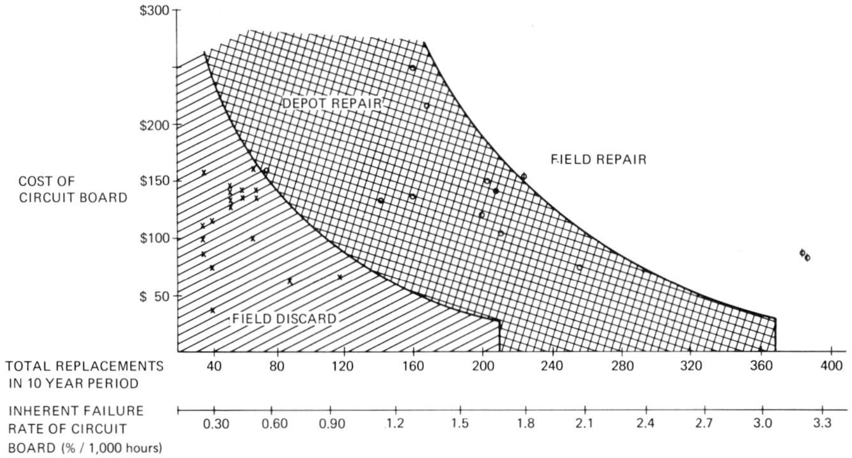

FIG. 7-8 Failure-rate-Cost Factors Related to Maintenance Levels.

dividually they may be reclassified as candidates for *depot repair.*

b. The unit cost of items designated for *field repair* should be decreased so that individually they may be reclassified as candidates for *depot repair.* Degrading of reliability should be held to a minimum.

c. The unit cost of the items designated for *depot repair* and *field repair* should be decreased so that they may be reclassified as candidates for *field discard.* Performance characteristics must be maintained within specified limits. Degrading of reliability should be held to a minimum.

A detailed investigation into the above-listed possibilities will likely result in the establishment of an optimum maintenance policy for each circuit board and for the items as a class.

Evaluation of Maintenance Concepts. This example illustrates the use of a maintainability prediction technique to (1) define the maintenance concept and (2) identify the primary maintainability-design requirements of an equipment item.

1. DEFINITION OF THE PROBLEM

   A communications test station (item of support equipment for a prime weapon system) contains a frequency converter comprised of a frequency unit, circuit board 1, circuit board 2, and circuit board 3. The converter must be repaired in less than 60 min. The problem is to identify what maintenance concept will ensure that the downtime limit will not be exceeded. In addition, what features must be inherent in the design to complement the maintenance concept?

2. PROPOSED ALTERNATIVE SOLUTIONS

   The trade-off study considers:

   a. Concept *A.* Frequency-converter repair is accomplished through the removal and replacement of circuit-board piece parts.

   b. Concept *B.* Frequency-converter repair is accomplished through the removal and replacement of circuit boards. Circuit boards are nonrepairable.

3. TRADE-OFF APPROACH

   a. The first step involves the preparation of an equipment functional diagram covering the entire test station down to the lowest level of item removal and replacement. In the example presented herein, the main question involves the maintenance concept associated with the frequency converter. Figure 7-9 illustrates this portion of the test station.

   b. The next step requires the identification of levels of test-station localization, isolation, and checkout, respectively. For example, the level of localization is determined by the depth of fault location possible without requiring the use of external test equipment.

   c. A maintainability prediction is then applied to both concept *A* and concept *B* to determine mean-time-to-repair (*MTTR*) factors.

4. DECISION

Based on the maintainability prediction of the two concepts and the resultant quantitative *MTTR* values, concept *B* is the best design approach from the standpoint of least downtime. Therefore, design of circuit boards 1, 2, and 3 should be such as to enable rapid fault isolation and removal and replacement.

Evaluation of Three Computers. This example outlines one procedure commonly used to aid in the selection of equipment. Emphasis is placed on the evaluation of the maintainability-design characteristics of the proposed equipment. In this case there are three equipment items under consideration.

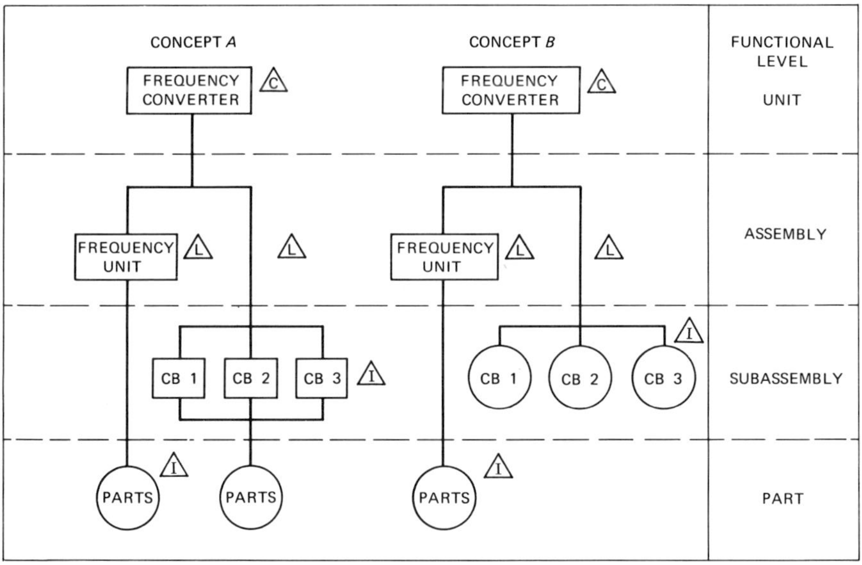

ITEM	ΣNλ	$\Sigma N\lambda R_p$
FREQ. UNIT	50.04	68.96
C. B. 1	36.22	65.46
C. B. 2	28.62	42.77
C. B. 3	38.57	49.67
TOTAL	153.45	226.86

$$MTTR = \frac{226.86}{153.45} = 1.48 \text{ Hr.}$$

ITEM	ΣNλ	$\Sigma N\lambda R_p$
FREQ. UNIT	50.04	68.96
C. B. 1	36.22	27.89
C. B. 2	28.62	22.04
C. B. 3	38.57	29.70
TOTAL	153.45	148.59

$$MTTR = \frac{148.59}{153.45} = 0.97 \text{ Hr.}$$

NOTES:

1. △C, △L, △I, represent levels of checkout, localization, and isolation respectively.
2. Quantitative values are obtained from detailed prediction worksheets.

FIG. 7-9 Functional Diagram with *MTTR*-prediction Data.

1. DEFINITION OF THE PROBLEM

A semiautomatic radar test station will incorporate a single rack computer for use in checkout and testing of the station and its associated items. Three computers have been selected that meet the performance requirements. The problem is to identify which

TABLE 7-10 Maintainability Checklist

MAINTAINABILITY CONSIDERATION	MAX VALUE	COMPUTER *A*	COMPUTER *B*	COMPUTER *C*
1. General	33			
a. *MTBM*	(25)	10	15	20
b. Warmup Time	(4)	4	2	3
c. Weight	(2)	1	2	2
d. Size	(2)	1	2	2
2. Handling	5	2	3	4
3. Equipment Racks	8	1	2	1
a. Slide-Mounted Drawers	(2)	1	2	1
b. Hinged Panels	(2)	1	2	1
c. Rack Layout	(2)	1	2	2
d. Cable Routing	(2)	2	1	1
4. Packaging	16			
a. Accessibility	(2)	2	1	0
b. Modularization	(2)	0	2	1
c. Plug-In Components	(2)	0	2	1
d. Component Stacking	(2)	2	1	1
e. Component Keying	(2)	0	1	1
f. Standardization	(2)	2	2	2
g. Interchangeability	(2)	2	2	2
h. Functional Grouping	(2)	1	1	1
5. Diagnostic Aids	8			
a. Built-In Self-Test	(2)	0	2	1
b. Test Points (Qty.)	(2)	0	2	1
c. Test Points (Location)	(2)	0	2	1
d. Meters/Displays/Lights	(2)	1	2	1
6. Adjustments	6			
a. Adjustment Points	(2)	0	2	1
b. Interaction Effects	(2)	0	2	1
c. Built-In Meter Jacks	(2)	0	2	0
7. Panel Displays/Controls	8			
a. Panel Layout/Labeling/ Illumination	(2)	1	1	2
b. Self-Indicating Fuses	(2)	0	0	2
c. Spare Fuses	(2)	0	0	2
d. Meters Externally Removable	(2)	0	2	0
8. Labeling	4			
a. External Labeling	(2)	2	2	1
b. Internal Labeling	(2)	0	2	1
9. Safety	4			
a. Personnel Safety	(2)	2	1	1
b. Equipment Safety	(2)	2	1	1
10. Support	8			
a. Tools/Test Equip.	(2)	1	2	1
b. Tech. Data	(2)	2	1	1
c. Personnel Qty.	(2)	2	1	1
d. Personnel Skill	(2)	2	1	1
TOTAL SCORE	100	47	71	66

of the three computers is superior from the standpoint of maintainability-design characteristics. Investment and sustaining costs are not to be considered in this trade-off.

2. PROPOSED ALTERNATIVE SOLUTIONS

As stated above, one of three proposed computers must be recommended based only on a trade-off of the maintainability-design characteristics of the items.

3. TRADE-OFF APPROACH

a. A detailed set of criteria (to include various maintainability characteristics in design) is prepared. The criteria allow for the assignment of quantitative factors.

b. A checklist, as illustrated in Table 7-10, is prepared from the detailed set of criteria.

c. Each of the three computers is evaluated against the detailed criteria. Quantitative factors are assigned and recorded on the form. For example, if there are 10 panels incorporated in the rack of equipment being evaluated and all are hinge mounted, the quantitative value against item 3*b* would be 2. If only 5 panels are hinge mounted, the quantitative value would be 50 percent or 1. Items not applicable in design of the equipment being evaluated are not penalized, but are given the maximum rating.

d. The total score for each configuration is determined through the addition of all quantitative values.

4. DECISION

Computer *B* is recommended.

QUESTIONS AND PROBLEMS

1. What is the definition of *trade-off*?
2. Of what value is a trade-off report to a manager? What purpose does it serve?
3. The trade-off technique is only applicable to military systems. True or false?
4. List five applications of trade-offs as related to maintainability functions.
5. Why must trade-off studies be concluded as early as possible in a program cycle?
6. Outline the trade-off approach. What steps are involved?
7. Define the term *life-cycle cost*.
8. What are the two major elements considered in determining the life-cycle cost of any item?
9. Technical (performance) and schedule requirements are never considered in a trade-off study. True or false?
10. A standard model for all types of trade-offs is feasible. True or false?
11. What approach would you take in performing a trade-off that involves the determination of life-cycle costs for two different system configurations?

12. What approach would you take in performing a trade-off to select one of several shop configurations?
13. The maintenance disposition of items is influenced by a number of factors. List three of these factors.
14. Given the following data, determine the life-cycle cost for each item. What item do you recommend? Why?

Item	Purchase cost	Operational cost	Modification cost
Equipment *A*	$20,000	$6,000/yr	0
Equipment *B*	$30,000	$3,000/yr	$19,700

a. The service life of either item is 10 years.
b. Delivery of both items is guaranteed for the same date.
c. It will take 30 days to modify equipment *B*.
d. After modification of equipment *B*, the item will have the same functional characteristics as equipment *A*.
e. Each item has a 2-year warranty unless modified by the buyer.

15. Recommend either of the following maintenance concepts based on the provided data:

Concept 1.—Discard all component-board assemblies upon failure.
Concept 2.—Repair all component-board assemblies upon failure.
Concept 3.—Discard component-board assemblies upon failure when the replacement cost is less than $100 each, and repair all other component boards.

a. Quantity data
Total number of assemblies: 10,000
Number of items under $100 unit cost: 5,000
Number of items over $100 unit cost: 5,000*

b. Failure-rate data
The items under $100 unit cost will fail on the average of every 36 months. The failure rates, respectively, are 0.00048 and 0.00016 per item; or 2.40 and 0.80 per class.

c. The service life of the system is 10 years. The system is operated 2,080 hr per year.

d. An investment of $1,000,000 is necessary for equipment to repair any or all of the subassemblies. The equipment life is 10 years, and it will have no scrap value at the end of that time.

e. Labor costs are $3.00 for each repair. Component spares cost $2.00 per repair.

f. Use figures obtained by slide rule computation.

* Average cost of these items is $300 per unit.

EIGHT

Design Liaison and Support Services

One objective of a maintainability-engineering organization is to provide the designer with a specialized source of information to (1) facilitate the design process and (2) identify the required inherent design support characteristics of the system or equipment. Design liaison and maintainability services play an important part in translating the available knowledge into usable form. To accomplish this, the maintainability engineer performs the following tasks:

1. Assesses the quantitative and qualitative support requirements or goals, policies, and general criteria, and generates specific design criteria for application to the design
2. Works with other members of the *design team* in performing detailed trade-off studies as related to a specific design approach
3. Assists in the evaluation and selection of vendor parts and assemblies
4. Verifies the feasibility of a design approach by utilizing models and mock-ups
5. Performs intermediate maintainability predictions and analyses of the proposed design approach
6. Performs a systematic review and evaluation of design data to

ensure compliance with the defined requirements and specified design objectives

7. Participates in formal design reviews
8. Provides specific design recommendations to correct design deficiencies or to enhance the support characteristics of the system or equipment

It should be emphasized that the successful completion of these tasks rests solely on the ability of the maintainability engineer to offer timely and constructive inputs to the design process. Although there are checking functions, the main value in performing these tasks is to facilitate the development of the design.

DESIGN CRITERIA

Application of General Criteria

While performing design-liaison tasks, the maintainability engineer will be called upon to recommend one detailed design approach from several alternatives. To do this effectively, the maintainability engineer must have knowledge of the conventional, proven maintainability criteria that exist in a number of documents. Three common references are:

1. ASD-TR-61-424, "Guide to Integrated System Design for Maintainability"

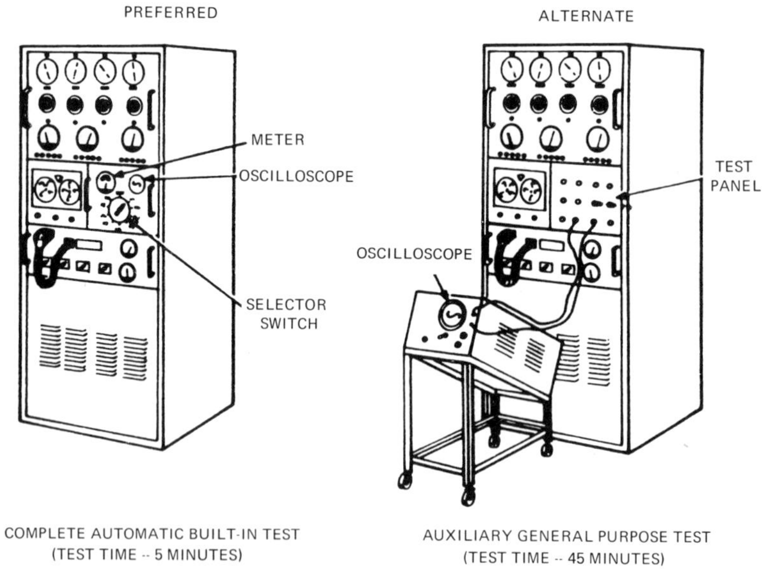

FIG. 8-1 Diagnostic Aids Criteria.

2. NAVSHIPS 94324, "Maintainability Design Criteria Handbook for Designers of Shipboard Electronic Equipment"

3. AMCP 706-134, "Engineering Design Handbook, Maintainability Guide for Design"

The types of information in these documents is illustrated in Figs. 8-1 through 8-9.

Based on the specific data presented to him, and with a knowledge of the preferred practices, the maintainability engineer can select the design approach that contains the greater amount of maintainability features, consistent with the requirements of the system or equipment.

Criteria Checklists

Other criteria of a general nature, but abbreviated, are usually recorded in company maintainability handbooks, checklists, or criteria listings (see Appendix A). The content of the handbooks may be information such as that in the references listed in the preceding paragraph. The checklists and criteria listings may be in either of the forms illustrated in Figs. 8-10 through 8-12. Whatever the format, this material serves the same purpose as the general criteria discussed in the above paragraph.

Specific Design Criteria

The maintainability engineer must provide the design engineer with information consistent with customer requirements, the maintenance con-

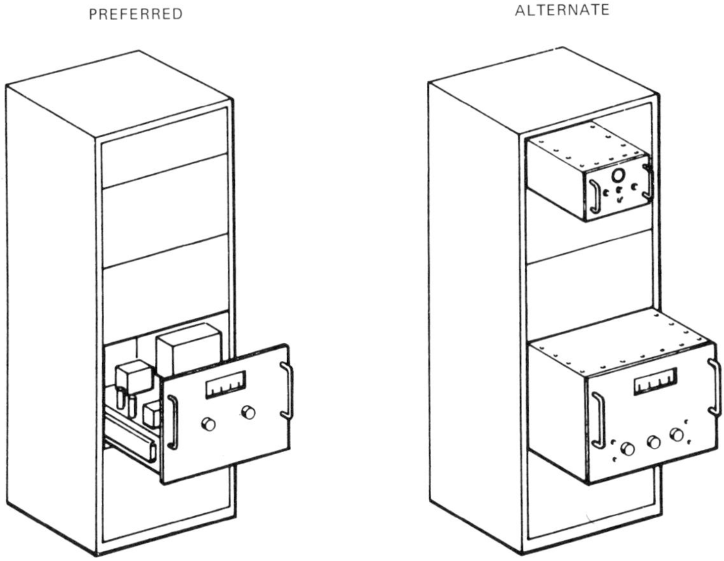

FIG. 8-2 Accessibility Criteria (Packaging).

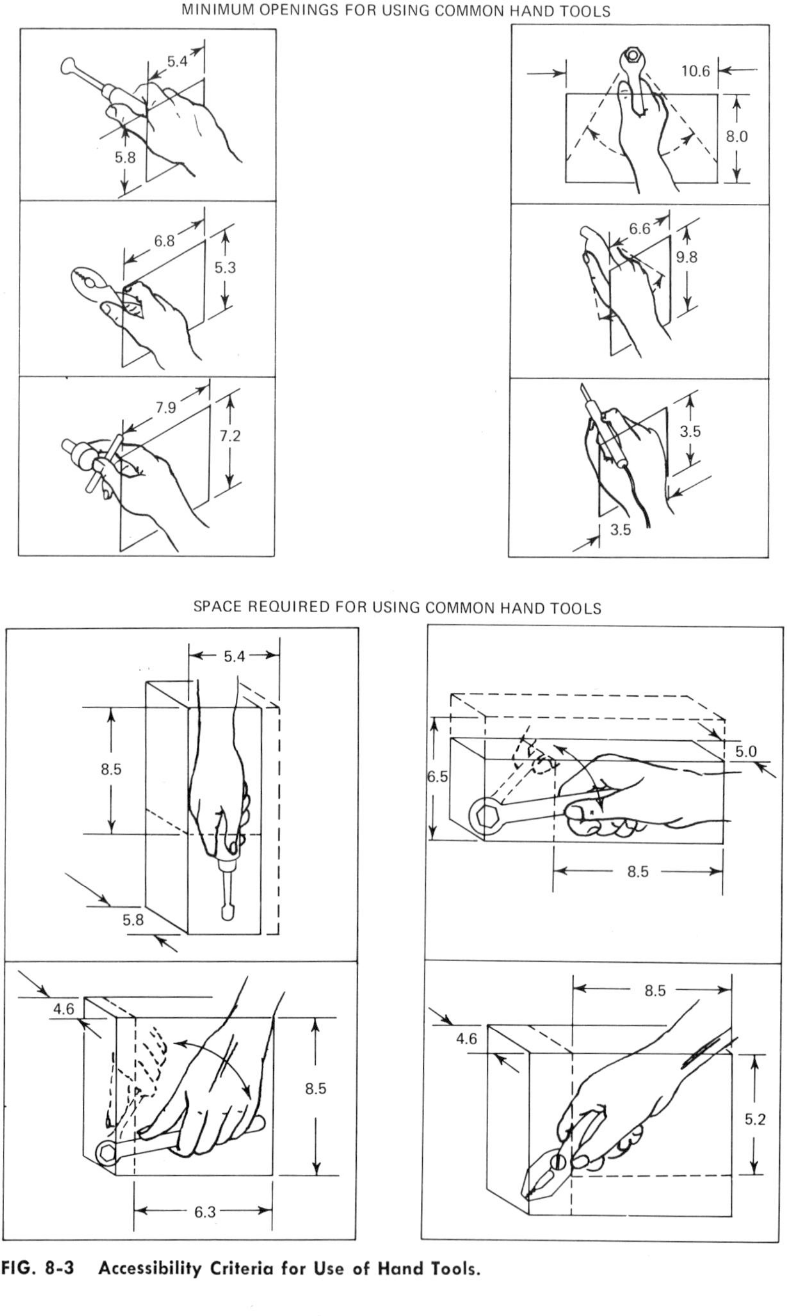

FIG. 8-3 Accessibility Criteria for Use of Hand Tools.

cept, and good maintainability principles. This information must be oriented to a specific system/equipment configuration, and the provided data should be inserted in the design specification. An example of this type of criteria is illustrated in Fig. 8-13.

TRADE-OFF STUDIES

The maintainability engineer, in conjunction with the designer, performs trade-off studies with the intent of selecting the best possible design approach. Alternative design configurations are evaluated in terms of performance, reliability, maintainability, logistics, cost, and other factors. The maintainability engineer supports the designer in evaluations of this type. Trade-off techniques are covered in Chap. 7 of this text.

SELECTION OF EQUIPMENT

When the system equipment is defined, the design engineer is faced with a "make or buy" decision. Influential in the decision are schedule requirements, costs, available manufacturing facilities, etc. In addition,

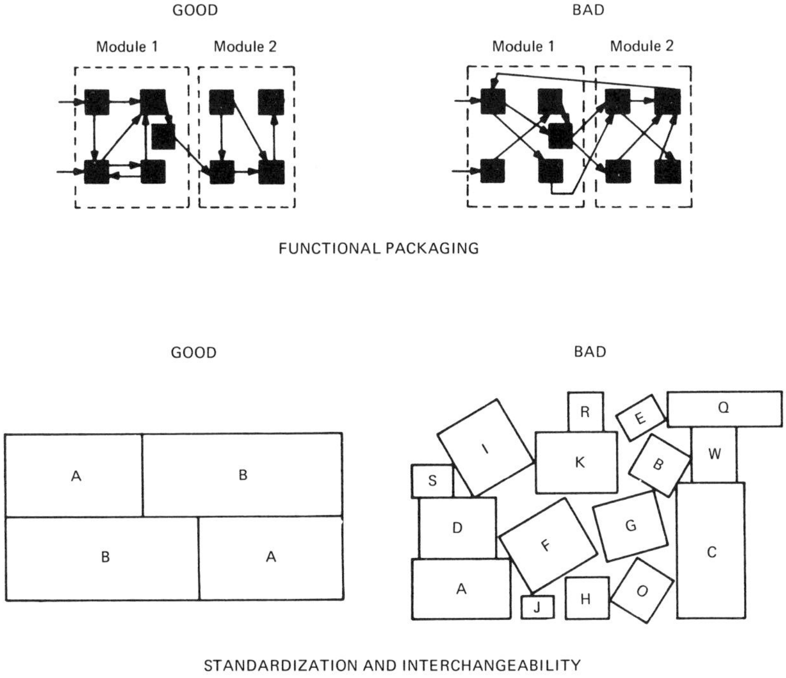

FIG. 8-4 Packaging Criteria (Function and Interchangeability).

the decision is based on the availability of qualified suppliers and equipment items meeting customer requirements. Assuming that the decision is to "buy," the maintainability engineer supports the designer relative to the accomplishment of a number of tasks associated with supplier activity.

Initially, supplier specifications and work statements are prepared. Maintainability-program tasks and design requirements are included in the specifications and work statements. Such design requirements include numerical *MTTR* allocations derived through the process covered in Chap. 6. Secondly, the maintainability engineer participates in the evaluation of various suppliers responding to the prime contractor's RFP. Such an evaluation may entail a visit to the supplier's facility,

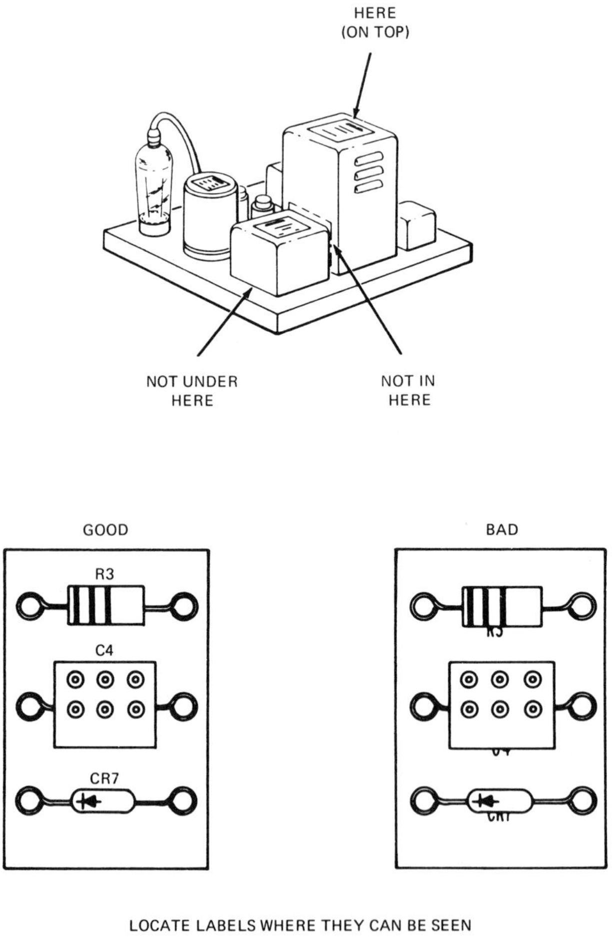

FIG. 8-5 Identification Criteria.

an evaluation of the supplier's proposal and capabilities, and an evaluation of the supplier's equipment.

Newly designed equipment. If an equipment item involving new design is to be subcontracted to a supplier, primary interest lies in the supplier's capability in the maintainability area (i.e., management recognition of the discipline, established policies and procedures, organizational structure and responsibilities, internal methods for meeting maintainability qualitative and quantitative requirements, and past performance relative to the degree of maintainability features incorporated in other systems or equipment). Assuming that several potential suppliers are being considered, the maintainability engineer assists in the formalized evaluation of each, leading to the ultimate selection of the best source.

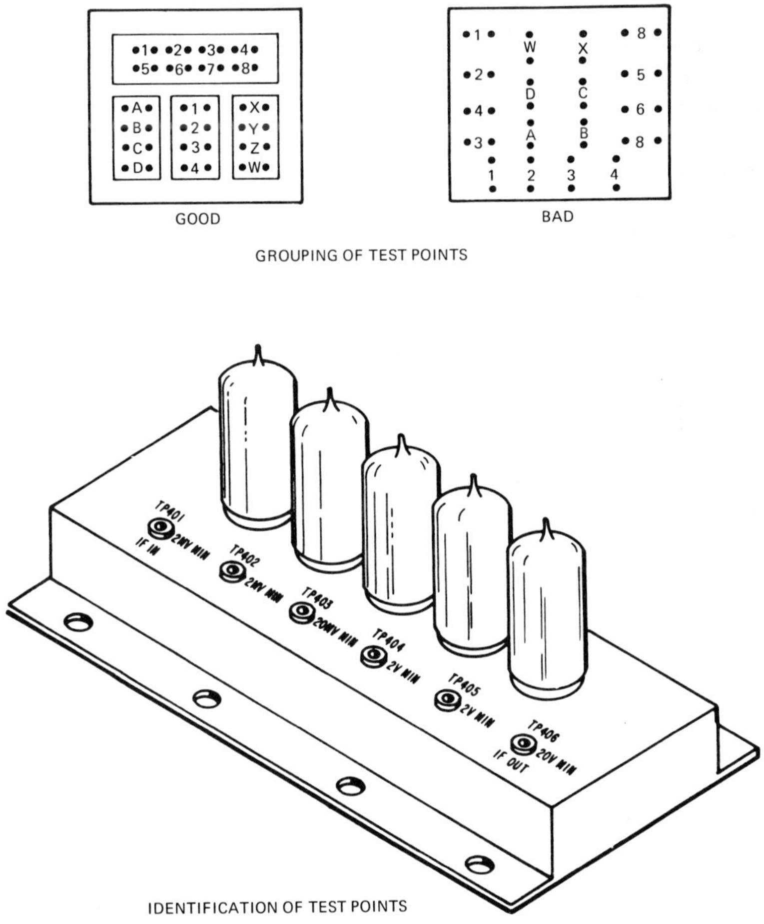

FIG. 8-6 Test-point Criteria (Identification and Grouping).

Off-the-shelf equipment. In the event that the equipment item to be selected is off the shelf (no new design is involved) and several possible sources are being considered, the maintainability engineer participates in the selection process by accomplishing a separate, independent evaluation of each supplier. Such evaluations are accomplished through a review of manufacturing data, technical manuals or handbooks, and/or the review of the actual equipment item. The best approach involves an evaluation of the actual production-item configuration. If at all possible, the prime contractor should obtain an item on a loan basis for formal testing and evaluation. The maintainability engineer reviews the equipment item for the incorporation of maintainability features.

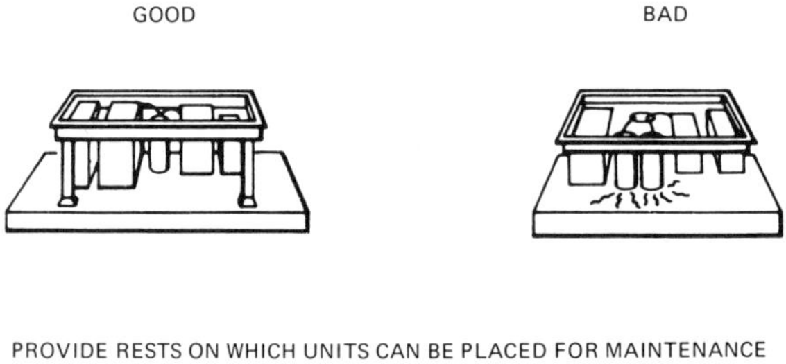

FIG. 8-7 Equipment Safety Criteria.

In completing the review process, the maintainability engineer accom-

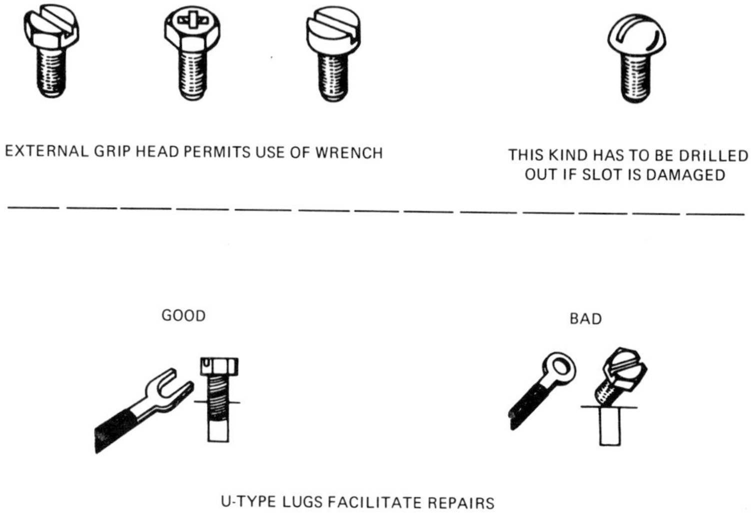

FIG. 8-8 Criteria for Fasteners.

plishes a prediction in accordance with one of the maintainability-prediction techniques covered in Chap. 10 of this text. In the event that a formalized maintainability prediction is not possible for any reason, an alternate method can be employed using a general checklist (similar to the one presented in Fig. 8-10) or an evaluation form (Fig. 8-14).

Referring to Fig. 8-14, the evaluation form covers seven basic areas relative to maintainability. In accomplishing the evaluation, each item on the form is given a numerical rating against specific checklist criteria developed by the maintainability engineer. An example of such criteria is shown in Fig. 8-15. Individual numerical values are then totaled.

Mounting Method	Maintainability Considerations	
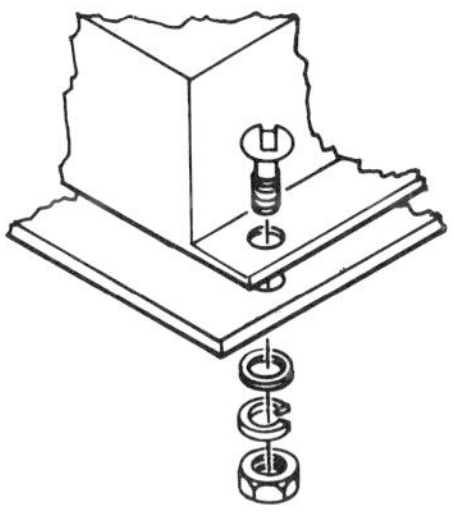	Description:	Horizontal surface mounted. Screws through clearance holes with nuts, flat washers, and lock washers.
	Advantages:	No significant advantages.
	Disadvantages:	Two-handed operation. Possible loss of nuts, washers, or screws. Alignment of holes difficult.
	Tools Required:	Screwdriver and wrench or "spintite,"
	Operating Time:	Approximately 0.8 minute per fastener.
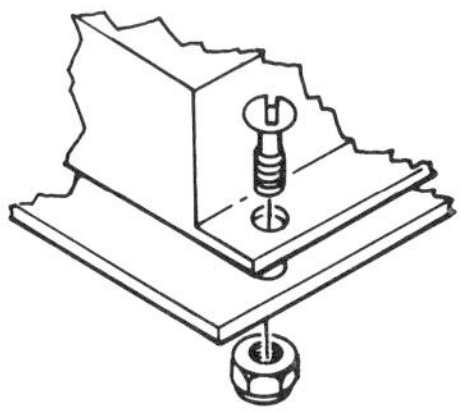	Description:	Horizontal surface mounted. Screws with stop nuts through clearance holes.
	Advantage:	No washer required.
	Disadvantages:	Stop nut difficult to turn. Two-handed operation. Alignment of holes difficult. Possible loss of nuts.
	Tools Required:	Screwdriver and wrench or "spintite."
	Operating Time:	Approximately 1.4 minutes per fastener.
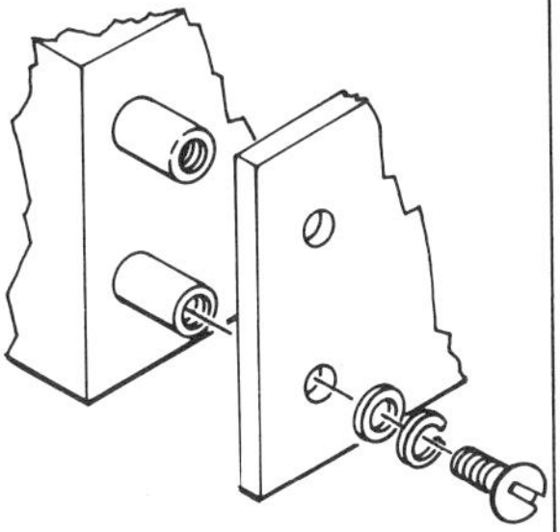	Description:	Vertical surface mounted. Screws with flat washers and lock washers into tapped holes.
	Advantage:	One-handed operation if part is supported.
	Disadvantages:	Parts must be supported to start nuts or screws. Alignment of holes difficult. Possible loss of washers or screws.
	Tools Required:	Screwdriver.
	Operating Time:	Approximately 0.8 minute per fastener.

FIG. 8-9 Mounting Criteria (Sheet 1 and Sheet 2).

The resultant value is entered in the provided (points) space at the top of the form. In support of the numerical rating for each item being evaluated, a summary is prepared using the space on the form under "remarks." The final rating of the item (good, fair, or poor) is based on a combination of the total numerical point count plus the factors listed under "remarks." Relative to point count, the following range can be employed as a guide:

Rating	Total score
Good	70 to 100 points
Fair	40 to 69 points
Poor	0 to 39 points

Mounting Method	Maintainability Considerations	
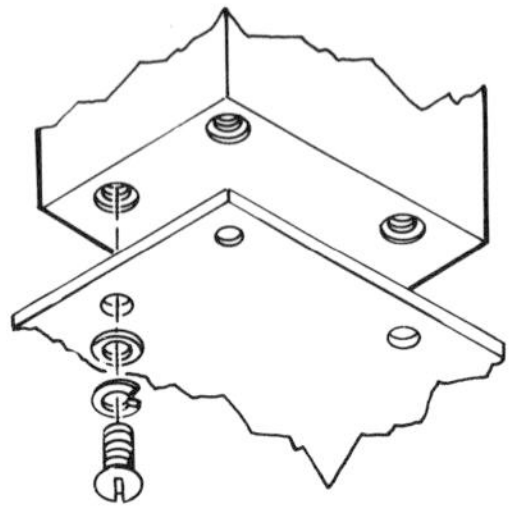	Description:	Horizontal surface mounted. Screws with flat washers, and lock washers into tapped hole in part.
	Advantage:	Can usually be performed with one hand if necessary.
	Disadvantages:	Possible loss of washers or screws. Alignment of holes difficult.
	Tools Required:	Screwdriver
	Operating Time:	Approximately 0.8 minute per fastener.
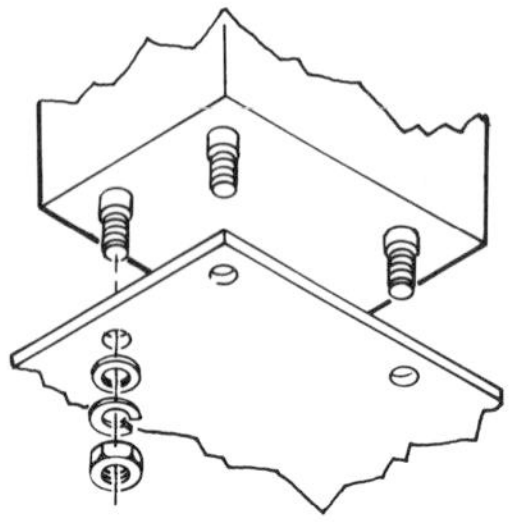	Description:	Horizontal surface mounted. Studs through clearance holes with nuts, flat washers, and lock washers.
	Advantage:	One-handed operation. Studs act as locating pins.
	Disadvantages:	Possible loss of nuts or washers. Part must be lifted.
	Tools Required:	Wrench or "spintite."
	Operating Time:	Approximately 0.7 minute per fastener.
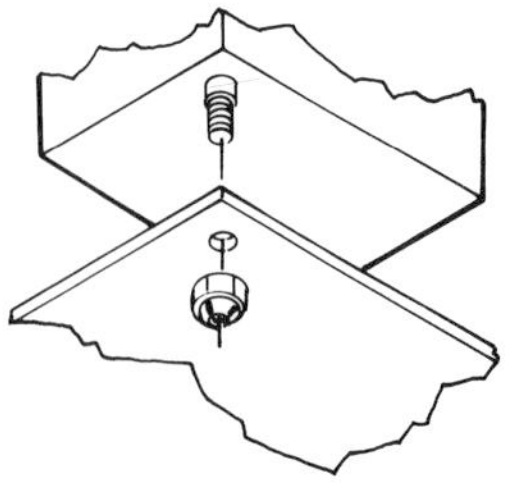	Description:	Horizontal surface mounted. Stop nuts on studs.
	Advantages:	No washer required. One-handed operation. Studs act as locating pins.
	Disadvantages:	Possible loss of nuts. Part must be lifted.
	Tools Required:	Wrench or "spintite."
	Operating Time:	Approximately 1.3 minutes per fastener.

FIG. 8-9 (continued)

MAINTAINABILITY CHECKLIST

GENERAL

1. STANDARDIZATION MAXIMIZED
2. COMPONENTS FUNCTIONALLY GROUPED
3. CONSOLE LAYOUT OPTIMIZED
4. COMPLEXITY MINIMIZED
5. SELF-TEST INCORPORATED
6. MAX. TIME TO REPAIR MINIMIZED
7. TOOLS & TEST EQUIP. MINIMIZED
8. LABELING MAXIMIZED
9. WEIGHT MINIMIZED
10. CALIBRATION REQUIREMENTS KNOWN
11. REPAIR/REPLACE PHILOSOPHY KNOWN
12. MAINT. PROCEDURES KNOWN
13. PERSONNEL REQUIREMENTS MINIMIZED
14. TRADE-OFFS DOCUMENTED

HANDLING

1. EQUIPMENT LIFTING MEANS EMPLOYED
2. EQUIPMENT BASE REINFORCED (FORK-LIFT APP.)
3. DRAWER/PANEL HANDLES EMPLOYED
4. ASSEMBLY HANDLES EMPLOYED
5. CONSOLE CASTORS EMPLOYED (AS APPLICABLE)
6. DAMAGE SUSCEPTIBILITY MINIMIZED
7. WEIGHT LABEL ON CONSOLE

EQUIPMENT RACKS-GENERAL

1. DRAWERS ON ROLL-OUT SLIDES
2. PANELS HINGED
3. IN-POSITION MAINTENANCE POSSIBLE
4. CABLES CONNECTED WITH DRAWERS EXTENDED
5. PERMANENT CABLE INLETS ON FRONT AVOIDED
6. HEAVIEST ITEMS ON BOTTOM
7. OPERATOR PANELS OPTIMUM POSITION
8. AIR INTAKE/EXHAUST PROVISIONS ADEQUATE

PACKAGING

1. PLUG-IN COMPONENTS EMPLOYED
2. COMPONENT STACKING AVOIDED
3. ACCESSIBILITY BASED ON REPLACEMENT FREQ.
4. WRONG INSTALLATION OF UNIT PREVENTED
5. MODULES & MOUNTING PLATES LABELED
6. GUIDES USED FOR MODULE INSTALLATION
7. INTERCHANGEABILITY INCORPORATED

ACCESSIBILITY

1. ACCESS DOORS PROVIDED
2. ACCESS DOORS SELF-SUPPORTED
3. ACCESS DOORS LABELED
4. ACCESS OPENINGS ADEQUATE IN SIZE
5. ACCESS FASTENERS MINIMIZED
6. SPECIAL TOOLS MINIMIZED
7. COMPONENT ACCESSIBILITY ADEQUATE
8. GUIDES FOR DANGEROUS ACCESSES CONSIDERED

FASTENERS

1. QUICK RELEASE FASTENERS EMPLOYED
2. FASTENERS STANDARDIZED
3. QUANTITY OF FASTENERS MINIMIZED
4. HEXAGONAL SOCKET-HEAD FASTENERS USED
5. CAPTIVE NUTS/SCREWS EMPLOYED
6. MINIMUM NUMBER OF TURNS REQUIRED

PANEL DISPLAYS/CONTROLS

1. CONTROLS STANDARDIZED
2. CONTROLS SEQUENTIALLY POSITIONED
3. CONTROLS PROPERLY SPACED
4. CONTROLS ADEQUATELY LABELED
5. CONTROLS ADJACENT TO APPLICABLE DISPLAY
6. RUGGEDIZED METERS EMPLOYED
7. METERS EXTERNALLY REMOVABLE
8. PANEL LIGHTING EMPLOYED
9. INDICATOR LIGHTS "PRESS-TO-TEST"
10. FUSE REQUIREMENTS SATISFIED
11. SPARE FUSES PROVIDED
12. WARNING LIGHTS EMPLOYED-CRITICAL FUNCTIONS
13. COLOR OF INDICATOR LIGHTS ADEQUATE
14. CONTROLS PLACED BY FREQUENCY OF USE

TEST POINTS

1. LOCATED ON FRONT PANEL
2. FUNCTIONALLY GROUPED
3. ADEQUATELY LABELED-NUMBER & SIGNAL VALUE
4. INTERNAL TEST POINTS ACCESSIBLE
5. DEGREE OF TEST INDICATED
6. ADEQUATELY PROTECTED
7. ADEQUATELY ILLUMINATED
8. LOCATED CLOSE TO APPLICABLE CONTROL

ADJUSTMENTS

1. ADJUSTMENT POINTS ACCESSIBLE
2. PERIODIC ADJUSTMENTS KNOWN
3. INTERACTION EFFECTS ELIMINATED
4. ADJUSTMENT LOCKING DEVICES PROVIDED
5. FACTORY ADJUSTMENTS SPECIFIED
6. ADJUSTMENT POINTS ADEQUATELY LABELED
7. FINE ADJUSTMENTS THROUGH LARGE MOVEMENTS
8. BUILT-IN JACKS FOR METER CALIBRATION
9. CLOCKWISE ADJUSTMENTS FOR INCREASING VALUES

PARTS/COMPONENTS

1. ARRANGED IN FAMILY GROUPS
2. ADEQUATELY LABELED
3. ADEQUATELY SPACED FOR TOOL ACCESS
4. INDIVIDUAL PARTS DIRECTLY ACCESSIBLE
5. DELICATE PARTS ADEQUATELY PROTECTED
6. NOT VULNERABLE TO EXCESSIVE SOLDER HEAT

FIG. 8-10 Maintainability Checklist (General).

In the event that the item maintainability requirements are not reflected in the numerical rating, the final factor may be influenced by the comments under "remarks."

UTILIZATION OF DESIGN AIDS

During the evolutionary design process, the design engineer is confronted with the task of defining the equipment-package size, weight,

CABLES
1. CABLES FABRICATED IN REMOVABLE SECTIONS
2. CABLES ROUTED TO AVOID SHARP BENDS
3. CABLES ROUTED TO AVOID PINCHING
4. PROTECTION FOR CABLES ROUTED THRU HOLES
5. CABLES IDENTIFIED
6. CABLE CLAMPING SUPPORT ADEQUATE
7. HANDHOLD/STEP PREVENTION CONSIDERED

CONNECTORS
1. QUICK DISCONNECT VARIETY
2. CONNECTOR SPACING ADEQUATE
3. LABELING ADEQUATE
4. CONNECTORS KEYED
5. CONNECTORS STANDARDIZED
6. SPARE PINS PROVIDED
7. MALE CONNECTORS CAPPED
8. RECEPTACLES "HOT" & PLUGS "COLD"
9. MOISTURE PREVENTION CONSIDERED

SERVICING/LUBRICATION
1. SERVICING REQUIREMENTS CONSIDERED
2. SERVICING POINTS ACCESSIBLE
3. SERVICING FREQUENCIES KNOWN

ENVIRONMENT
1. TEMPERATURE/HUMIDITY RANGES CONSIDERED
2. ILLUMINATION ADEQUATE
3. TRANSPORTABILITY CONDITIONS CONSIDERED
4. MOBILITY CONDITIONS CONSIDERED
5. STORAGE CONDITIONS CONSIDERED

SAFETY
1. ELECTRICAL OUTLETS/JUNCTION BOXES LABELED
2. INTERLOCKS EMPLOYED
3. FUSE/CIRCUIT BREAKER PROTECTION ADEQUATE
4. WARNING DECALS ADEQUATE
5. GUARDS/SAFETY COVERS--HIGH POTENTIALS
6. PROTRUDING DEVICES ELIMINATED
7. EXTERNAL METAL PARTS ADEQUATELY GROUNDED
8. DRAWER/PANEL/STRUCTURE EDGES ROUNDED
9. TOOL USE CONSIDERED

RELIABILITY
1. ALLOCATED MTBF KNOWN
2. FAIL-SAFE PROVISIONS INCORPORATED
3. CRITICAL/SERVICE LIFE CONSIDERED
4. WEAR-IN/WEAR-OUT CYCLES CONSIDERED
5. FAILURES TRACEABLE BY TEST

WHEN REVIEWING LAYOUTS/DRAWINGS, THIS CHECKLIST MAY PROVE BENEFICIAL IN COVERING VARIOUS DESIGN FEATURES APPLICABLE TO MAINTAINABILITY. THE ITEMS OR CATEGORIES LISTED ARE IN MOST CASES BACKED UP WITH MORE DETAILED QUESTIONS BASED ON SPECIFIC CRITERIA. THE LIST IS SO DESIGNED THAT THE ANSWER TO EACH ITEM (AS APPLICABLE) SHOULD BE "YES".

REFERENCE DOCUMENTS SPECIFYING GOOD MAINTAINABILITY CRITERIA INCLUDE:

1. Navships 94324, Maintainability Design Criteria Handbook for Designers of Shipboard Electronic Equipment
2. ASD-TR-61-424, Guide to Integrated System Design for Maintainability
3. General Dynamics/Electronics Maintainability Handbook

FOR FURTHER INFORMATION, CONTACT YOUR MAINTAINABILITY ENGINEER

FIG. 8-10 (continued)

internal layout (placement of assemblies/modules), access points, etc. As a normal function, the maintainability engineer works on a day-to-day liaison basis with the mechanical or packaging engineer with the intent of developing the best design approach. The process of determining an optimum equipment-packaging scheme during these early stages is often quite difficult when relying solely on two-dimensional layout drawings. All assemblies, subassemblies, modules, test points, accesses, etc., are not shown in their true perspective. As an aid to the design engineer, the maintainability engineer can assist through the development and utilization of three-dimensional scale-models or mock-ups. Such models or mock-ups can be developed to any desirable scale and to varying degrees of detail depending on the level of emphasis required. In addition, mock-ups can be developed on a relatively inexpensive basis and in a short period of time when employing the right materials and personnel services (industrial-design, human-factors, and/or model-shop personnel are usually well oriented to this area of endeavor and should be utilized to the maximum degree possible). The intent of a mock-up is to provide realistic three-dimensional simulation of a proposed final equipment-design configuration at an early stage of a program prior to the development of formal design data or actual prototype hardware. The values of a mock-up are numerous:

1. It provides the design engineer with the opportunity of experiment-

1. Provide a signal that shows when test equipment is warmed up, or if this is not feasible, indicate warm-up time required next to the turn-on switch.
2. Provide quick, simple check to indicate when out of calibration or not functioning.
3. Provide for direct reading displays not requiring multiplication or other transformation of values.
4. If transformation of display value cannot be avoided, provide conversion table mounted on equipment.
5. If mounted conversion table is not feasible, provide transformation factor adjacent to each switch position or display requiring use of the factor.
6. When more than one scale must be in technician's view, differentiate by labeling and color coding to respective control positions.
7. Provide selector switches in lieu of a number of plug-in connections.
8. Provide circuit breakers or fuses to safeguard against damage if wrong switch or jack position is used.
9. Provide some device, such as warning light or automatic disconnect, which operates when test equipment lid is closed and written warning notice to ensure equipment turn-off when testing is completed.
10. Prominently display on outer surface a statement of purpose (use) for tester and any special precautions.
11. Label every item which technician must recognize, read, or manipulate.
12. Label outer case and all removable parts (loose items such as cords, plugs, probes, etc.) with their official nomenclature.
13. When adapters are required, be sure they are provided as a removable accessory.
14. Provide fasteners or holders in storage compartments for all accessories.
15. Recess handles on outer or carrying cases of portable test equipment for convenient and space-saving storage.

FIG. 8-11 Statements of Maintainability Criteria (General) Example.

MAINTAINABILITY
INFORMATION
BULLETIN

NO: 2.1
DATE: 5/8/63

SUBJECT:

ACCESSIBILITY

Opening Dimensions	Dimension* (in inches) A	Dimension* (in inches) B	Maintenance Task
	4.8	5.0	Grasping small objects (less than 2 1/2" diameter)
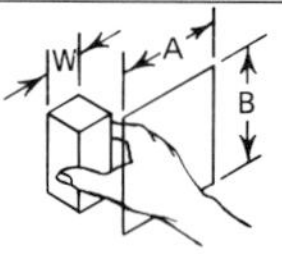	W + 1.75	5.0 †	Grasping large objects (more than 2 1/2" wide)
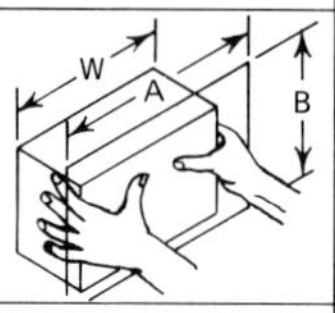	W + 3.0	5.0 †	Grasping large objects with two hands, with hands extended through openings up to fingers
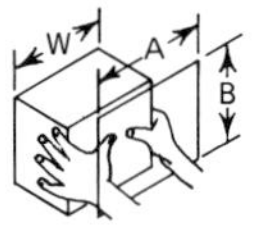	W + 6.0	5.0 †	Grasping large objects with two hands, with hands extended through openings up to wrists
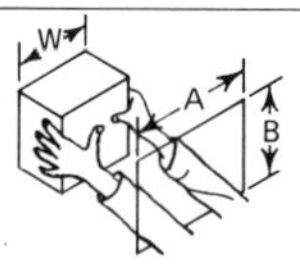	W + 6.0	5.0 †	Grasping large objects with two hands, with arms extended through openings up to elbows

Minimum Rectangular Openings for Grasping Parts

* The dimensions are extrapolated from data presented in: *Guide to Design of Electronic Equipment for Maintainability*. WADC Technical Report 56-218, Wright Air Development Center, Wright-Patterson Air Force Base, Ohio, 1956; and *Anthropometry of One-Handed Maintenance Actions*, Technical Report NAVTRADEVCEN 330-1-3, U. S. Navy Training Device Center, Port Washington, New York, 1960.

† Or sufficient to clear port
(Data Extracted From NAVSHIPS 94324)

PREPARED Ben S. Blanchard APPROVED John J. Smith

FIG. 8-12 Maintainability Information Bulletin. *(Data extracted from NAVSHIPS 94324.)*

4.1 Maintainability Design Criteria:

4.1.1 The design characteristics of the MPPG shall be such to permit fault isolation, repair, and test within a time frame of thirty minutes. Maintenance tasks shall not require the use of special equipment aside from what is normally supplied as part of AGERD 6811. A high impedance voltmeter and an accurate (1 part in 10^6) countertimer are not considered special equipment.

4.1.2 Design characteristics shall be consistent with operator skills and technician skills normally associated with AFSC-3 (Repair tasks) and AFSC-5 (Troubleshooting, test, and MPPG operation) classifications.

4.1.3 Self-test provisions shall be considered in the design of the MPPG.

4.1.4 Circuit functions should be logically grouped to permit rapid fault isolation. Physical packaging of specific circuit elements should be by functional grouping wherever feasible to enhance repair tasks.

4.1.5 Plug-in type modules and parts should be utilized where feasible in terms of operation requirements and cost.

4.1.6 Provisions shall be incorporated to ensure that parts are mounted (installed) properly during the repair tasks. Guides and location keys should be provided on plug-in items.

4.1.7 Reference designators and part identification data shall be provided wherever practical. Test points shall be marked and coded. Color coding shall be utilized where practical to aid in distinguishing parts.

4.1.9 Controls and indicators should be functionally grouped, sequentially positioned, and adequately labeled.

4.1.10 Front panel mounted test points shall be provided, as a minimum for the following functions:

a. Power supply output

b. Delay sync output

c. Pulse sync output

d. Test pulse output

e. Main pulse output

f. Start sync input

g. Stop sync input

4.1.11 Test points shall be provided on each major plug-in module to determine input and output functions.

4.1.12 A template containing adequate markings shall be provided as overlay to the wire-wrap connections.

4.1.13 Standard parts should be utilized where possible. The variety of parts should be minimized.

4.1.14 The MPPG shall have provisions to extend the item beyond the cabinet facing. Slides shall be considered. Access to all internal parts and circuits should be possible with the MPPG in the extended position, except where internal covers are included by necessity.

4.1.15 Covers shall be mounted with quick-disconnect fasteners.

4.1.16 Servicing requirements of the MPPG should be held to a minimum. Sealed "LIFE" bearings should be used where possible.

FIG. 8-13 Maintainability Criteria (Specified) (Sheet 1 and Sheet 2).

PRELIMINARY MAINTAINABILITY EQUIPMENT EVALUATION

NO. M227

EQUIPMENT NAME Sweep Oscillator MANUFACTURER Green Engineering

TYPE 38149-001 MODEL B R. Krest SERIAL NO. 321-09

POINTS 57 ☐ GOOD ☒ FAIR ☐ POOR ENGINEER R. Krest DATE 3-6-64

ADEQUACY, ACCESSIBILITY, ETC. DETERMINED ON FOLLOWING ITEMS AS APPLICABLE (0 TO 4 POINTS OBTAINABLE DEPENDING UPON THE DEGREE OF ACCEPTABILITY. 100 POINTS POSSIBLE).

1. GENERAL	PTS
1. WARMUP TIME	4
2. LABELING	2
3. MAINT. TOOLS/EQUIP.	1
2. CIRCUITRY	
1. COMPLEXITY	3
2. FUNCTIONAL PKGING	3
3. TEST POINTS	
1. AVAILABILITY	0
2. LABELING	0
3. LOCATION/GROUPING	0

4. PACKAGING	PTS
1. FAST VARIETY/TYPE	3
2. PARTS STANDARDIZED	3
3. CONNECTORS	4
4. WIRING/CABLING	4
5. PRINTED CIRCUITS	2
6. PROBE/SOLD/IRON/OTHER TOOL SPACE	3
7. PLUG-IN-MODULES	0
5. MAINT. AIDS	
1. MALF. EQUIP. KNOWN	2
2. SELF CHK/CALIB.	1

6. HUMAN FACTORS	PTS
1. INST. POS. MAINT	2
2. DISPLAYS	3
3. CONTROLS	2
4. WEIGHT/HANDLING	2
5. DAMAGE SUSCEPTIBLE	3
6. PERSONNEL TRAINING	2
7. SAFETY	
1. EQUIPMENT	4
2. PERSONNEL	4

REMARKS:

1) Warmup time is approximately 2 minutes.

2) Labeling inadequate - not completely labeled with reference designations.

3) Packaging - P/C boards were used where possible, but some of them are almost impossible to remove. P/C boards are wired-in to chassis - not plug-in type.

4) Special tools are needed for P/C boards removal.

5) No test points are designed into the equipment. However, there are a few BN connectors and terminal points that can be used as test points.

6) Equipment, when rackmounted, must be removed from the rack for fuse replacement (fuse is located on the rear surface of the chassis). No spare fuse is provided.

FIG. 8-14 Equipment Evaluation (Example).

ing with different packaging layouts prior to the preparation of formal design drawings.

2. It provides the design engineer with the opportunity of experimenting with different packaging sizes prior to the preparation of formal design drawings.

3. It provides the maintainability engineer with the opportunity to accomplish a more effective review of a proposed design configuration relative to the incorporation of desired maintainability features.

Category/Item (Refer to Figure 8-14, Item 4.2.)	Numerical Rating (Points)
4.2 Parts Standardized a. All equipment parts are of the standard AN, NAS, or MIL-STD variety and are likely to be available in government supply.	4
b. Seventy-five percent of all equipment parts are of the standard AN, NAS, or MIL-STD variety and are likely to be available in government supply. Twenty-five percent are commercially available, but are not likely to be available in government supply.	3
c. Fifty percent of all equipment parts are of the standard AN, NAS, or MIL-STD variety and are likely to be available in government supply.	2
d. Seventy-five percent of all equipment parts are of the standard AN, NAS, or MIL-STD variety or are commercially available. Twenty-five percent are nonstandard or peculiar parts not likely to be readily available in either government supply or commercial supply.	1
e. Fifty percent of all equipment parts are of the standard AN, NAS, or MIL-STD variety or are commercially available. Fifty percent are nonstandard or peculiar parts not likely to be readily available in either government supply or commercial supply.	0

FIG. 8-15 Equipment-evaluation Criteria (Rating Data).

4. It provides the maintainability engineer with a tool for use in the accomplishment of maintainability predictions and the development of maintenance-analysis data. Simulated maintenance operations accomplished on the mock-up (e.g., removal of outside access panels, removal and replacement of major assemblies) are often possible, providing detailed maintenance task-time data.

5. It provides the design engineer with an excellent tool for conveying his final design approach during a formal design review.

In general, the mock-up approach not only is beneficial for design improvement, optimization, and verification purposes, but also serves as an excellent sales aid. Examples of several mock-ups which have actually been employed for the above purposes are presented in Figs. 8-16 and 8-17.

FIG. 8-16 Mock-up of Control Center (Operational Shelter). (*General Dynamics Corporation.*)

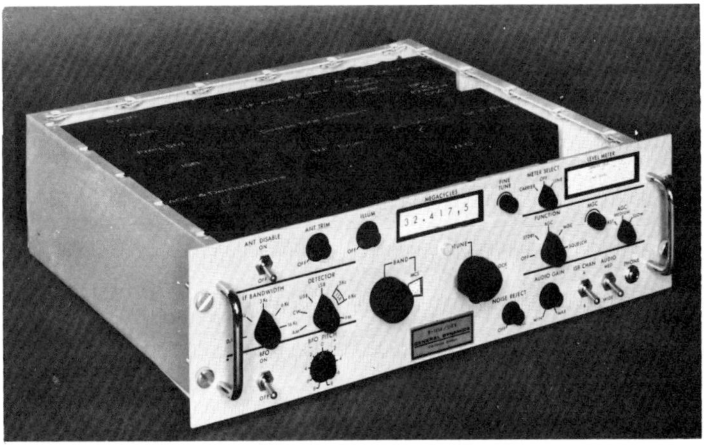

FIG. 8-17 Equipment Drawer Mock-up (Module Layout Study). (*General Dynamics Corporation.*)

DESIGN REVIEW AND ANALYSIS

Drawing Review

As equipment design progresses and engineering layouts, detailed drawings, lists of material, etc., are generated, the maintainability engineer provides a day-to-day review and drawing sign-off function. The purpose of the review is to ensure maximum consideration of maintainability features in design. The review is conducted by evaluating individual design features. Approval results in drawing sign-off. Disapproval results in the preparation of recommendations for product improvement, which are in turn submitted to the cognizant design engineer for consideration. After review of the maintainability recommendation, the design engineer makes a decision as to whether the recommendation should be incorporated or rejected. Incorporation results in a drawing change and subsequent second review by the maintainability engineer. Rejection generally results in a meeting between design engineering and maintainability personnel. The reasons for the rejection are discussed. In instances where a rejection results in significant compromises of the maintainability design, such a rejection is documented in the form of trade-off report which becomes a part of the data package required for formal design reviews. The overall informal data-review sequence is illustrated in Fig. 8-18.

Design Analysis

In instances where specific quantitative maintainability-design requirements are imposed, the design engineer may (on a periodic basis) wish

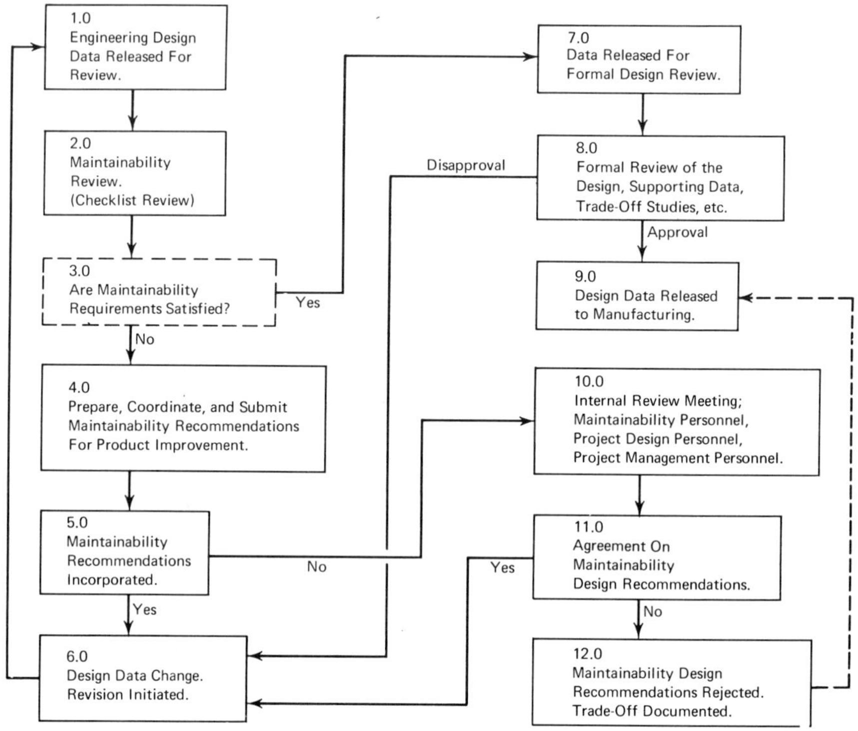

FIG. 8-18 The Sequence of Review of Design Data.

to evaluate his design against those specified requirements. Such an evaluation is best accomplished through a maintainability prediction, the maintenance analysis, or a maintainability demonstration, depending on the program phase. Through these methods of evaluation, the maintainability engineer is able to provide the design engineer with a quantitative index measure of the design. Comparison of this value against the specified requirement will provide either an indication that the design is progressing satisfactorily, or that a problem exists and that corrective action is required. In addition, the evaluation process will localize the problem area for the designer. The prediction, maintenance analysis, and demonstration methods and techniques are covered in subsequent chapters of this text.

RECOMMENDATIONS FOR PRODUCT IMPROVEMENT

In submitting recommendations for product improvement to the design engineer, several approaches may be used. The most effective approach is through the verbal conveyance of ideas and suggestions on a day-to-

GENERAL DYNAMICS
Electronics Division

DESIGN ASSURANCE CORRECTIVE ACTION REQUEST

DATE ORIGINATED	CONTROL NUMBER
9/6/66	M105

TO John S. Stone AREA 54 ROOM 18E EXT. 5398

ORIGINATOR Joe D. Smith AREA 54 ROOM 18E EXT. 5345

REQUEST AFFECTS: ☐ RELIABILITY ☒ MAINTAINABILITY ☐ VALUE ENGINEERING ☐ HUMAN FACTORS ☐ EVALUATION TEST ENG.

PROGRAM F-118 BNS Test Station Des & Dev

ITEM NAME IRU Switching Panel ITEM NO. ____

DWG/PART NO. A 11016-0011 REV/EFF. A

D.A. ENDORSEMENTS:

SECTION SUPERVISOR Sam H. Brown DATE 9/7/66

PROJECT D.A. ADM. Robert A. Anderson DATE 9/7/66

NO.	PROBLEM DESCRIPTION	CORRECTIVE ACTION REQUESTED	DISPOSITION SHOULD BE MADE AND THE DACAR COMPLETED AND RETURNED TO DESIGN ASSURANCE WITHIN FIVE DAYS OF RECEIPT.
1	The 36-Position, G-Wafter Switch on the IRU Switching Panel Requires an extensive amount of time for switch replacement and/or repair (54 wires to switch and 61 wires between wafers).	Mount Switch on a separate bracket which in turn should be mounted on the front panel. Route wires (54) from switch thru 60-PIN MS Connector. Corrective maintenance will involve removal of 4 screws and a connector break.	☒ ACCEPTED INCORPORATE IN: Problem No. 1 1. Proposed switch design accepted 9/10/66, and will be incorporated in S/N 2 and on. S/N 1 will not be retrofitted.
2	Access to the numerous small switches, terminals, etc., mounted behind the IRU switching panel requires removal of 24 countersunk screws. An additional 20 countersunk screws are used to fasten a dust cover which must be removed for tool access.	Provide one-quarter (¼) turn fasteners in place of countersunk screws to facilitate access Reduce quantity of fasteners to 8 on the panel and 6 on the dust cover	☐ REJECTED REASON:
			☒ PARTIAL ACCEPTANCE Problem No. 2 Use of ¼ turn fasteners accepted 9/12/66. S/N 2 and on. Reduction in Qty. of fasteners rejected due to R.F.I problems. SIGNATURE ____ TITLE Sr. Design Eng. DATE 9/14/66

FORM 1116 E 9/66

FIG. 8-19 Formal Maintainability Recommendation (Example). (*General Dynamics Corporation.*)

MAINTAINABILITY RECOMMENDATION (Example *A*)	MAINTAINABILITY RECOMMENDATION (Example *B*)
PROBLEM:	PROBLEM:
The 36 position, 6 wafer switch on the Test Control Unit (A11471-00) requires an excessive amount of time for switch replacement and/or repair (54 wires to switch and 61 wires between wafers).	Troubleshooting the Serializer is difficult due to lack of test points on front panel or lack of tip jacks internally. In addition, heavy use of the terminal points presently provided will likely cause excessive maintenance-induced faults.
RECOMMENDATION:	RECOMMENDATION:
Mount switch on a separate bracket which in turn should be mounted on the front panel. Route wires (54) from switch through 60 pin MS connector. Corrective maintenance will involve removal of 4 screws (required for mounting switch assembly on panel) and a connector break in lieu of individual wire replacement with the switch installed on the panel.	Add the following test point coverage to the Serializer front panel: a. Tape Levels 1 - 7 b. Clear Pulse c. Enter Pulse d. Clock Pulse e. Three (3) Common Inputs
RESULTING EFFECTS:	RESULTING EFFECTS:
Average Time Saving/Maintenance Action - - 1 hour,22 minutes	Time Saving/Maintenance Action - - 7 min
Estimated Quantity of Maintenance Actions Covering 30 Equipments for 10 Years - - 390	Estimated Quantity of Maintenance Actions Covering 30 Equipments for 10 Years - - 1770
Estimated Cost Saving (Labor) - $2,949	Estimated Total Cost Saving - - $1,143
Estimated Additional Material Expenditure - - $1,920	
Estimated Total Net Cost Saving - - $1,029	
Increase in Equipment Availability (A_a) - - 2.4 percent	

FIG. 8-20 Formal Maintainability Recommendation (Alternate Form) (Examples).

day basis prior to the generation of formalized design data. In presenting such ideas and recommendations, the maintainability engineer must employ tact and diplomacy. It must be remembered that the ultimate responsibility for the design lies with the design engineer, and the equipment characteristics are the results of *his* decisions. If a good professional relationship exists between the design engineer and the maintainability engineer, and the exchange of ideas is handled on an informal basis, the probability of acceptance of submitted maintainability recommendations is relatively high. Hopefully, the majority of maintainability recommendations will be handled in this manner. A second approach involves the generation and submittal of maintainability recommendations through more formal means. At a later stage in the design process, when engineering layouts and detailed design drawings are being reviewed by the maintainability engineer, or during the prototype fabrication and test phase, when problem areas are detected, maintainability recommendations resulting in desired design changes are submitted in writing to the design engineer. At this point in time, major design changes may result in program schedule delays and additional costs. Therefore, it becomes necessary to handle such recommendations in a formal manner. Examples of the format that may be used to record recommendations are illustrated in Figs. 8-19 and 8-20.

QUESTIONS AND PROBLEMS

1. For what purpose does the maintainability engineer develop general design criteria? Specific criteria?
2. Why does the maintainability engineer work with other members of the design team in performing major trade-offs?
3. Of what value is a maintainability evaluation of vendor parts and assemblies?
4. List three reasons why a mock-up is of value to the designer or the maintainability engineer.
5. In your own words, outline the purpose of intermediate maintainability predictions.
6. What is the reason for reviewing design data (drawings, documents) and obtaining sign-off by the maintainability engineer?
7. The main purpose in performing maintainability design-liaison and support services is to facilitate the design process. True or false?
8. List one reference document that contains general maintainability criteria.
9. Specific maintainability criteria should be inserted in the system/equipment design specification. Why?
10. The submittal of a maintainability-design recommendation by formal means is the most effective manner to obtain a positive response. True or false? Why?

NINE

Design Characteristics

The purpose of this chapter is to illustrate a variety of maintainability design characteristics as they exist in present products. The examples are typical results of the efforts applied by maintainability engineers while performing design-liaison work.

The selected illustrations concern some of the maintainability-design characteristics of automobiles, electronic equipment, and fluid systems. It is believed by the authors that these examples provide contrast as regards the types of items subject to maintainability-design influence. Equally good examples are evident in some household appliances, television sets, production machinery, aircraft, atomic power plants, ships, missiles, and a great number of other items, products, or installations.

Regardless of the choice of illustrations, the reader may take issue with one or more of the examples based on his experiences. This type of response, however, tends to support the need for more and better maintainability-design features in all commercial and military products.

AUTOMOBILE MAINTAINABILITY-DESIGN CHARACTERISTICS[1]

Engine and Compartment

Recent innovations and progressive design developments have resulted in the following maintainability-design characteristics for the automobile engine and items in the engine compartment. (See Figs. 9-1 and 9-2.)

1. New spark-plug design (longer life, facilitates cleaning and reconditioning)
2. New alternator design (modified bearing housings for longer life, permanent lubrication feature)
3. New battery design (one piece, sealed unit minimizes loss of electrolyte, and protects against terminal corrosion)
4. Central placement of engine in the compartment, unrestricted access to maintenance significant items:

a. Cooling-system elements
b. Alternator
c. Spark plugs
d. Oil filter
e. Coil and capacitor
f. Distributor
g. Brake master cylinder
h. Steering case
i. Washer motor and pump
j. Electric-circuit parts and wiring
k. Carburetor
l. Air and fuel filters
m. Oil-filler inlet (servicing point)
n. Battery
o. Oil measuring stick

5. Major redesign of the engine to make fullest use of common components between several types of engines (reduction in type and quantity of supply parts)

Transmission

Figures 9-3 and 9-4 illustrate the design progress made on automatic transmissions over a period of 10 years. Note the decrease in complexity of the 1966 model (Fig. 9-4) as compared with the 1956 model. Aside from the significant reduction in spare-part requirements due to the decrease in design complexity, the later model contains parts that have longer life characteristics. As regards servicing needs, the 1966 model does not require periodic changes of the transmission oil.

Axle-shaft Design

Figure 9-5 illustrates the difference between the design characteristics of the 1964 model and 1965 model axle shafts. The features of the

[1] The illustrations and information on automobile maintainability-design characteristics were supplied by Chrysler Corporation.

FIG. 9-1 Automobile Engine. (*Chrysler Corporation.*)

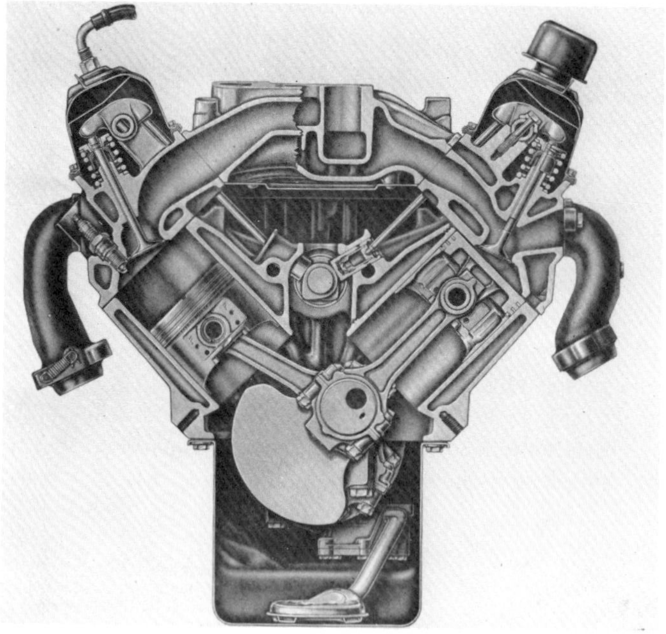

FIG. 9-2 Cross Section of V-8 Engine. (*Chrysler Corporation.*)

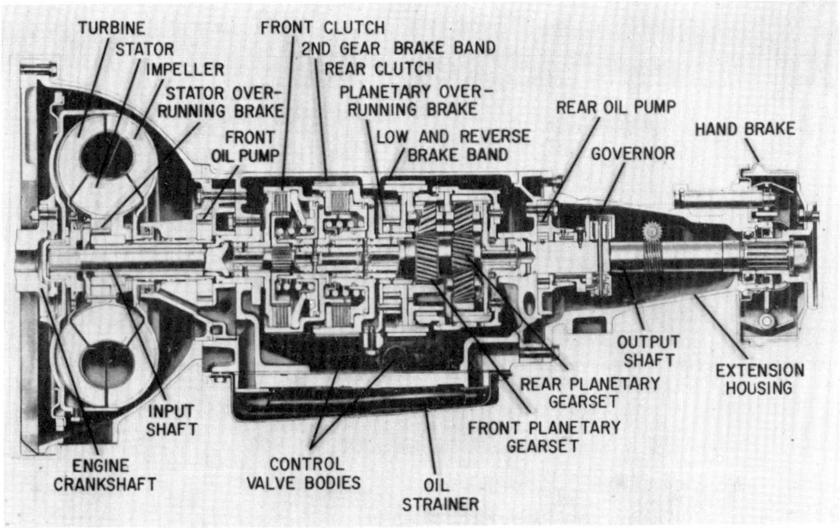

FIG. 9-3 Automatic Transmission, 1956 Model. *(Chrysler Corporation.)*

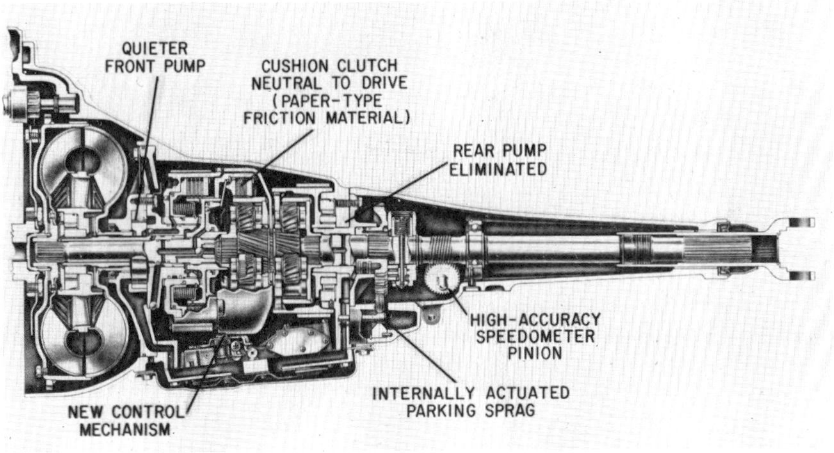

FIG. 9-4 Automatic Transmission, 1966 Model. *(Chrysler Corporation.)*

1965 design are:

1. The new shaft is extruded, rather than forged, resulting in better grain structure. It is resistant to deflection and fatigue (longer life).

2. The physical design of the new shaft permits fast and easy removal of brake drums for periodic inspection.

3. The single-threaded adjustment feature controls end play, and simplifies assembly procedures.

Selector Linkage

Figure 9-6 illustrates the new design of the drive gear selector linkage. This simple push-pull rod replaces a two-cable method previously used. Servicing needs have been eliminated in the new design.

Transmission Shaft and Flange

The addition of a heavy-duty bearing extended the life of the transmission assembly (see Fig. 9-7). The change in shaft configuration (addition of flange) permitted an increase in thrust forces without damage to the assembly. As a result, the torque capacity could be increased for better performance of the vehicle.

ELECTRONIC EQUIPMENT DESIGN CHARACTERISTICS[2]

Figures 9-8 through 9-11 illustrate some typical maintainability-design features in electronic equipment. The characteristics of design are explained in the following key.

[2] The illustrations and information on electronic equipment maintainability-design characteristics were supplied by the Electronics Division of General Dynamics Corporation.

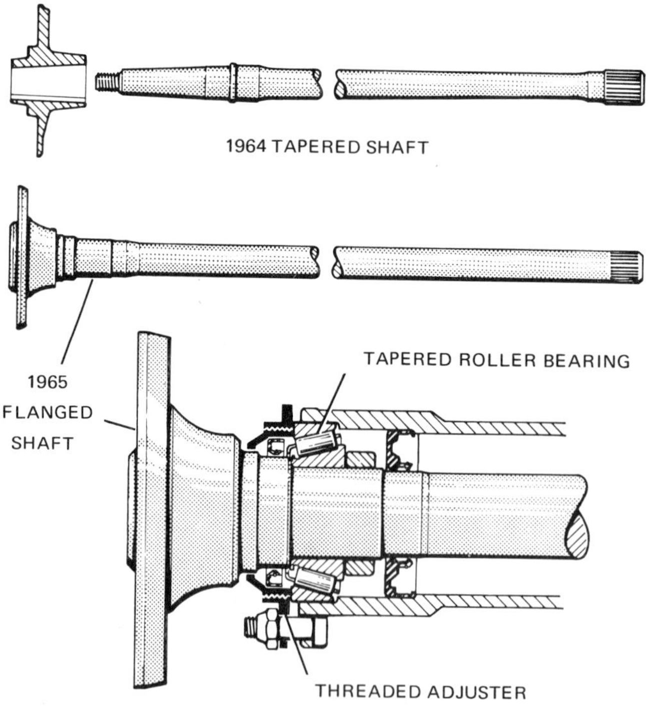

FIG. 9-5 Rear Axle-shaft Comparison. (*Chrysler Corporation.*)

Associate each number call-out on the figure with the comparable number in the key.

① Modular construction. Plug-in boards facilitate repair. Repair time is minimized.

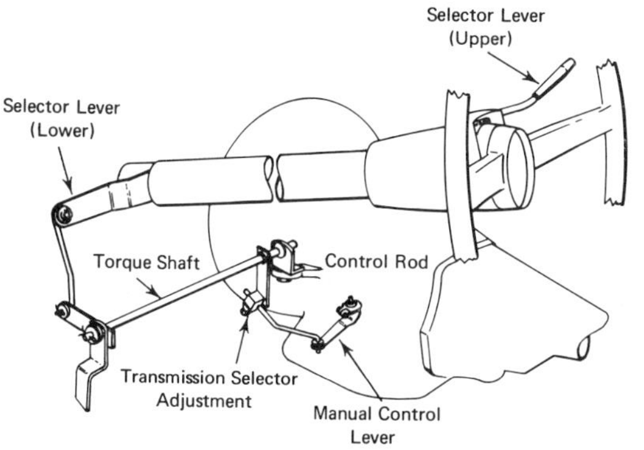

Fig. 9-6 Column-mounted Selector Linkage. *(Chrysler Corporation.)*

② Test points provide for positive fault isolation to a board assembly.

③ Cover-attaching hardware is of the quick-disconnect type. Access time is minimized.

④ Test points provide for positive fault isolation to a subassembly.

⑤ Adequate marking aids in troubleshooting and repair. The equipment elements can be related to schematics, and parts can be fully identified for replacement purposes.

⑥ Connectors are recessed to avoid damage when mating the drawer to the rack.

⑦ Hinged-panel assembly provides full access to mounted parts.

⑧ Parts are functionally grouped to reduce troubleshooting time.

⑨ Expansion capability is considered in the initial design. Provisions

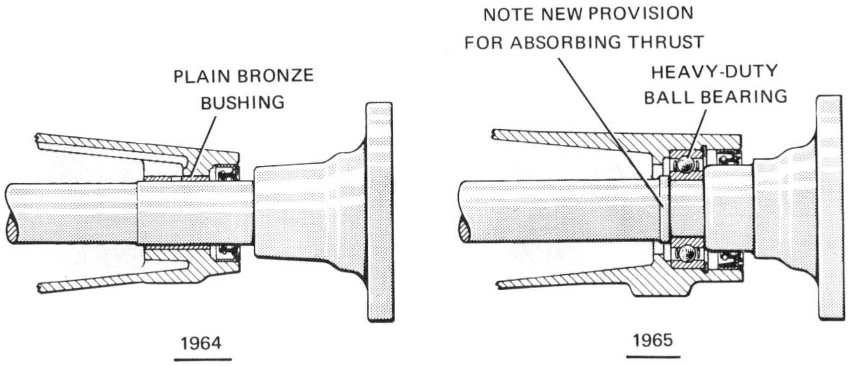

FIG. 9-7 Transmission Output Shaft and Flange Comparison. *(Chrysler Corporation.)*

are made to extend the functional capability of the equipment without causing redesign of the entire drawer.

⑩ Spare fuses are provided with the assembly.

⑪ Fuses are of the indicator type.

⑫ Fault indicator is mounted on front panel.

⑬ Handles are provided to remove (or extend) the panel, and to facilitate handling of the panel assembly.

FLUID SYSTEM DESIGN CHARACTERISTICS[3]

Portable-standard Test Kit

The use of a portable standard to test or calibrate fluid (pneumatic, hydraulic) systems can be equally applied to large installations (e.g., industrial plant-fixed facilities, missile underground-site facilities) and

[3] The illustration data and information on fluid system maintainability-design characteristics were supplied by Mr. Paul H. Hayden.

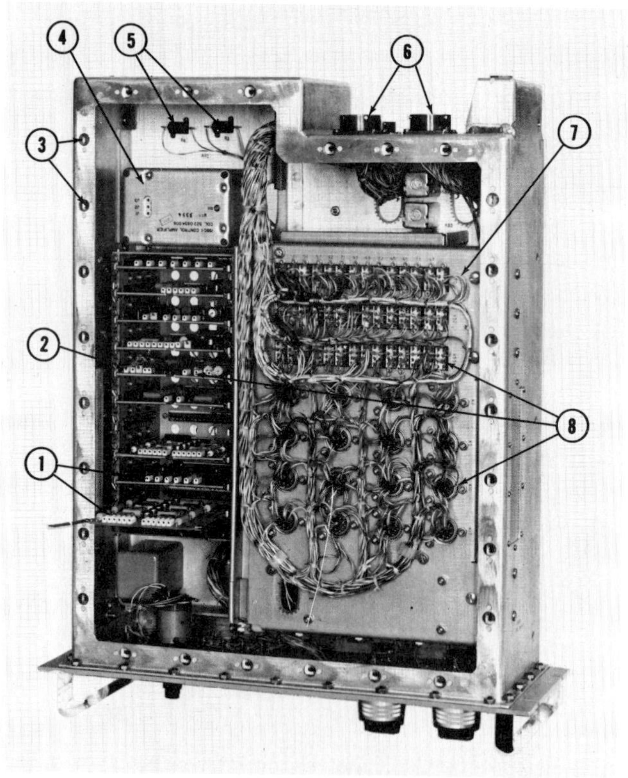

FIG. 9-8 Electronic Equipment Drawer (Example No. 1). *(General Dynamics Corporation.)*

conventional test equipment (e.g., hydraulic test stand). The maintainability-design features illustrated in Figs. 9-12 and 9-13 are applicable to either case. The principles involved are:

1. The use of a portable standard permits "in-place" calibration of fixed line gages. With this approach, it is *not necessary* to remove and replace the line gages for calibration purposes. The advantages of this are:
   a. System downtime is reduced or eliminated.
   b. Labor, tools, and equipment requirements for removal and replacement of gages are eliminated.
   c. Secondary support facilities and equipment requirements to calibrate gages are minimized through use of the portable standard.
   d. The in-line test reduces the possibility of system contamination.
   e. The possibility of equipment damage through remove-and-replace maintenance actions is reduced by utilizing the in-line method of test.
2. The portable-standard test kit is comprised of a "master gage" and

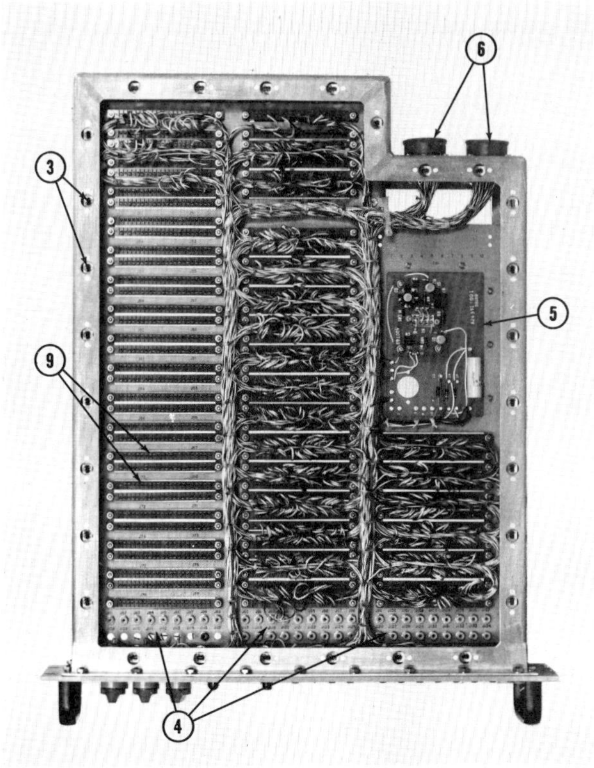

FIG. 9-9 Electronic Equipment Drawer (Example No 21).
(General Dynamics Corporation.)

a regulated pressure source. The system gage to be calibrated is isolated from the system pressure source. The *standard pressure source* is inserted into the system and applied to the gage under test. The reading on the system gage is then compared to that on the master gage. Comparisons can be made over the full range of the system gage.

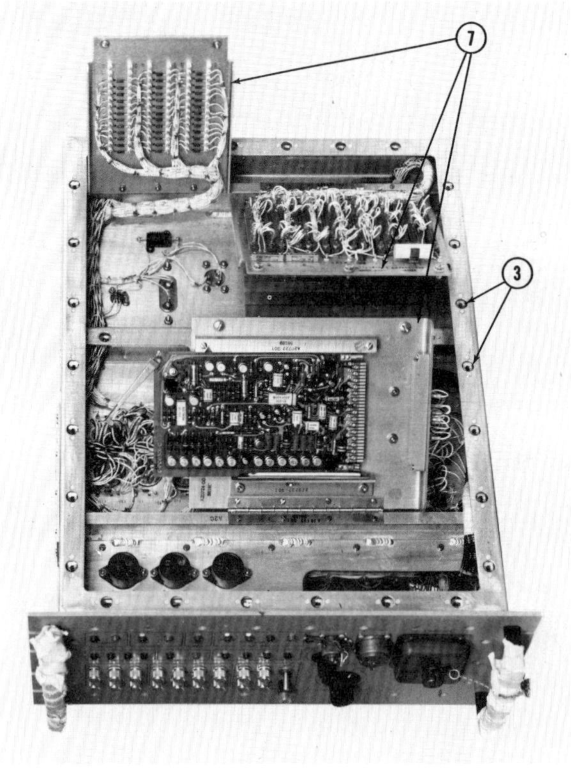

FIG. 9-10 Electronic Equipment Drawer (Example No. 3). (*General Dynamics Corporation.*)

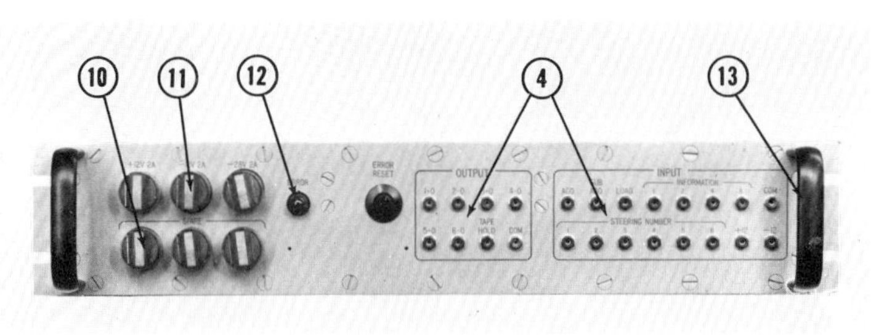

FIG. 9-11 Electronic Equipment Panel. (*General Dynamics Corporation.*)

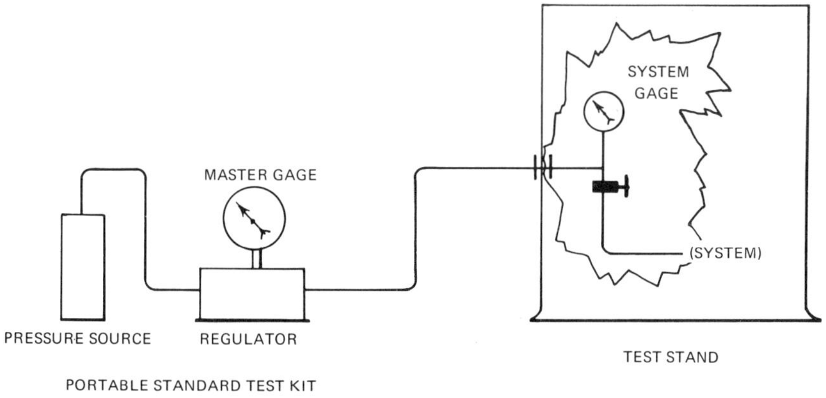

FIG. 9-12 Portable Standard Applied to System Test.

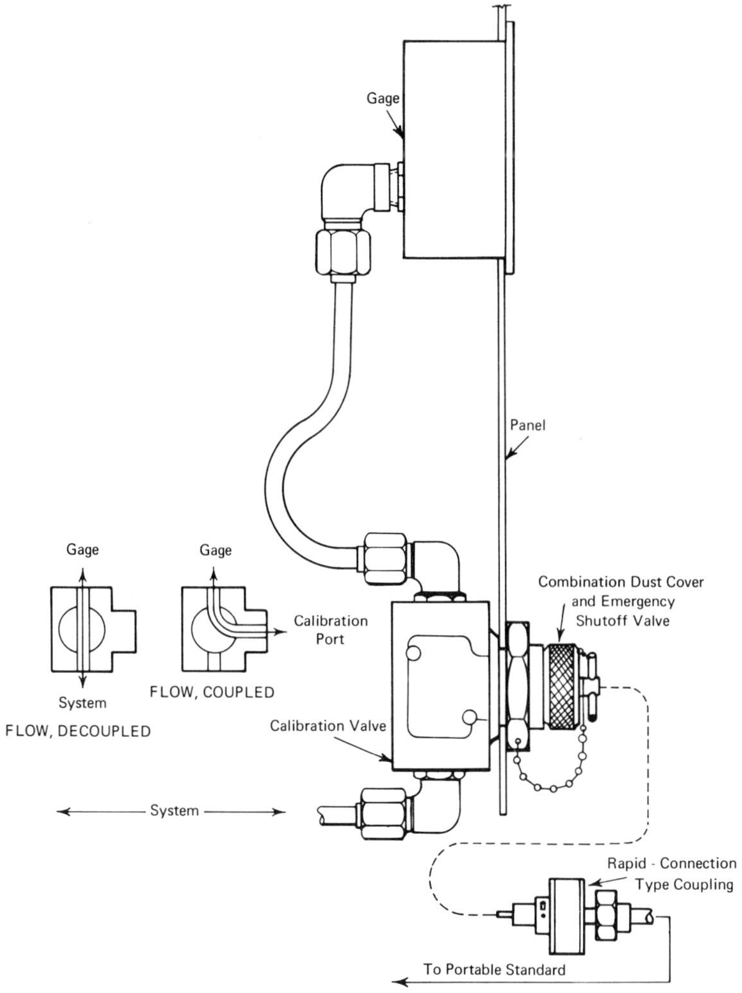

FIG. 9-13 Interface of Portable Standard with System Gage. *(Courtesy of Paul H. Hayden.)*

Line-pressure Checks

Maintenance tests of sealed pressure lines can be accomplished with the use of a portable system gage (see Figs. 9-14 and 9-15). The advantages of using a portable system gage are:

1. Periodic line-pressure checks can be performed to determine line loss or pressure drops without the need of having a permanently installed line gage.

Installed line gages require periodic calibration, which imposes additional maintenance requirements.

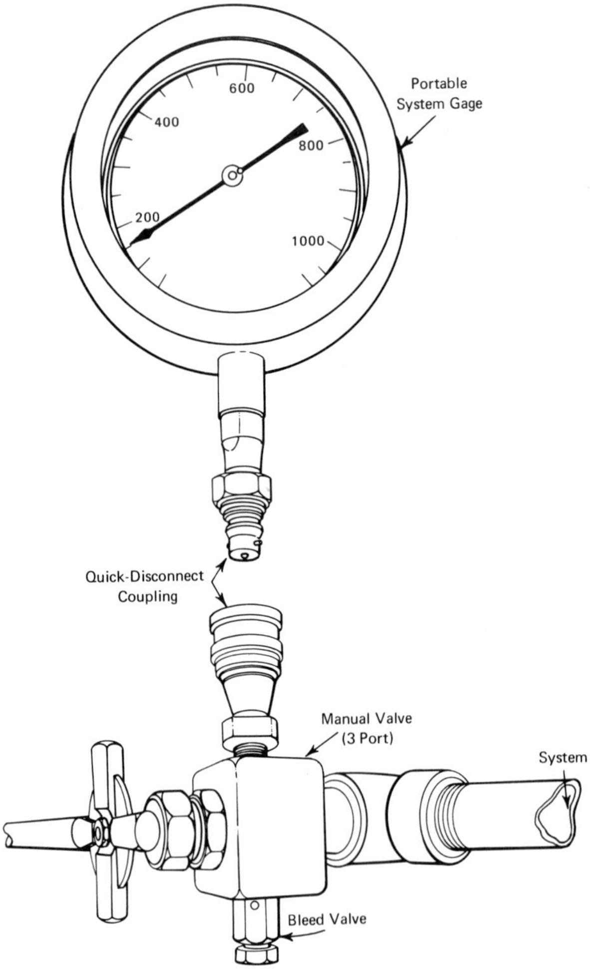

FIG. 9-14 Line Pressure Test with Use of Portable Gage (Method No. 1). *(Courtesy of Paul H. Hayden.)*

2. Investment costs for line gages are eliminated by adopting the in-line method of test.

3. The probability of contaminating the system line is reduced by utilizing the in-line method of test.

QUESTIONS AND PROBLEMS

1. List three possible maintainability-design features of a home television set.
2. What are some of the benefits that will be realized by a washing-machine manufacturer who incorporates good maintainability features into the design of his products?
3. List some of the benefits to the consumer when a manufacturer produces maintainable products.
4. Given the following information, make maintainability-design recommendations.

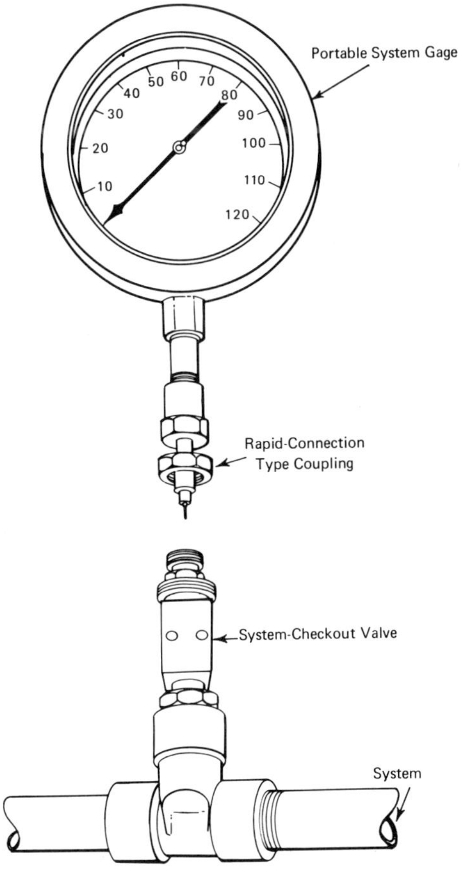

FIG. 9-15 Line Pressure Test with Use of Portable Gage (Method No. 2). (*Courtesy of Paul H. Hayden.*)

a. PARTS LIST

ITEM	PART NO.	QUANTITY
Test Kit		
Case	A10320	1
Front Panel	A10321	2
Meter	A10322	2
Meter	A10323	1
Screw	A10321-A	5
Screw	A10321-B	6
Screw	A10320-A	10
Screw	A10320-D	4
Handle	A10324	1
Flange	A10325	3

b. FUNCTIONAL DESCRIPTION

The test kit is used to monitor three missile pneumatic line pressures of 80, 160, and 200 psi.

c. PHYSICAL DESCRIPTION

The test kit dimensions are 36 in. long, 24 in. high, and 18 in. deep. It has a handle for carrying the test kit from one missile location to another The total weight of the test kit (case assembly and accessories) is 70 lb.

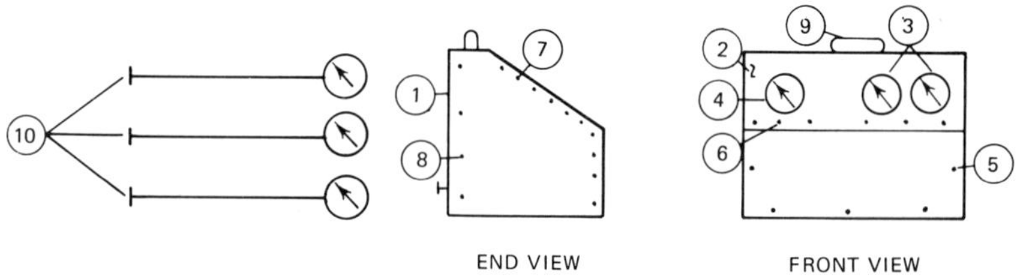

TEN

Prediction Techniques

As illustrated in Fig. 10-1, maintainability predictions are accomplished immediately after definition of the basic system. This is the earliest point in time when sufficient engineering data is available to perform a meaningful quantitative evaluation of design characteristics in terms of performance and supportability. On the other hand, the system-design process at this point in time has not advanced beyond the stage where predictions can influence the course of the design.

Predictions can provide an indication of design compliance with specified quantitative requirements (e.g., $MTTR$, ERT, $\bar{M}$) and can serve to establish a level of confidence for successfully completing maintainability demonstration. Predictions are applicable to all programs and all types of systems and equipment; however, they are particularly pertinent in programs where monetary incentives and high contractor risks are based on the successful demonstration of a given quantitative requirement.

The effectiveness of maintainability prediction as an evaluation tool depends on the technique and accuracy of input data. This in turn is based upon the applied knowledge and insight of the analyst. Above all, the prediction process should be undertaken with full objectivity on the part of the analyst.

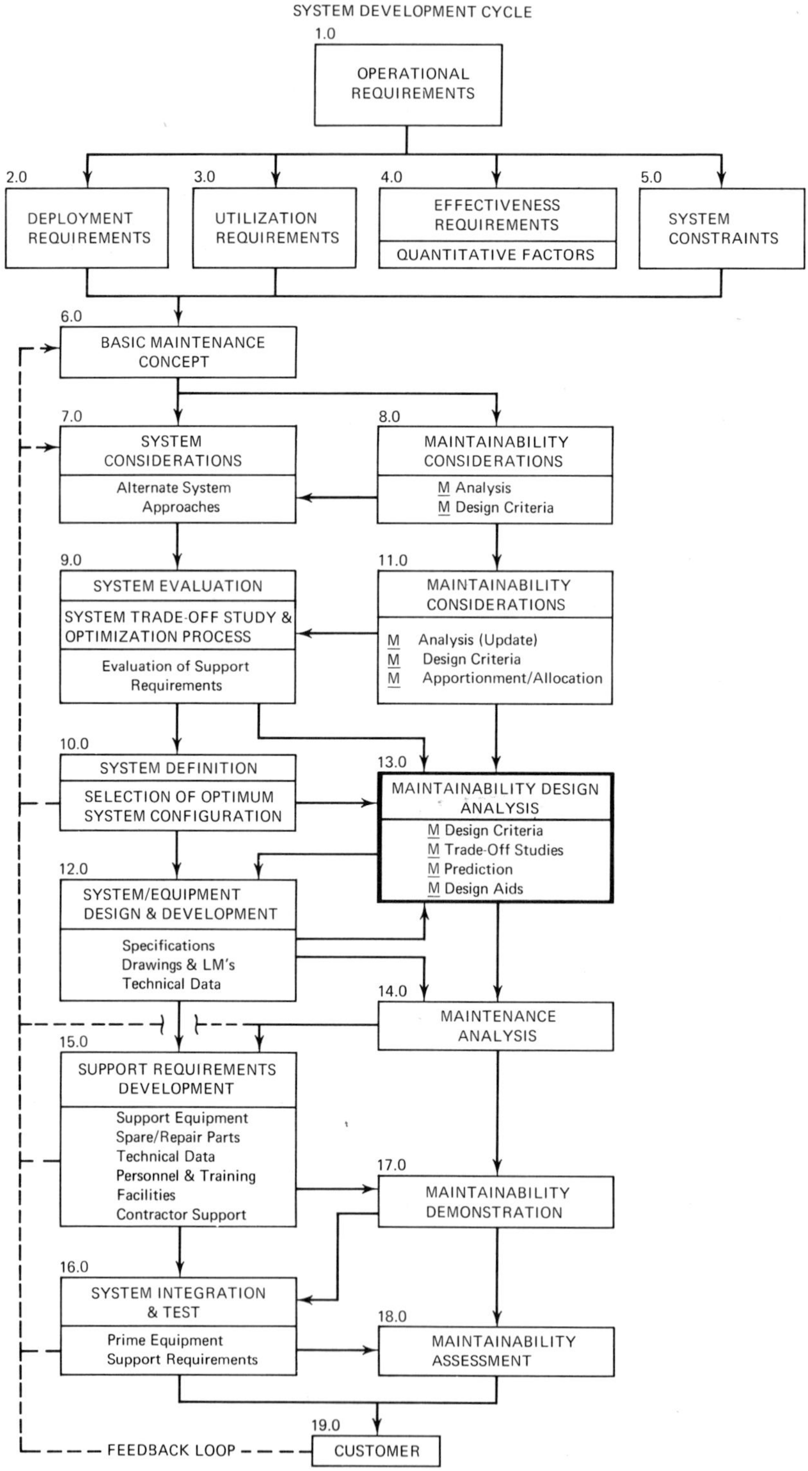

FIG. 10-1 System Development Cycle.

There are a great number of maintainability-prediction techniques presently being used by industry and government organizations. The procedures used vary depending upon the specific need for measurement, differences in imposed requirements, peculiarities of the equipment being measured, and because of individual or company preferences. The purpose of this chapter is not to detail all possible methods, but instead to outline several of the most commonly used prediction techniques.

PREDICTION TECHNIQUES

The techniques discussed herein are interpretations of procedures II and III as presented in MIL-HDBK-472, "Maintainability Prediction Handbook." Some of the information presented in the paragraphs below was extracted directly from the military handbook.

Corrective-maintenance-time Prediction; Technique No. 1

This technique involves the detailed functional-level breakdown of equipment, establishing the functional level at which maintenance features are effective, and performing the prediction of specific maintenance tasks (localization, isolation, remove and replace, etc.) based on equipment-design features and item failure rates. The steps involved in accomplishing a prediction using this technique are:

1. Determine the functional-level breakdown of the system or equipment item. This is accomplished by dividing the system into its various physical subdivisions beginning with the highest subdivision and continuing down to the lowest item that will be replaced during corrective maintenance. The various functional levels are defined below:

PART—one piece or two or more pieces joined together which are not normally subject to disassembly without destruction of designed use (e.g., tubes, resistors, transformers).

STAGE—a combination of two or more parts which form a portion of a subassembly. In electronics, a stage is usually considered as a tube or transistor, together with its directly associated parts (e.g., amplifier stage, detection stage).

SUBASSEMBLY—two or more parts or stages which form a portion of an assembly and which are replaceable as a whole but containing a part or parts which are individually replaceable (e.g., i-f strip, circuit board with mounted parts).

ASSEMBLY—a number of parts, stages, or subassemblies which, when joined together, perform a specific function and are replaceable as a whole (e.g., frequency amplifier).

UNIT—any combination of parts, subassemblies, and assemblies mounted together normally capable of independent operation in a variety of situations (e.g., power supply, converter). A unit is normally accessible and can be removed without disassembly or removal of other equipment items.

GROUP—a collection of units, assemblies, or subassemblies which is a subdivision of the equipment or system, but which is not capable of performing a complete operational function (e.g., antenna group, indicator group, receiver group, transmitter group).

EQUIPMENT (SET)—one or more units and assemblies, subassemblies, and parts connected or associated together (including all necessary cabling and accessories) which perform an operational function (e.g., radio receiving set, radar set).

SUBSYSTEM—a combination of equipment, groups, etc., which perform an operational function within a system. Subsystems form the major subdivision of a system (e.g., UHF communications as a subsystem of an aircraft).

SYSTEM—a combination of two or more subsystems or equipment sets, generally physically separated when in operation (e.g., airplane with all of its electrical, electronic, hydraulic, and pneumatic subsystems).

Employing the above descriptions as a guide, a functional-level diagram is prepared similar to that presented in Fig. 10-2. The functional-level diagram should complement the maintenance concept and the system/equipment breakdown derived through the maintainability analysis described in Chap. 6. Referring to Fig. 10-2, a hypothetical subsystem is presented showing the various subdivisions. Each branch of the diagram is terminated with a circle which indicates the item or items that will be replaced to correct malfunctions existing in that branch. For instance, if a malfunction exists in the bearing computer unit, the malfunction will be corrected by removing and replacing circuit board 1A4, 1A5, 1A6, or 1A7.

The connecting lines on the functional-level diagram indicate physical relationships and not electrical connections.

2. The functional levels at which localization, isolation, access, alignment, and checkout occur should be determined based on the overall characteristics of design. These levels are assigned in accordance with the description below:

LEVEL OF LOCALIZATION—the functional level to which a failure can be located without employing accessory support equipment. Referring to Fig. 10-2, this level is designated by △L. In other words, with built-in means, a malfunction can be identified to the transmitter group within equipment *A*.

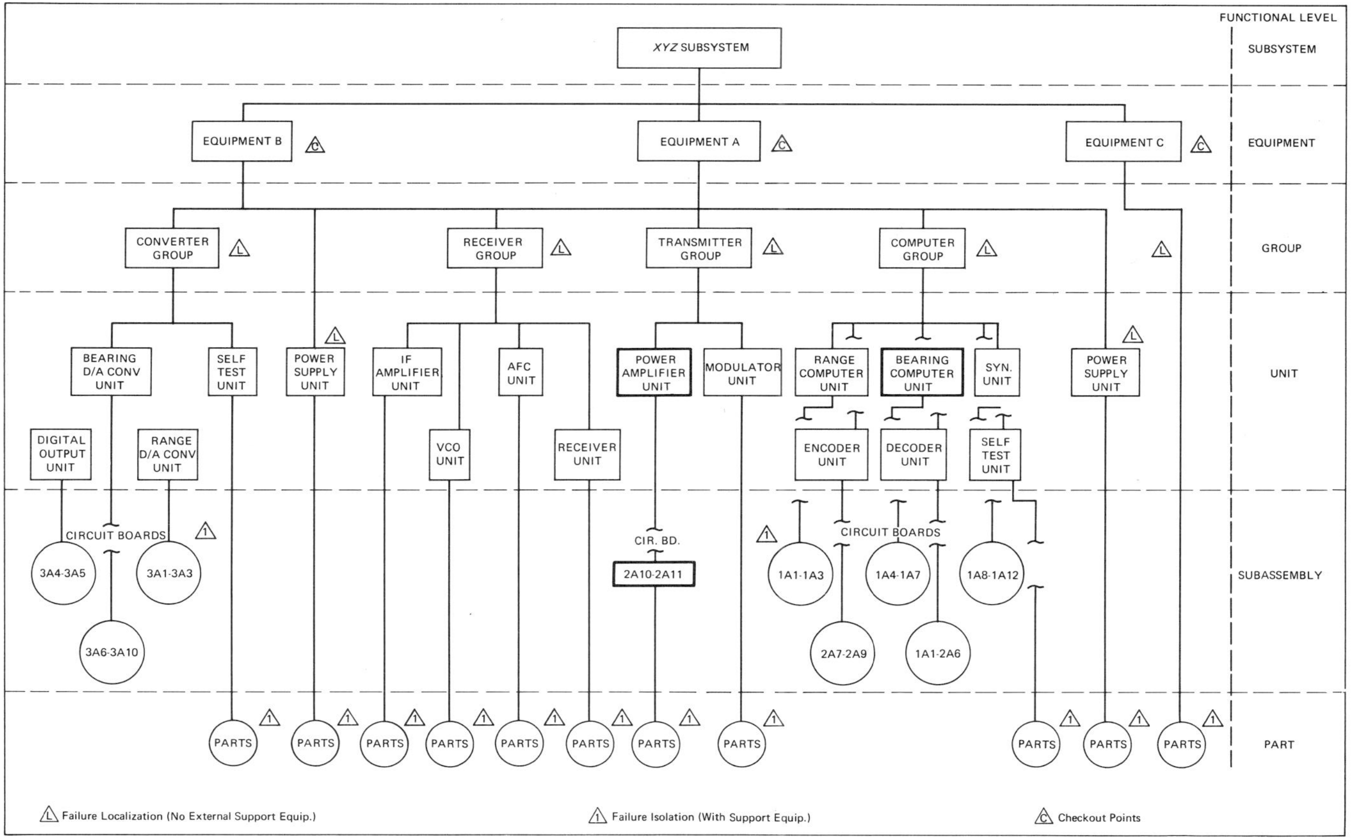

FIG. 10-2 Functional-level Diagram of the *XYZ* Subsystem.

LEVEL OF ISOLATION—the functional level to which a failure can be located using accessory support equipment at designated test points. This level is designated by △1. Referring to Fig. 10-2, fault isolation can be accomplished to the circuit board or part level.

LEVEL OF ACCESS—the access level for a replaceable item is that level to which disassembly must be accomplished in order to gain access to the item that is to be replaced and from which reassembly must be accomplished after item replacement. This can be determined directly from the functional level diagram as the level of the first rectangular block above the replacement item. For example, replacement of circuit board 1A4 in the bearing computer (Fig. 10-2) requires access to the unit level.

LEVEL OF ALIGNMENT—the functional level at which alignment must be done following replacement of a variable item such as a potentiometer, or following replacement of subassemblies, assemblies, or units containing sensitive mechanisms requiring alignment or adjustment. This level is designated by △A.

LEVEL OF CHECKOUT—the functional level at which system/equipment operation can be verified using self-test or other testing facilities. This level is designated by △C.

3. The next step in the prediction process requires the use of a *worksheet for maintainability prediction* (see Fig. 10-3). A worksheet is prepared on each functional-level item (except piece parts). Referring to Fig. 10-3, two examples are presented. Example *A* covers the branch in Fig. 10-2 containing the power amplifier unit consisting of circuit boards 2A10 and 2A11. The circuit boards are deemed repairable, and piece parts are removed and replaced to correct malfunctions. Note that maintenance task times are included for each part category. Example *B* covers the branch in Fig. 10-2 containing the bearing computer unit consisting of circuit boards 1A4, 1A5, 1A6, and 1A7. In this instance, the circuit boards are coded as being throwaway items, and individual maintenance times covering specific part categories are not applicable.

Completion of the worksheet for maintainability prediction is accomplished as follows:

a. Block 1—Item: Enter the item nomenclature.

b. Block 2—Method of repair: The type of item being replaced to correct the malfunction is indicated by a checkmark in the appropriate block. Parts are replaced in Example *A*.

c. Block 3—Part category: List each part. Items can be combined

1. ITEM *CIRCUIT BOARD NO. 1A6*
2. METHOD OF REPAIR Replace - Parts ☐ Subassembly ☒ Assembly ☐ Unit ☐ SHEET *12* OF *66*

3. Part Category	4 N	5 * λ	6 N·λ	8 LOC	9 ISO	10 ACC	11 ALIN	12 CHEC	13 INT	14 Rp	15 NλRp	17 LOG Rp	18 NλLOG Rp
				MAINTENANCE TASK TIMES									
CIRCUIT BOARD	–	–	–	0.045	0.265	0.228	0	0.108	0.015	0.661	20.385	-0.1798	-5.545
CAPACITORS													
ELECTROLYTIC	2	2.48	4.96										
MICA	1	0.46	0.46										
PAPER	2	0.50	1.00										
CONNECTORS	1	0.06	0.06										
DIODES	6	2.98	17.88										
RESISTORS													
COMPOSITION	3	0.36	1.08										
CARBON FILM	2	0.39	0.78										
WIREWOUND	1	1.40	1.40										
TRANSFORMER	1	1.16	1.16										
TRANSISTOR	2	1.03	2.06										
* FAILURES/10⁶ HRS		7 SUM	30.84							16. SUM	20.385	sum 19	-5.545

Example *B*. Prediction of Nonreparable Subassembly.

1. ITEM *CIRCUIT BOARD NO. 2A10*
2. METHOD OF REPAIR Replace - Parts ☒ Subassembly ☐ Assembly ☐ Unit ☐ SHEET *1* OF *66*

3. Part Category	4 N	5 * λ	6 N·λ	8 LOC	9 ISO	10 ACC	11 ALIN	12 CHEC	13 INT	14 Rp	15 NλRp	17 LOG Rp	18 NλLOG Rp
				MAINTENANCE TASK TIMES									
CAPACITORS	2	2.48	4.96	0.106	0.772	0.427	0	0.108	0.081	1.494	7.410	0.1744	0.865
ELECTROLYTIC	2	0.46	0.92	0.106	0.772	0.427	0	0.108	0.081	1.494	1.374	0.1744	0.160
MICA	1	0.50	0.50	0.106	0.772	0.427	0	0.108	0.081	1.494	0.747	0.1744	0.087
PAPER	3	0.28	0.84	0.106	0.772	0.427	0	0.108	0.081	1.494	1.255	0.1744	0.146
R.F. COILS	1	0.06	0.06	0.106	0.772	0	0	0.108	0.272	1.258	0.075	0.0997	0.006
CONNECTORS	4	2.98	11.92	0.106	0.772	0.427	0	0.108	0.081	1.494	17.808	0.1744	2.079
DIODES	2	1.16	2.32	0.106	0.772	0.427	0	0.108	0.142	1.555	3.608	0.1917	0.445
TRANSFORMERS													
RESISTORS	3	0.36	1.08	0.106	0.772	0.427	0	0.108	0.081	1.494	1.614	0.1744	0.188
COMPOSITION	2	0.39	0.78	0.106	0.772	0.427	0	0.108	0.081	1.494	1.165	0.1744	0.136
CARBON FILM	1	1.40	1.40	0.106	0.772	0.427	0	0.108	0.081	1.494	2.092	0.1744	0.244
WIREWOUND	4	2.86	11.44	0.106	0.772	0.427	0	0.108	0.081	1.494	17.091	0.1744	1.995
RELAYS													
* FAILURES/10⁻⁶ HRS		7. SUM	36.22							16. SUM	54.293	SUM 19	6.351

Example *A*. Prediction of Reparable Subassembly.

FIG. 10-3 Worksheet for Maintainability Prediction.

TABLE 10-1 Average Maintenance Times (Functional Level-Time Matrix) (Sheet 1)

TUBES[a]1

AVERAGE TIME[d] (HOURS) TO PERFORM CORRECTIVE MAINTENANCE TASKS FOR ELIMINATING EQUIPMENT OR SYSTEM MALFUNCTION CAUSED BY TUBE FAILURE

FUNCTIONAL LEVELS									CORRECTIVE MAINTENANCE TASKS					
FOR DETERMINING LOCALIZATION AND ISOLATION TIME USE COLUMN BEGINNING WITH THE FUNCTIONAL LEVEL THROUGH WHICH FAILURE IS REMOVED[f]									DIAGNOSIS		REPLACEMENT[b]		TEST	
9	8	7	6	5	4	3	2	1	Localization	Isolation[c] 1	Disassembly	Reassembly	Alignment	Checkout[g]
System	Subsystem	Equipment	Group	Unit	Assembly	Subassembly	Stage	Part	0.021e	0.265e	0.313	0.438	0.156	0.175
None	System	Subsystem	Equipment	Group	Unit	Assembly	Subassembly	Stage	0.029	0.315	0.231	0.362	0.077	0.167
	None	System	Subsystem	Equipment	Group	Unit	Assembly	Subassembly	0.037	0.394	0.165	0.262	0.045	0.154
		None	System	Subsystem	Equipment	Group	Unit	Assembly	0.045	0.506	0.122	0.191	0.030	0.149
			None	System	Subsystem	Equipment	Group	Unit	0.053	0.636	0.094	0.134	0.021	0.138
				None	System	Subsystem	Equipment	Group	0.063	0.776	0.071	0.090	0.015	0.12
					None	System	Subsystem	Equipment	0.075	0.907	0.049	0.061	0.010	0.108
						None	System	Subsystem	0.090	1.090	0.032	0.037	0.007	0.091
							None	System	0.107	1.305	0.016	0.017	0.003	0.062
								None	0.125	1.546	0.000	0.000	0.000	0.000

PARTS[a]2

AVERAGE TIME[d] (HOURS) TO PERFORM CORRECTIVE MAINTENANCE TASKS FOR ELIMINATING EQUIPMENT OR SYSTEM MALFUNCTION CAUSED BY PART FAILURE

FUNCTIONAL LEVELS									CORRECTIVE MAINTENANCE TASKS					
FOR DETERMINING LOCALIZATION AND ISOLATION TIME USE COLUMN BEGINNING WITH THE FUNCTIONAL LEVEL THROUGH WHICH FAILURE IS REMOVED[f]									DIAGNOSIS		REPLACEMENT[b]		TEST	
9	8	7	6	5	4	3	2	1	Localization	Isolation[c] 2	Disassembly	Reassembly	Alignment	Checkout[g]
System	Subsystem	Equipment	Group	Unit	Assembly	Subassembly	Stage	Part	0.021e	0.772[e]	1.281	1.223	0.156	0.175
None	System	Subsystem	Equipment	Group	Unit	Assembly	Subassembly	Stage	0.039	1.179	0.328	0.561	0.077	0.167
	None	System	Subsystem	Equipment	Group	Unit	Assembly	Subassembly	0.056	1.417	0.165	0.262	0.045	0.158
		None	System	Subsystem	Equipment	Group	Unit	Assembly	0.073	1.569	0.122	0.191	0.030	0.149
			None	System	Subsystem	Equipment	Group	Unit	0.089	1.700	0.094	0.134	0.021	0.138
				None	System	Subsystem	Equipment	Group	0.106	1.821	0.071	0.090	0.015	0.124
					None	System	Subsystem	Equipment	0.121	1.924	0.049	0.061	0.010	0.108
						None	System	Subsystem	0.136	2.022	0.032	0.037	0.007	0.091
							None	System	0.150	2.100	0.016	0.017	0.003	0.062
								None	0.165	2.172	0.000	0.000	0.000	0.000

into a part type (e.g., combine all composition resistors and include as one part type).

d. *Block 4—Quantity N:* Enter the quantity of each part type.

e. *Block 5—Failure rate* λ*:* After each part type, indicate the applicable failure rate. If more than one part-numbered item is included in the part type, the listed failure rate will be an average value

TABLE 10-1 Average Maintenance Times (Applications, Notes, and Definitions) (Sheet 2)

(see superscript letters on sheet 1)

a.

1. The average task times in the TUBES chart should be applied to front panel fuses, encased plug-in modular subassemblies, plug-in crystals, and tubes (excluding special purpose tubes).

2. The average task times in the PARTS chart should be applied to fuses not on the front panel, transistors, special purpose tubes (Magnetrons, Klystrons, etc.), wired in nonsealed modules, and all electronic parts (resistors, capacitors, inductors, transformers, etc.).

b. The total replacement time should include disassembly time, interchange time and reassembly time. However, since an average interchange time applicable to all situations is difficult to obtain, it has not been included in the tabulated average task time. Interchange time must be determined from Table 10-2.

c.

1. Specific techniques are used to isolate tube failures. The designer must consider these techniques in determining R_p for tubes.

   (a) Test points are normally not used to isolate tube failures.

   (b) The most common method of isolating tube failures during corrective maintenance is by substitution of a known good tube for the suspected tube. Although isolation features may be effective at the unit or assembly level, tube isolation can be accomplished at the tube or stage level through substitution means without using test points.

   (c) Some tube failures (open filaments, broken envelope, etc.) can be located visually; however, this is true in only a very small percentage of cases.

2. Because the task durations were derived from field data, this isolation time includes the tube checks, etc., the electronic technician usually makes before deciding to use test points and point-to-point checks. This average time also accounts for the point-to-point checks which would usually be required to isolate a failed part.

d. The average time intervals in this chart do not include administrative time. Administrative time consists of time expended in part procurement (time spent by maintenance personnel in obtaining replacement items) and in nontechnical military routine (inspection, eating, other military duties, etc.).

e. When equipment maintenance features enable localization to the functional level through which failure is being removed (the top row of FUNCTIONAL LEVELS columns), do not use the value shown in the ISOLATION column at this functional level, instead use *0.000 hours.*

f. To determine the time for the disassembly, reassembly, alignment and checkout tasks, the designer should only use Column 1 of FUNCTIONAL LEVELS in the appropriate row at which the task is performed.

g. To properly approximate the checkout time, enter column 1 at the functional level at which the checkout is being made and multiply by the number of operational modes affected by the replaced functional level.

(combined). Failure rates are obtained from reliability-prediction reports.

f. *Block 6—Part-type failure rate* $N\lambda$: Determine the reliability of each part type by multiplying the failure rate λ in block 5 by the quantity N of parts in block 4.

g. *Block 7—Item failure rate:* To obtain the overall reliability factor for the item covered by the worksheet, add the figures in block 6.

h. *Block 8—Maintenance time for localization (LOC):* Maintenance times should be determined from the most reliable data source available to the maintainability engineer. Very likely, the contractor has established a data bank containing experience data of systems already in operation. Data may stem from maintenance-analysis reports, special studies involving utilization of design aids (mock-ups), and/or demonstrations. Reviews and comparison of such data, applied to the system/equipment being evaluated through this prediction, may result in realistic estimates of maintenance times.

TABLE 10-2 Maintenance Interchange Task Times

TUBES AND FUSES	
PART TYPE	AVERAGE TIME (HOURS)
Plug-in tubes	0.015
Wired tubes (4 wires)	0.149
Wired tubes (more than 4 wires)	0.149 + 0.034 per wire over 4
All tubes with shield with clamp with cap	ADD 0.007 ADD 0.027 ADD 0.007
Plug-in fuses	0.010
Screw-in fuses	0.015
All fuses with screw cap	0.014
PARTS OTHER THAN TUBES AND FUSES	
PART TYPE	AVERAGE TIME (HOURS)
Parts with 2 wires or 2 tabs to be soldered	0.081
Parts with more than 2 wires or 2 tabs to be soldered with clamp	0.081 + 0.034 per wire over 2 ADD 0.027
Parts attached with screws, nuts and washers	ADD 0.022 for each screw, nut and washer combination

NOTE: For attachments not conforming to the above type determine the interchange time by using work factor data, Table 10-3.

- This is the preferred source of data for completing the maintenance-task times (blocks 8 through 13 of the *worksheet for maintainability prediction*).
- In the event that maintenance-task times are not available through other sources, data may be obtained from such charts as those presented in Tables 10-1 to 10-4. Table 10-1 covers localization, isolation, disassembly, and reassembly, or access, alignment, and checkout. Tables 10-2 to 10-4 pertain to removal and replacement or interchange of parts.

To obtain localization time using Table 10-1, select the functional-level column headed by the type of replacement that will be

TABLE 10-3 Task-element Time Data

ELEMENT DESCRIPTION	Element Time (Hours)
1. PLUG-INS (includes handling)	
Pin-type tubes, plug-in parts, etc.	0.0075
Tube cap or shield	0.0035
Fuse	
Insert into horizontal holder	0.0050
Insert into vertical holder	0.0075
2. WIRING AND SOLDERING	
Wire - wrapping and splicing	
Bare copper wire (1) End	0.0150
(2) Ends	0.0237
Jumper wire and cable leads (1) End	0.0134
(2) Ends	0.0265
Part with axial leads (includes part handling)	
(1) End	0.0178
(2) Ends	0.0289
Soldering	
Per joint	0.0058
3. REPLACEMENT WITH HARDWARE	
Replace screw into tapped hole	0.0093
Replace screw through clearance hole	0.0023
Replace washer	0.0018
Replace nut	0.0071
Replace stop nut	0.0210
Replace set screw	0.0075
Apply glyptol screw	0.0018
4. PART HANDLING	
Pull up part and position in chassis for assembly	0.0025
5. PRINTED CIRCUIT WIRING	
Replace (insert)	0.0033/End
Solder	0.0056/End

TABLE 10-4 Determination of Task Time with Use of Element Time Factors (Example)

The following example is for a resistor with two tab terminals which are attached to the chassis by two screws, two nuts, and two washers.

DETAILED INTERCHANGE STEPS	ELEMENT DESCRIPTION	ELEMENT TIME (Hours)	MULTIPLIED BY	NUMBER TIMES PERFORMED	=	TIME FOR DETAILED INTERCHANGE STEPS (Hours)
Unsolder joint	Soldering (per joint)	0.0058	x	2	=	0.0116
Disconnect lead	Wire (jumper Wire - 1 end)	0.0134	x	2	=	0.0268
Remove nut	Remove nut	0.0071	x	2	=	0.0142
Remove washer	Remove washer	0.0018	x	2	=	0.0036
Remove screw	Remove screw through clearance hole	0.0023	x	2	=	0.0046
Remove failed resistor	Pull up part and position	0.0025	x	1	=	0.0025
Position new resistor	Pull up part and position	0.0025	x	1	=	0.0025
Replace screw	Replace screws through clearance hole	0.0023	x	2	=	0.0046
Replace washer	Replace washer	0.0018	x	2	=	0.0142
Replace nut	Replace nut	0.0071	x	2	=	0.0268
Connect lead	Wire (jumper wire - 1 end)	0.0134	x	2	=	0.0116
Solder joint	Soldering (per joint)	0.0058				
			Hours of Maintenance Per Part Failure		=	0.1266

accomplished. For example, replacement of a part would indicate column 1, replacement of a stage would indicate column 2, etc. Relative to Example *A*, power-amplifier design is such that parts are replaced on circuit board 2A10; therefore, column 1 is the correct column in this instance.

The appropriate corrective-maintenance task time is found by determining the functional level at which maintenance features are effective. In this instance, Fig. 10-2 indicates that LOC for the branch containing the power amplifier unit and circuit board 2A10 is effective to the group functional level as designated by △L. Referring to Table 10-1, the maintenance task time is determined by following column 1 on the parts chart down to the group level and extracting the appropriate quantitative value under localization. The value 0.106 hr is entered in block 8 on the worksheet presented in Example *A* (Fig. 10-3). It is assumed that the localization value for each part on circuit board 2A10 is the same; therefore, the value 0.106 is entered after each part type.

i. Block 9—Maintenance time for isolation (ISO): Determination of the maintenance time for fault isolation follows the same approach as employed for localization in block 8. In the absence of actual experience data, Table 10-1 may be used. Referring to the table, select the appropriate functional-level column for IOS for the branch containing the power-amplifier unit and circuit board 2A10. Column 1, the part-level row, intersects with the "isolation" column which contains a maintenance task-time value of 0.772 hr. This value is entered for each part type in block 9 on the worksheet presented in Example *A* (Fig. 10-3).

j. Block 10—Maintenance time for access (ACC): Maintenance time for access includes the disassembly time to gain access to the failed item and the reassembly time after item removal and replacement. Both the disassembly and reassembly maintenance times may be determined from Table 10-1. Referring to the table, select column 1 and proceed to the applicable functional level. Referring to Fig. 10-2, access to the parts on circuit board 2A10 is to the subassembly level. The appropriate level is indicated by the first rectangular box in Fig. 10-2 above the replacement item. Thus, the appropriate disassembly time is 0.165 hr and the reassembly time is 0.262 hr. Access time is the sum of the two values, or 0.427 hr. This value is entered against each part in block 10 on the worksheet presented in Example *A* (Fig. 10-3).

k. Block 11—Maintenance time for alignment (ALIN): The maintenance time for alignment or adjustment subsequent to item removal, replacement, and reassembly is determined by the same method as presented above. Select column 1 of Table 10-1 and proceed to the applicable functional level. Relative to Example A, design is such that alignment or adjustment is not required; therefore, the value entered in block 11 of the worksheet is zero.

l. *Block 12—Maintenance time for checkout (CHEC):* The maintenance time involves checkout of the item after part replacement or repair has been accomplished (to verify proper function of the item). Referring to Table 10-1, select column 1 and proceed to the equipment level. The predicted value of 0.108 hr is entered in block 12 of the worksheet.

m. *Block 13—Maintenance time for interchange (INT):* The maintenance time for interchange involves the time to remove and replace the malfunctioning item. Table 10-2 contains time information for most replaceable parts. For items having mounting methods not conforming to the description given in Table 10-2, the interchange times can be determined from the data presented in Table 10-3.

- In determining the interchange time from Table 10-3, each detailed step involved in removing an item that has malfunctioned and in positioning and attaching a new item must be assigned a specific element of time. The interchange time for an item is equal to the sum of all time intervals. An example of determining interchange time for a typical part is illustrated in Table 10-4.

n. *Block 14—Maintenance repair time R_p:* To determine the repair time required for performing a corrective-maintenance task, add the values recorded on the worksheet in columns 8 through 13. The repair time R_p is equivalent to the active-maintenance time per corrective-maintenance task M_{ct}.

o. *Block 15—Weighted maintenance factor ($N\lambda R_p$):* Determine the weighted factor by multiplying the value in block 6 by the repair time in block 14. The resultant value is entered in block 15.

p. *Block 16—Summation of weighted factors $\Sigma N\lambda R_p$:* Add the individual values in block 15.

q. *Block 17—Logarithm of repair time log R_p:* Enter the logarithm of each recorded value of R_p in block 17. When a value is less than one, the logarithm should be expressed as a negative number. For instance, $\log 0.25 = 9.3979 - 10 = -0.6021$.

r. *Block 18—Product of log R_p and $N\lambda$:* The value entered in block 18 is obtained by multiplying the value in block 6 by that in block 17.

s. *Block 19—Total value of log R_p and $N\lambda$:* The value entered in block 19 is the summation of the individual values in block 18.

Referring to Fig. 10-3, two examples are presented. Example *A*, covering maintenance through the replacement of piece parts on circuit board 2A10, is illustrated by the data included in the above steps. Example *B*, on the other hand, covers maintenance through the replacement of circuit board 1A6, a plug-in item in the bearing computer unit shown in Fig. 10-2. The maintainability-prediction worksheet for Example *B*, also presented in Fig. 10-3, is discussed briefly in the following steps:

a. Blocks 1—7: Determination of data for these blocks is accomplished by following the same procedures as defined above. In this instance, however, the item checked in block 2 is subassembly, as the circuit board itself is replaced for corrective maintenance.

b. Blocks 8—13: Since the circuit board is the item that is replaced in accomplishing corrective maintenance and not individual parts mounted on the circuit board, only a single set of values is entered. These values are obtained using Tables 10-1 to 10-4 as applicable.

c. Blocks 14—19: Determination of data for these blocks is accomplished by following the same procedure as defined for Example *A*. In this instance, however, the value indicated in block 15 is the product of the values in blocks 7 and 14.

4. When individual prediction worksheets (Fig. 10-3) have been prepared on each replaceable item (other than piece parts), the values presented in blocks 7, 16, and 19 of each worksheet are recorded on a form similar to that presented in Table 10-5. Based on the tabulated data, the *MTTR* of the *XYZ* subsystem can now be calculated as follows:

$$MTTR = \frac{\Sigma N\lambda R_p}{\Sigma N\lambda} = \frac{\text{block 7 value}}{\text{block 6 value}} = \frac{3548.99}{1982.68}$$
$$= 1.79 \text{ hr or } 1 \text{ hr } 47 \text{ min} \quad (1)$$

The geometric mean-time-to-repair $MTTR_G$, which is the median equipment repair time *ERT* when the repair times are log-normally dis-

TABLE 10-5 Prediction Summary Worksheet

(1) Work Sheet No.	(2) Item Designation	(3) W. S. Block 7	(4) W. S. Block 16	(5) W. S. Block 19
1	Ckt. Bd. 2A10	36.22	54.293	6.351
2				
12	Ckt. Bd. 1A6	30.84	20.385	−5.545
GRAND TOTAL		(6) 1982.68	(7) 3548.997	(8) −11.892

tributed, is calculated as follows:

$$MTTR_G = \text{antilog} \frac{\Sigma N\lambda \log R_p}{\Sigma N\lambda} = \text{antilog} \frac{\text{block 8 value}}{\text{block 6 value}}$$

$$= \text{antilog} \frac{-11.892}{1982.68} = \text{antilog}\,(-0.00599) \quad (2)$$

or

$$\text{antilog } 9.99401 - 10 = 0.986 \text{ hr or } 59 \text{ min}$$

Preventive-maintenance-time Prediction; Technique No. 2

This prediction technique involves the determination of preventive-maintenance task requirements, task frequencies, and task times. A data format similar to the one presented in Table 10-6 may be used for the recording of the data. The completion of the preventive-maintenance summary is accomplished by performing the following:

1. *Block 1—Description of preventive-maintenance task:* Determine all of the preventive-maintenance tasks that must be performed during the service life of the product. A detailed description of each task should be documented in a separate report. A description of each task that cannot be performed while the system or equipment is operating is listed in block 1.

2. *Block 2—Frequency of task* (f)*:* The frequency of occurrence of each task listed in block 1 should be determined and entered in block 2.

3. *Block 3—Preventive-maintenance time* (M_{pt})*:* The maintenance man-hours required to perform each task should be listed in block 3 after the applicable task.

4. *Block 4—Maintenance time-frequency product* (fM_{pt})*:* The value

TABLE 10-6 Preventive-maintenance Requirements Summary

(1) Description of Preventive Maintenance Task	(2) Frequency (f)	(3) Preventive Maintenance Time (M_{pt})	(4) $(f)\,M_{pt}$
1. Lubricate Gear Train	961	0.5	480.5
2. Calibrate Board 1A6	1940	0.2	388.0
3. Clean Air Filter (F3)	961	0.1	96.1
28. Align Resolver	1940	0.1	194.0
29. Ripple Check of PS3	961	0.2	192.2
TOTALS	(5) Σf		(6) ΣfM_{pt}

entered in block 4 is derived by multiplying the maintenance-task time M_{pt} in block 3 by the task frequency f in block 2.

5. *Blocks 5 and 6—Totals:* The individual values in blocks 2 and 3 are added and entered in blocks 5 and 6, respectively. These values are then used to determine the mean preventive-maintenance time ($\bar{M}_{pt}$) as follows:

$$\bar{M}_{pt} = \frac{\Sigma f M_{pt}}{\Sigma f} \tag{3}$$

Corrective-maintenance-time Prediction; Technique No. 3

This technique involves a method of performing a maintainability prediction utilizing the basic principles of random sampling. The technique is primarily oriented to ground electronic equipment.

General. The procedures described herein outline the methods used to select random samples of replaceable items from the total complement of components comprising the system, to subdivide the sample into discrete classes of items, and to conduct an analysis for every replaceable item in the sample.

Typical replaceable items are transistors, tubes, capacitors, etc. The underlying philosophy of this technique is that system failures are principally due to the malfunction of replaceable items, and therefore, the time to replace these items is a measure of downtime. The duration of downtime is assumed to be a function of specific design parameters which relate to (1) the physical configuration of the system, (2) the facilities provided for maintenance, and (3) the degree of maintenance skills required to perform the maintenance tasks. Further, it is assumed that because of a basic uniformity of design, a random selection of replaceable items by class will provide a sample of maintenance tasks that is representative of the system-maintenance task times.

The values of the task times are determined by using three checklists. Checklist A is used for scoring physical-design factors, checklist B scores design dictates-facilities, and checklist C is used to score design dictates-maintenance skills. The theory is that by using these checklists, which include scoring values and criteria, variations due to individual appraisers are minimized, and the resulting scores can then be correlated with downtime. A regression equation is provided to determine downtime from the numerical scores in the checklists.

Procedural steps. The steps involved in a prediction process are presented below:

1. Determine a sample size by using the equation

$$N = \left[C_x \frac{Z}{k} \right] \tag{4}$$

where N = sample
C_x = coefficient of variation
Z = confidence level
k = desired accuracy

and

$$C_x = \frac{\sigma}{\bar{M}_{ct}} \tag{5}$$

where σ = standard deviation of sample data
$\bar{M}_{ct}$ = mean repair time

Given Z = 95% or 1.645
k = 10%
σ = 17 min
$\bar{M}_{ct}$ = 65 min

and substituting these data into the equation, the sample size for this case can be determined.

$$N = \left[0.262 \frac{1.645}{0.10}\right]^2 = (4.31)^2 = 18.57 \text{ (assume 20)}$$

It should be recognized that the value of C_x must usually be approximated in order to establish the magnitude of the sample size N. Although it is desirable to derive specific values for σ and $\bar{M}_{ct}$, this is not always possible. In cases where ground electronic equipment is involved, a C_x value of 1.07 may be used.

2. Determine the sample allocation to the part classes. The procedure for this is as follows:

Assume an equipment with part classes as shown in column 1 of Table 10-7.

a. Determine the quantity of parts in each class and enter the value in column 2 of Table 10-7.

b. Determine the average failure rate for each part category and enter the value in column 3. Failure-rate information is obtained from reliability-prediction reports.

c. Determine the expected quantity of failures per 10^6 hr for each part class by multiplying the quantity value Q in column 2 by the failure rate λ in column 3. The resultant value is entered in column 4.

d. Based on the values entered in column 4, establish the percent contribution for each part class and enter the value in column 5.

e. Determine the quantity of failures per part class based on a sample size N. Multiply the line entry in column 5 by the sample number.

TABLE 10-7 Part-class Contribution to the Sample (Example)

(1) Part Class	(2) Quantity (Q)	(3) Part Failure Rate (λ)*	(4) Expected Qty. Of Failures $f = (Q\lambda)$	(5) Contribution To Total Expected Failures (%)	(6) Quantity of Failures Per Class When $N = 20$	(7) Actual Sample Used
Capacitors	706	0.107	75.5	10.8	2.16	2
Diodes	458	0.105	48.0	6.8	1.36	1
Transistors	430	0.221	95.0	13.6	2.72	3
Int. Circuits	460	0.140	64.4	9.2	1.84	2
Connectors	89	0.140	12.4	1.7	0.34	0
Relays	3	0.600	1.8	0.5	0.10	0
Inductors	149	0.086	12.8	2.0	0.40	0
Transformers	65	0.333	21.6	3.0	0.60	1
Tubes	5	25.000	125.0	17.9	3.58	4
Resistors	1,090	0.179	195.1	28.0	5.60	6
Miscellaneous	252	0.181	45.6	6.5	1.30	1
TOTAL	3,112		697.2	100%		20

f. The values in column 6 are then rounded off to yield the values in column 7.

3. When the quantity of samples per part class has been established (as shown in column 7 of Table 10-7), the next step is to randomly select specific parts to be analyzed. Table 10-8 illustrates a listing of random selections.

4. Complete a task-prediction form on each item randomly selected. To accomplish the task predictions, the maintainability engineer should have design data (schematic diagrams and layouts). In addition, he should have some concept of the methods of testing and the employment of maintenance acids, and he should be thoroughly familiar with the functional operation of the system or equipment. With this information, the maintainability engineer proceeds to complete the *maintainability prediction form* (see Fig. 10-4).

The maintenance-analysis section of the form is divided into two columns. The left column is used to record maintenance steps or the sequence of maintenance actions. Scoring comments associated with each step are entered in the right column.

The section of the form identified as *checklist scores* is completed with the aid of criteria listings:

a. Checklist A: Used to score specific maintenance tasks that are a function of physical-design variables such as packaging, access characteristics, and test-point displays.

b. Checklist B: Used to score "design dictates-facilities." These

TABLE 10-8 Task Listing as Basis for Random Selection of Tasks for Analysis

(1) Part Class	(2) Quantity	(3) Designation	(4) Location	(5) Remarks	(6) Task No.
Capacitors	2	C1	52.3 MHz Osc. Can, IF Amp	Replace Osc. Can	1
		C16	VCO Osc. Driver	Replace Piece Part	2
Diodes	1	CR3	Unit A Power Supply Chassis	Replace Piece Part	3
Transistors	3	Q5	Bearing Computer Card – A11	Replace Card	4
		Q14	Digital Output Card – 3A5	Replace Card	5
		Q1	Modulator Chassis	Repair to Piece Part	6
Int. Circuits	2	Z3	Bearing Computer Card – A12	Replace Card	7
		Z6	Frequency Synthesizer – A16	Replace Card	8
Transformers	1	T6	Unit A Power Supply Chassis	Replace Piece Part	9
Tubes	4	V2	Power Amplifier Assembly	Replace Assembly	10
		V3	Power Amplifier Assembly	Replace Assembly	11
		V4	Power Amplifier Assembly	Replace Assembly	12
		V6	Power Amplifier Assembly	Replace Assembly	13
Resistors	6	R2	IF Amplifier Assembly	Replace Piece Part	14
		R3	VCO Tripler Assembly	Replace Piece Part	15
		R33	Digital Output Card – 3AJ	Replace Piece Part	16
		R45	Decoder Card – A18	Replace Piece Part	17
		R5	Modulator Chassis	Replace Piece Part	18
		R1	Unit A Power Supply	Replace Piece Part	19
Miscellaneous	1	S1	Switch On Front Panel	Replace Piece Part	20

include such variables as external test equipment and personnel assistance.

c. *Checklist C:* Used to evaluate the personnel requirements relating to physical, mental, and attitude characteristics. Typical of criteria are visual acuity, logical analysis, memory, endurance, and energy.

These checklists are presented in Appendix B of this text.

Referring to the checklists, the scoring for each item ranges from 0 to 4. Intermediate values of 1, 2, and 3 are provided for some questions where the nature of the characteristic being assessed may take on varying magnitudes. This is contrasted to the yes-no situation. The questions have been framed in a manner that permits general application across equipment lines.

Situations may arise where insufficient information is available to score a particular checklist question, or where a question is not applicable to a particular task. If insufficient data are available, the average ques-

Equip. Unit A Unit/Part Resistor R3 Task No. 15

Ass'y VCO/Tripler By J. Doe Date 11/15/66

Primary function failed unit/part Provide proper degenerative feedback in Tripler Assy. emitter circuit.

Mode of failure Opened resistor

Malfunction symptoms XYZ System indicates a NO-GO condition during confidence check.

Maintenance Analysis

Maintenance Steps	Scoring Comments
1. The equipment malfunction is initially indicated by a no lock-on condition in both the T/R mode and the A/A mode of operation. Initiating interruptive self-test switch on the Unit C front panel isolates the trouble to Unit A.	Isolation of fault to Unit A can be accomplished using only the built-in self-test features and a 50-ohm dummy load attached to the Antenna output of Unit A.
(Continued on Sheet 2)	(Continued on Sheet 2)

Checklist Scores

	1	2	3	4	5	6	7	8	9	10	11	12	13	14	15	Total
A	4	4	3	2	0	2	2	2	4	0	2	4	4	0	4	37
B	1	4	2	4	2	4	2									19
C	3	2	0	2	1	4	4	2	3	2						23

FIG. 10-4 Maintainability-prediction Form (Sheet 1).

MAINTENANCE ANALYSIS CONTINUATION SHEET

Equip. Unit A   Part Resistor, R3Task No. 15

Maintenance Steps	Scoring Comments
2. In the Field Electronics Shop, the faulty Unit A is connected to the System Test Station for fault isolation. The Unit is then prepared for fault isolation within the Synthesizer-VCO-AFC Signal Loop.	Isolation of fault to Synthesizer-VCO-AFC Signal Loop is accomplished using the System Test Station control panel self-test readout lights. Digital word code on readout lights correlated with readings listed in Report No. 101A5 provide fault location.
3. VCO signal outputs at VCO Connector are observed and checked against correct values. No 1025 MH_z sine wave output indicates trouble is in VCO.	Signal Generator SG-215 and Oscilloscope TKX-545 used to check VCO signals on subass'y. Connector. Oscilloscope and Signal Generator set-up and adjustments required. Proper reading listed in Report No. 101A5.
4. By removing the VCO output R. F. cables connected to the VCO Tripler ass'y., the correct driver output from the VCO subassembly can be observed. Presence of the correct signal indicates trouble is probably in the Tripler Subassembly.	Location of VCO module requires the use of a module extender. Dense packaging within the VCO requires high manual dexterity in performing repair services on this subassembly. Some delay to be expected in repair action due to part location and necessity to use care in part removal to avoid heat damage or solder contamination to adjacent parts.
5. Stage-by-stage probing of L. O. Tripler board indicates an open resistor in Q1 emitter circuit.	

FIG. 10-4 Maintainability-prediction Form (Sheet 2).

tion score for the checklist should be inserted. For example, if 14 questions were scored and the total score was 42, the average score is 3. Inserting this value for the unevaluated item, the final score is 45. For other situations (e.g., not applicable), a score of 4 should be used. The reasoning here is that if a particular item does not apply, it is not detrimental to maintenance.

To illustrate the scoring process, the scores for the sample maintenance-analysis tasks are shown in Fig. 10-4 (sheet 1). The score for each checklist question is obtained by referring to the scoring comments in the maintenance analysis. Each task was reviewed for items that pertain to each question, and the equipment was then scored in accordance with the criteria.

5. The last step in this prediction process is to calculate the corrective-maintenance downtime for each task. This is accomplished by in-

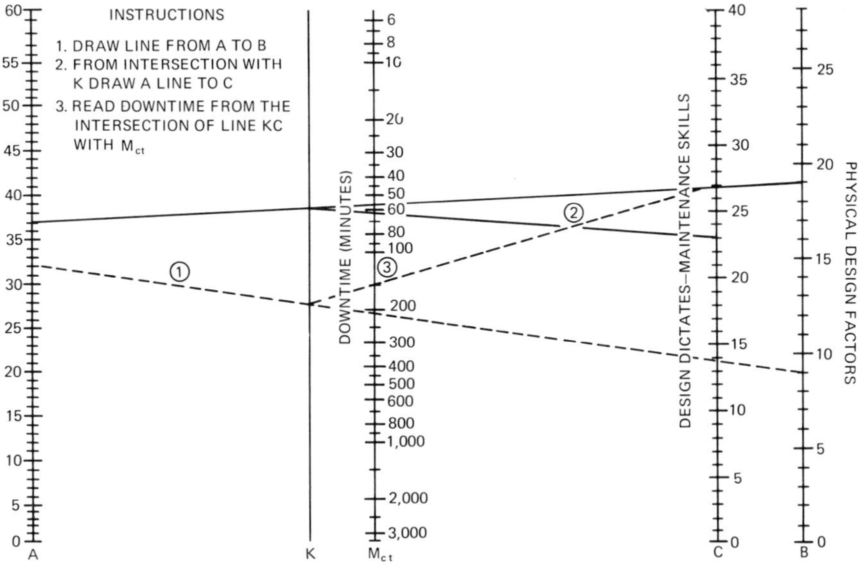

FIG. 10-5 Downtime Nomograph.

serting the checklist scores (indicated in Fig. 10-4) in the equation below.

$$M_{ct_i} = \text{antilog}\ (3.54651 - 0.02512A - 0.03055B - 0.01093C) \quad (6)$$

In the example illustrated, the scores for checklists *A*, *B*, and *C* are 37, 19, and 23 respectively. Substituting these values in the equation results in the following:

$$M_{ct_i} = \text{antilog}\ [3.54651 - (0.02512)(37) - (0.03055)(19) - (0.01093)(23)]$$

$$M_{ct_i} = \text{antilog}\ (1.78543) = 61 \text{ min}$$

To facilitate this calculation, a nomograph is presented in Fig. 10-5. The use of the nomograph permits the determination of downtime directly.

When the above procedure is followed for all tasks (Table 10-8), computation of the mean corrective maintenance time ($\bar{M}_{ct}$) is possible. The M_{ct_i} values for each task, as entered in Table 10-9, are added and entered in the following equation:

$$\bar{M}_{ct} = \frac{\sum_{1}^{N} M_{ct_i}}{N} = \frac{877}{20} = 43.8 \text{ min} \quad (7)$$

N = sample size of corrective-maintenance tasks.

Further

$$\overline{\log M_{ct}} = \frac{\sum_{1}^{N} \log M_{cti}}{N} = \frac{31.733}{20} = 1.586 \tag{8}$$

It is now possible to calculate the maximum maintenance time M_{max}.

$$M_{max} = \text{antilog}\,(\overline{\log M_{ct}} + 1.645\sigma_{\log M_{ct}}) \tag{9}$$

where $\sigma_{\log M_{ct}}$ is the standard deviation with respect to $\overline{M}_{ct}$.

The 95% confidence level (factor of 1.645) is assumed.

$$\sigma_{\log M_{ct}} = \sqrt{\frac{\sum_{1}^{N} (\log M_{cti})^2 - (\sum_{1}^{N} \log M_{cti})^2/N}{N-1}} \tag{10}$$

Therefore

$$\sigma_{\log M_{ct}} = \sqrt{\frac{51.587 - (31.733)^2/20}{19}}$$

And

$$\sigma_{\log M_{ct}} = 0.3525$$

Substituting the values in Eq. (9)

$$M_{max} = \text{antilog}\,[1.586 + 1.645(0.3525)]$$

Therefore

$$M_{max} = \text{antilog}(2.1658) = 146 \text{ min}$$

TABLE 10-9 Maintenance-task Time Summary

Task No.	M_{ct}	Log M_{ct}	(Log M_{ct})^2
1	54	1.732	2.999
2	67	1.826	3.334
3	46	1.662	2.762
4	22	1,342	1.800
5	25	1.397	1.951
6	73	1.863	3.470
7	20	1.301	1.692
8	31	1.491	2.223
9	69	1.838	3.378
10	26	1.414	1.999
11	27	1.431	2.047
12	27	1.431	2.047
13	24	1.380	1.904
14	64	1.806	3.261
15	61	1.791	3.207
16	23	1.361	1.852
17	21	1.322	1.774
18	75	1.875	3.515
19	62	1.792	3.211
20	60	1.778	3.161
TOTAL	877	31.733	51.587

QUESTIONS AND PROBLEMS

1. What is the purpose of maintainability prediction? Define benefits!
2. When are maintainability predictions accomplished?
3. The results of a maintainability-prediction process will be the same regardless of who may perform the prediction. True or false? Why?
4. There is only one valid prediction technique. True or false? Why?
5. The maintenance concept for the equipment or system has no bearing on the method or extent of prediction. True or false? Why?
6. The maintainability-design features of the equipment or system do not influence the prediction results. True or false? Why?
7. What purpose do the failure rates of the equipment/system items serve in the prediction process? Why are they used?
8. What are the maintenance tasks or functions that you would consider in performing a maintainability prediction?
9. Given that the sum of all failure rates is $2{,}000 \times 10^{-6}$ and that the sum of all time-weighted factors is $100{,}000 \times 10^{6}$ (task time in minutes), what is the *MTTR* value for the item?
10. Illustrate or describe how the values of $2{,}000 \times 10^{-6}$ and $100{,}000 \times 10^{-6}$ were obtained in the above problem.
11. Calculate the sample size when:
    a. The standard deviation of the sample data is 10 min.
    b. The mean repair time is 40 min.
    c. The confidence level is 95%.
    d. The desired accuracy is 10%.
12. Calculate the mean repair time when:
    a. The standard deviation of the sample data is 20 min.
    b. The coefficient of variation is .250.
13. Assuming the checklist scores for *A*, *B*, and *C* are 30, 10, and 20, respectively, calculate the corrective-maintenance time for the item.

ELEVEN

Maintenance-policy Definition

In the broadest sense, a maintenance policy is a statement or illustration that conveys all of the regulations, directives, procedures, and resources necessary to service and maintain a system or equipment item. As such, the maintenance policy is an expression of the adopted support concept based on the results of trade-offs, the maintainability analysis, and the maintenance analysis. The maintenance policy sets forth the maintenance-task requirements, the frequency of maintenance needs, the level of maintenance support, facility requirements, replacement-part needs (spares), support equipment and tool requirements, the required number and type of maintenance personnel, and the definition of maintenance procedures. The maintenance policy, therefore, is a summary of the proposed system/equipment-support program.

The maintenance policy may be documented in the form of a narrative statement, or illustrated in a manner similar to that outlined in Fig. 11-1.

POLICY APPLICATION

As stated above, the maintenance policy is a summary of all pertinent facts contained in trade-off reports, the maintainability-analysis records, the maintenance-analysis reports, and other documents reflecting the support needs or requirements of the system or equipment. This summary data, by virtue of being in a composite form, can be applied by a number of people to evaluate the support characteristics of the system or equipment to aid in fulfilling their particular job responsibilities. Some specific applications of the maintenance policy (document or illustration) are as follows.

Design engineer. The design engineer utilizes the maintenance policy to assess the overall support requirements of the equipment item that he is responsible for. In doing so, he can reflect on alternate design approaches, and perhaps improve the design, to overcome some of the undesirable factors inherent in the present design.

Systems engineer. To the systems engineer, the maintenance policy serves to identify the requirements for support equipment, facilities, and personnel needs (type and number). In addition, the policy data permit an overall evaluation of the system-support needs to assess alternate means of performing and sustaining the system functions.

Technical-publications writer. The maintenance policy provides procedural criteria which form the basis for detailed maintenance procedures that are documented in the system/equipment technical manuals. The policy serves as a guide to the writer in the generation of detailed maintenance procedures.

Spares analyst. The spares analyst uses the maintenance policy to identify the spares-allocation needs (type and quantity of items and their physical location).

Maintainability engineer. The maintainability engineer uses the maintenance policy as an aid in communicating with the designer, the systems engineer, the technical writer, and the spares analyst, and with management and the customer. The policy can be used to highlight the attributes of the design as regards the adopted support plan.

Management. Project-management personnel use the maintenance policy as a reference in measuring the compatibility of spares allocations, technical procedures, the procurement of tools and support equipment, and other products and services to ensure consistency of approach toward a common goal.

Customer. The customer uses the maintenance policy as basis for short- and long-range planning and to evaluate the qualitative characteristics of the system or equipment.

Planning can focus on training of personnel, the establishment of new calibration standards, or similar activities that become evident from reviewing the maintenance policy.

The system can be measured in qualitative terms (from the standpoint of support) by evaluating the need for the types and quantities of men, material, and facilities. These factors can be equated with the operational needs of the system to verify the feasibility of the support plan.

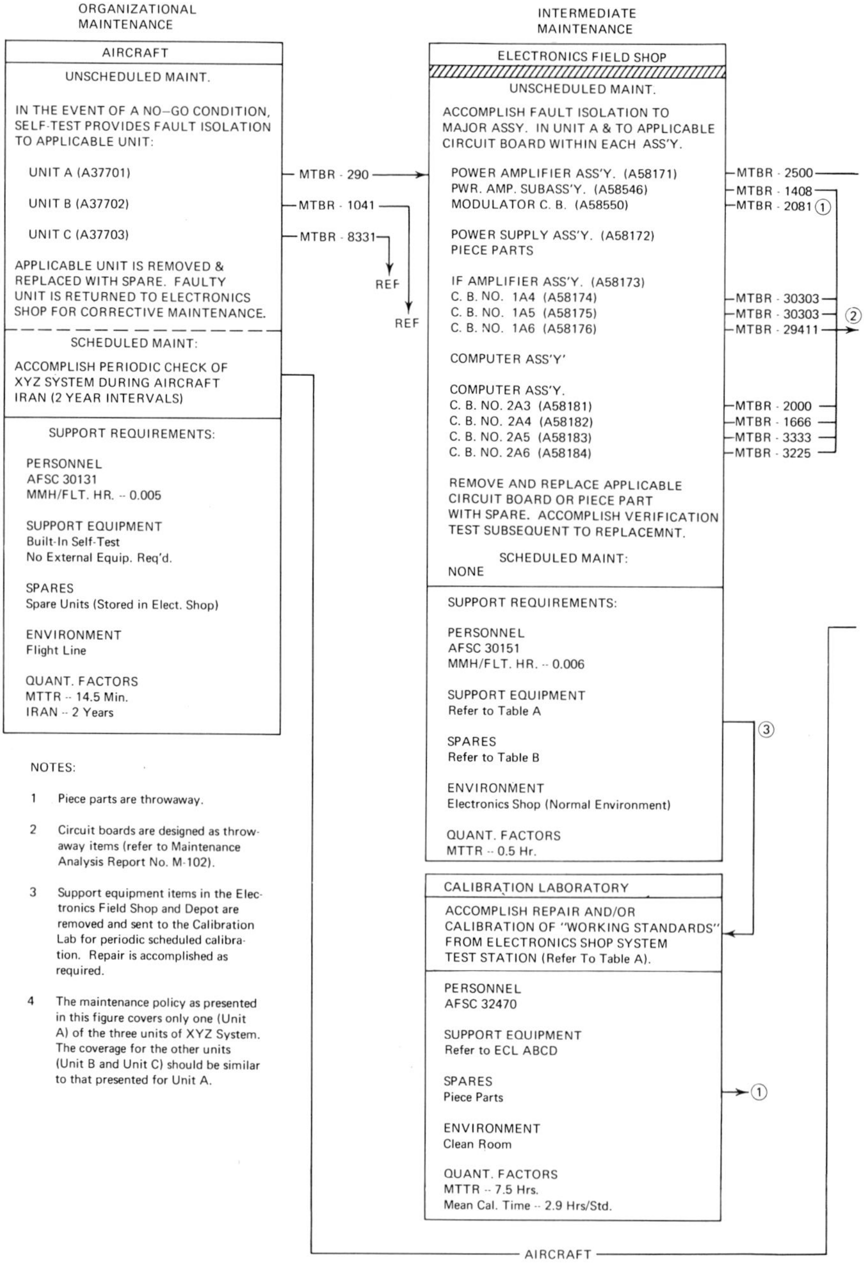

FIG. 11-1 Illustrated Maintenance Policy (Example).

DEPOT MAINTENANCE

DEPOT/SPECIALIZED REPAIR ACTIVITY

UNSCHEDULED MAINT:

ACCOMPLISH FAULT ISOLATION & REPAIR OF POWER AMPLIFIER SUBASS'Y. (A58546). REMOVE & REPLACE APPLICABLE CIRCUIT BOARD:

- C. B. NO. 3A1 (A58185) — MTBR - 13333
- C. B. NO. 3A2 (A58186) — MTBR - 13333
- C. B. NO. 3A3 (A58187) — MTBR - 10000
- C. B. NO. 3A4 (A58188) — MTBR - 6666

→ ②

ACCOMPLISH VERIFICATION TEST SUBSEQUENT TO REPLACEMENT

PERSONNEL
Existing Capability

SUPPORT EQUIPMENT
Refer to Table C

SPARES
Refer to Table B

ENVIRONMENT
Standard Depot

QUANT. FACTORS
MTTR -- 2.3 Hrs.

→ ③

DEPOT/IRAN

ACCOMPLISH CHECKOUT AND REPAIR OF SYSTEM WHEN AIRCRAFT IS SCHEDULED THRU IRAN FUNCTION.

PERSONNEL:
Existing Capability

SUPPORT EQUIP:
Same As Electronics Shop

SPARES:
Refer to Table B

ENVIRONMENT
Hangar

QUANT. FACTORS:
Checkout -- 5 Min.
MTTR -- 14.5 Min.

→ ③

TABLE A -- SUPPORT EQUIPMENT (ELECTRONICS SHOP)

ITEM	CALIBRATION	
	180 Days	360 Days
SYSTEM TEST STATION TO INCLUDE:		
D. C. Voltmeter, GX1001 or equiv.	X	-
A. C. Voltmeter, GR2010 or equiv.	X	-
Oscilloscope, XY1002A or equiv.	X	
Electronic Counter, MC8017 or equiv.	X	-
Frequency Converter Plug-In, QR8018 or equiv.	-	-
Special Test Set (Stimulus & Monitor)	-	-
Test Panel (Meters)	-	X
Misc. Loads, Transformers, Etc.	-	-
Assembly & Circuit Board Adapters	-	-
Tool Kit, MS10002	-	-

TABLE C -- SUPPORT EQUIPMENT (DEPOT)

ITEM	CALIBRATION	
	180 Days	360 Days
Signal Generator, UR70172 or equiv.	X	-
Spectrum Analyzer, TR10175 or equiv.	X	-
Pulse Generator, PG20172B or equiv.	-	X
Oscilloscope, XY1002A or equiv.	X	-
Power Supplies (PS2001, PS2002, PS2003)	-	-
Attenuators, Terminations, Directional Couplers (Refer to Report No. 10-1).	-	X
Test Panel (Meters)	-	X
Assembly & Circuit Board Adapters	-	-
Unit A (A38801)	-	-
Tool Kits, MS10002 & MS10003		

* Based on 273,600 Aircraft Flight Hours, and Utilization of 1.1 XYZ System Operation/Aircraft Flight Hour or 300,960 Hours of XYZ System Operation for a 10-year Period (Load Per Electronics Shop). Electronics Field Shop turnaround time - 30 days.

† Depot Level Spares (Supporting Two Electronics Field Shops -- Based on 562,320 Hours of Operation for a 10-year Period). Depot turnaround time - 60 days.

TABLE B -- SPARE PART REQUIREMENTS

ITEM	QUANTITY*	ITEM	QUANTITY*
Unit A (A37701)	9	C. B. No. 2A5 (A58183)	90
Unit B (A37702)	5	C. B. No. 2A6 (A58184)	93
Unit C (A37703)	4	Piece Parts	10 ea.
Power Amplifier Subass'y. (A58546)	15	Unit A (A37701)†	3
Modulator C.B. (A58550)	210	Unit B (A37702)+	2
C. B. No. 1A4 (A58174)	11	Unit C (A37703)†	1
C. B. No. 1A5 (A58175)	11	C. B. No. 3A1 (A58185) †	42
C. B. No. 1A6 (A58176)	12	C. B. No. 3A2 (A58186)†	42
C. B. No. 2A3 (A58181)	150	C. B. No. 3A3 (A58187)†	56
C. B. No. 2A4 (A58182)	180	C. B. No. 3A4 (A58188)†	85

FORMS OF POLICY DOCUMENTATION

Illustration

Figure 11-1 outlines one form of illustrating a maintenance policy. The format for policy illustration will vary from case to case, depending upon the complexity and type of the system or equipment and upon the degree of information that the maintainability engineer wishes to convey. As a general rule, however, the illustrated policy should highlight:

1. The division of maintenance tasks or functions by level of maintenance (e.g., organizational, intermediate, depot)
2. The type of maintenance performed at each level of support (e.g., removal and replacement of subassembly no. A37701)
3. The type of skills required to perform the maintenance tasks (e.g., AFSC 30131)
4. The support-equipment requirements (e.g., signal generator UR70172)
5. The type of spares required to support the maintenance actions
6. Special facility or environmental requirements
7. Traceability of support between the echelons of maintenance (e.g., item A58546 is repaired at depot, is replaced within assembly A58171 at intermediate level, and A58171 is replaced within unit *A* at the organizational level)

An illustration sheet may contain tables to relate details of policy that would otherwise detract from the main purpose of the illustrated material (see Fig. 11-1, Tables A to C). Also, an illustration sheet may be supported by narrative statements of policy if they are required to clarify details not evident in the illustration.

Narrative Statement

The common form for conveying maintenance policy is by narrative statement (see Fig. 11-2, for example). The narrative statement should contain the same type of information as discussed above for illustrated matter.

QUESTIONS AND PROBLEMS

1. What is a *maintenance policy*?
2. What documents and reports are used to define the maintenance policy?
3. How do the results of the maintainability analysis influence the maintenance policy?
4. Of what value is the maintenance policy to the following people?
   a. The design engineer

MAINTENANCE CONCEPT

VIDEO TEST STATION NOMENCLATURE	J. H. JONES PREPARED BY	30 MARCH 1967 DATE	A4811200 CONTROL NO
123456-1 DESIGNATION PART NUMBER	R. A. SMITH APPROVED BY	4 APRIL 1967 DATE	F-118 AGE MODEL

2 System Assembly Function THE PURPOSE OF THE TEST STATION IS TO PROVIDE INTERMEDIATE LEVEL MAINTENANCE SUPPORT FOR VARIOUS LRU'S IN THE ATTACK RADAR, RADAR ALTIMETER, TERRAIN FOLLOWING RADAR, AND ELECTRONIC COUNTERMEASURES SYSTEMS. THE TEST STATION CONTAINS PROGRAMMABLE CIRCUITS THAT PROVIDE STIMULI TO AND MONITOR THE RESPONSE OF COMPONENTS OF THE F-118 AIRCRAFT SYSTEM. THE TEST STATION CONTAINS A SEMI-AUTOMATIC AND A MANUAL OPERATIONAL CAPABILITY. DURING SEMI-AUTOMATIC OPERATION, THE TEST STATION CONTROL FUNCTIONS ARE EXECUTED ON COMMAND OF PROGRAMMED TEST INFORMATION TRANSFERRED FROM AN EXTERNAL CENTRAL PROCESSOR AND CONTROLLER (CENPAC). MANUAL OPERATION OF THE TEST STATION IS ACCOMPLISHED IN EITHER A KEYBOARD MODE OR A MANUAL MODE. THE KEYBOARD MODE IS USED TO SIMULATE COMMANDS NORMALLY RECEIVED FROM CENPAC FOR EXECUTING THE NORMAL PROGRAMMABLE FUNCTIONS WITHIN A PROGRAMMABLE UNIT OF THE TEST STATION. THE MANUAL MODE IS SIMILAR TO THE KEYBOARD MODE EXCEPT THAT THE FRONT PANEL CONTROLS ON CERTAIN PROGRAMMABLE UNITS ARE ENABLED. THIS MODE ALLOWS OPERATION OF THE PROGRAMMABLE UNITS USING FRONT PANEL CONTROLS TO ESTABLISH SPECIFIC CONDITIONS WITHIN THE UNIT.

3. Maintenance Concept UNSCHEDULED MAINTENANCE: WHEN A MALFUNCTION IS INDICATED, TROUBLESHOOTING IS ACCOMPLISHED TO ISOLATE FAULT TO A PARTICULAR TRU. STATION REPAIR IS THEN UNDERTAKEN TO REMOVE-REPLACE OR REPAIR IN-PLACE. FAULT ISOLATION IS AIDED BY PANEL INDICATORS, MAINTENANCE TAPE, MANUAL-KEYBOARD, AND/OR MANUAL-LOCAL MODES OF OPERATION FOR THE SEMI-AUTOMATIC SYSTEM. REPAIR IN-PLACE SHALL CONSIST OF EXTENDING THE TRU ON ITS SLIDE MOUNTS OR REMOVING THE TRU TO THE STATION WORK AREA FOR ACCESS. REPAIR ACTIONS WILL CONSIST OF SUB-ASSEMBLY OR PIECE PART REMOVAL-REPLACEMENT. PIECE PARTS, SUCH AS RESISTORS, SWITCHES, DIODES, RELAYS, ETC. ARE CONSIDERED THROW AWAY ITEMS. TROUBLESHOOTING IS PERFORMED BY V32251 PERSONNEL, AND REPAIR ACTION PERFORMED BY V32231 PERSONNEL. SCHEDULED MAINTENANCE: A MONTHLY SERVICE IS REQUIRED TO CLEAN AIR FILTERS IN THE BLOWER ASSEMBLIES (A11566), LRU COOLER (A31550), AND MICRO LOGIC POWER SUPPLY (A30021). THE FOLLOWING ITEMS WILL BE SENT TO PMEL FOR CALIBRATION AND/OR REPAIR AT INTERVALS AS INDICATED. 90 DAYS, DVM (A14001). 180 DAYS, OSCILLOSCOPE (A14026); FREQ. COUNTER (A14945); PROGRAMMABLE VARIABLE RATIO TRANSFORMER (A14020); AND PHASE SENSITIVE CONVERTER (A14008). THE FOLLOWING ITEMS WILL BE VERIFIED ON STATION AT 180 DAY INTERVALS: PROGRAMMABLE PULSE GENERATOR (A14463), FUNCTION GENERATOR SWITCHING (A20981), DIGITAL ANALYZER (A14859), TIMER COUNTER (A14850), SERVO COMMAND GENERATOR (A20677). THE FOLLOWING ITEMS WILL BE CALIBRATED ON STATION AT 180 DAY INTERVALS: POWER DISTRIBUTION PANEL (A14446), TRANSFORMER VARIABLE POWER (A14567), SERVO COMMAND GENERATOR (A20677). M2 IN VAR. POWER TRANSFORMER ASSY. WILL BE CALIBRATED ON STATION AT 360 DAY INTERVALS.

4. Reason for Maintenance Concept THE EXTENSIVE USE OF PLUG-IN MODULES AND SUBASSEMBLIES MAKES THE REPAIR IN-PLACE CONCEPT MORE FEASIBLE. THIS CONCEPT RESULTS IN A SIGNIFICANT COST REDUCTION BY REDUCING THE REQUIREMENT FOR SPARE TRU PROVISIONING, BY MINIMIZING PERSONNEL TRAINING REQUIREMENTS, AND BY DECREASING STATION DOWN TIME. THE AIDS FOR FAULT ISOLATION ALSO ENHANCE IN-PLACE REPAIR CONCEPT.

FIG. 11-2 Maintenance Policy—Narrative Statement (Example).

b. The systems engineer
c. The technical-publications writer
d. The spares analyst
e. The customer
f. The ILS manager
g. The maintainability engineer

5. What are the two basic forms of maintenance-policy documentation? Which is the common form?
6. At a minimum, what should the maintenance policy convey to the reader?
7. Tables and narrative statements should never accompany an illustrated maintenance policy. True or false?
8. Using Fig. 11-1 as a reference, answer the following.

a. The illustration outlines the complete maintenance policy for each of the items designated unit *A*, unit *B*, and unit *C*. True or false?
b. What maintenance tasks are performed on unit *A* at the organizational level of maintenance?
c. List (by name) the major subassemblies that comprise unit *A*.
d. List (by designator) the items that comprise A58546.
e. What test equipment is required at the intermediate level of maintenance? What is the purpose of the test equipment?
f. Circuit boards are repaired. True or false?
g. What items on unit *A* are calibrated at the intermediate level of maintenance?
h. What spare parts are required to support corrective maintenance accomplished within the electronic field shop? What spares are required to support maintenance accomplished at the aircraft?

TWELVE

Design Review

Design review, as discussed in this chapter, refers to the *formal design reviews* conducted at critical points throughout the definition and acquisition phases of a program. The formal design review, serving as a check function on the day-to-day design participation and liaison activity (discussed in Chap. 8), is intended to ensure that all facets of design (to include maintainability principles) are being incorporated or adequately considered.

The formal design review constitutes a coordinated activity (resulting in a meeting or series of meetings) intended to reflect the interests of the design engineer, the technical-discipline support areas (such as maintainability, reliability, human factors), manufacturing engineering and support, quality assurance, program management, etc. The subject of design review encompasses a wide scope of interests and disciplines. It is basically agreed that holding separate formal reviews for maintainability, other formal reviews for reliability, human factors, etc., would result in confusion, waste of man-hours and cost, schedule delays, and most likely would result in little or no accomplishment. Thus, the intent is to formally and logically review the proposed design from the *total-system* standpoint in the most effective and economical manner through a combined integrated review effort.

The formal design review serves a number of purposes.

1. It provides a formalized audit (check) of proposed system/equipment design with respect to contractual and specification requirements. Major problem areas are discussed, and corrective action is taken.

2. It provides a common baseline for all project personnel. The designer is provided the opportunity to explain and justify his design approach, and the various technical disciplines (e.g., maintainability) are provided the opportunity to hear the design engineer's problems. This tends to facilitate a better understanding among design and support personnel.

3. It provides a means of solving interface problems, and promotes the assurance that the system elements will be compatible. For example, major interface problems between engineering and manufacturing relative to the producibility of the system being designed are often undetected until design data are released and production commences. The results of major problems discovered at that time are often quite costly. Another example includes the common problem of compatibility, or lack thereof, between prime-equipment and support-equipment design. Such interface problems are often undetected at an earlier point in time due to a wide variance in organizational interests and activity. A design review prevents these major problems.

4. It provides a formalized record of what design decisions were made and the reasons for making them. Trade-off–study reports are noted and are available to support design decisions. Compromises to maintainability are documented and included in the trade-off–study reports.

5. It promotes a greater probability of mature design as well as the incorporation of the latest techniques. Group review may lead to new ideas, possibly resulting in simplified processes and subsequent cost savings.

In summary, it may be stated that the formal design review, when appropriately scheduled and conducted in an effective manner, causes a significant reduction in the contractor's risk relative to meeting contract and specification requirements and often results in improvement to the contractor's methods of operation.

SCHEDULING OF DESIGN REVIEWS

Formal design reviews are generally scheduled prior to each major evolutionary step in the design process. In other words, prior to proceeding to the next phase of design (e.g., from concept development to systems definition through the preparation of detailed specifications), it is often advantageous to schedule a formal design review to ensure that all areas have been adequately considered.

Although the quantity and type of design reviews established vary from program to program, four basic types are readily identifiable and common to most major programs. They include the conceptual review, the system-design review, the preliminary design review (PDR), and the critical design review (CDR). These are discussed in the following paragraphs.

The time phasing of the various design-review sessions is illustrated in Fig. 12-1.

Conceptual Design Review

The conceptual design review may be scheduled during the early portion of the definition phase (preferably not more than four to six weeks after contract award), when initial system requirements and concepts are defined. The purpose is to examine management and technical approaches relative to system development with respect to customer requirements. The conceptual design review should be conducted prior to the system-definition process resulting in the generation of system specifications.

Relative to maintainability requirements, the conceptual design review should cover the basic-maintenance concept, maintainability-analysis data (qualitative and quantitative design requirements and allocations), and major trade-offs where maintainability interests are included.

System-design Review

System-design reviews are generally scheduled during the definition phase and/or the early portion of the acquisition phase when preliminary system/equipment layouts and specifications have been prepared (prior to their formal release). There may be one or more formal reviews scheduled depending on the size of system and the number of defined end items. The purpose is to review individual specifications and supporting data to ascertain (1) whether the documentation reflects all customer requirements, and (2) whether the documentation reflects the best overall design approach (relative to economical considerations) within the constraints imposed by the customer requirements

In line with the system-design reviews, maintainability interests are concentrated on reviewing the adequacy of maintainability-analysis data coverage (the results of functional flows, definition of qualitative and quantitative design criteria, trade-off studies, etc.) incorporated in the system/equipment specifications.

Preliminary Design Review (PDR)

Preliminary design reviews are scheduled during the acquisition phase when design layouts, preliminary mechanical and electrical drawings, functional and logic diagrams, engineering breadboards, and design aids or mock-ups are available. The design process at this point has identi-

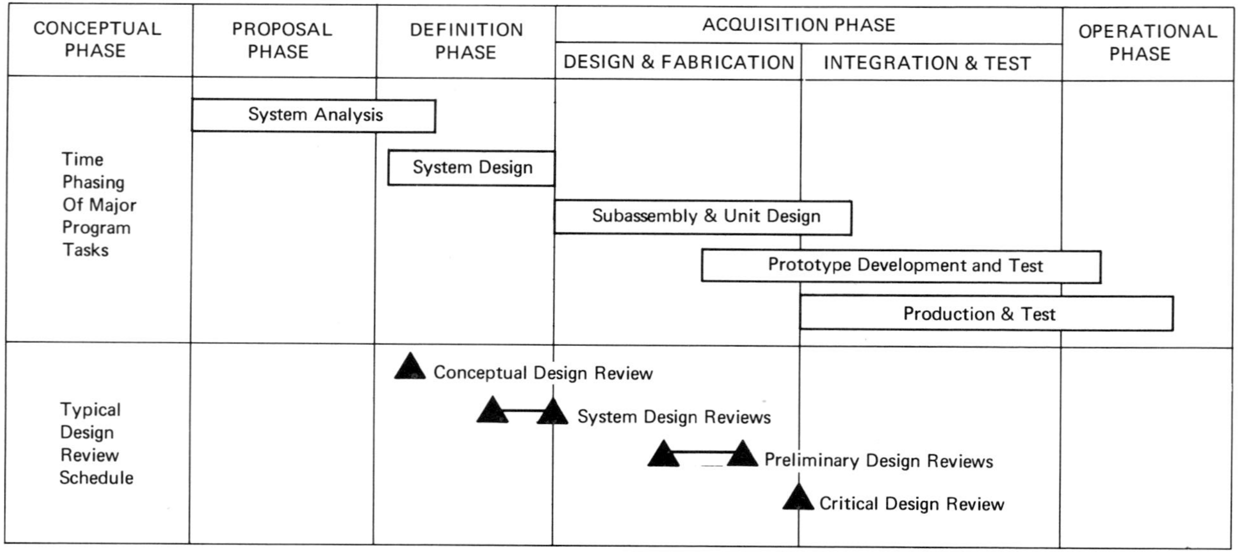

FIG. 12-1 Design Review Schedule in Relationship to Program Phases.

fied detailed design constraints, additional requirements, major problem areas, etc. Such reviews are conducted prior to proceeding with the finalization of detailed design.

During the early design process, the maintainability engineer is involved with the design liaison, prediction, and maintenance-analysis tasks (refer to Chaps. 8, 10 and 13). These tasks provide design support as well as the prediction and early assessment of the proposed design configuration against customer requirements. Major problem areas are readily identified, and are reviewed as part of the formal design-review process to seek solutions.

Critical Design Review (CDR)

Critical design reviews are scheduled during the acquisition phase, after detailed design has been completed but prior to the release of firm design data (e.g., detailed drawings, lists of material) to manufacturing for production purposes. Such reviews should be conducted to verify the adequacy and producibility of the basic design. Design is essentially "frozen" at this point, and manufacturing methods, schedules, and costs are reevaluated for final approval.

The critical design review covers all design effort accomplished subsequent to the completion of the preliminary design review. In addition, it covers those changes resulting from recommendations for corrective action stemming from the PDR. Maintainability interests at this point are essentially the same as for the PDR. Major problem areas (covered by outstanding maintainability recommendations and problem-identification reports) are reviewed, and resulting action is documented.

REQUIREMENTS FOR DESIGN REVIEWS

The success of the formal design review is dependent on the extent of planning, organization, and data preparation accomplished prior to the review itself. Such activity involves an extensive amount of coordination relative to the establishment of:

1. A specific date of the review.
2. A location or facility where the review is to be held.
3. An agenda for the review including the specific areas of design and/or support coverage, and the basic objectives of the review.
4. A design-review board (representing the organizational elements and disciplines affected by the review). Maintainability representation is included. Individual organization responsibilities should be identified. Customer participation may or may not be required. If required, the customer should be notified of the review (time and place) and

review agenda. In certain instances, the customer may specify the time and place of the review.

5. Equipment (hardware) requirements for the review. Equipment prototypes, breadboards, and/or mock-ups may be required for the review.

6. Design-data requirements for the review. This may include all applicable specifications, functional flows, trade-off–study reports, detailed drawings, lists of material, maintainability problem-identification reports, logistics data, etc.

7. Funding requirements. Planning is necessary relative to identifying the means for providing funds for conducting the design review.

8. Reporting requirements and the methods for accomplishing the necessary follow-up action stemming from design-review recommendations. Outstanding maintainability recommendations which are significant relative to compliance with contract requirements are included.

Design-review Board

One of the most significant aspects in the design-review planning process is the establishment of a design-review board. The function of the design-review board is to conduct the formal review plus determine the extent of any recommended follow-up action that may be required subsequent to completion of review meetings. The design-review board is the ultimate authority relative to all formal design-review activity.

The design-review board may be composed of permanent members, or organized on a temporary basis to review a specific design, or a combination of both. In any event, the design-review board must include top talent (specialists) representing such areas as design engineering, reliability, maintainability, configuration management, value engineering, components engineering, logistics, manufacturing engineering, and quality assurance. The design-review board is chaired by an impartial individual who is knowledgeable of overall system-engineering design and development processes.

The design-review board is the mechanism to ensure that all facets of design (e.g., maintainability, reliability) are being adequately considered. In addition, the design-review board has the authority to effect corrective action and to direct the initiation of design changes as required to meet contractual requirements or to improve total product effectiveness.

THE CONDUCT OF DESIGN REVIEWS

There are several basic principles to be applied in conducting a meeting of any type. These principles are equally applicable in conducting a successful formal design-review meeting and can be summarized as follows.

1. Design-review meetings must be firmly controlled by the design-review-board chairman.

2. Design-review meetings should be brief and to the point and must not be allowed to drift away from the topics on the agenda.

3. Design-review-meeting attendance should be limited to those having a direct interest in a given problem to be presented at the review. Only those specialists who can contribute directly to the attainment of the review objectives are permitted to participate actively. Specialists who participate should be authorized both to speak for and make decisions concerning their area of specialty.

4. Design reviews must make provisions for corrective actions to be identified, recorded, and subsequently monitored. Specific responsibilities for follow-up action must be designated by the design-review-board chairman.

QUESTIONS AND PROBLEMS

1. List three reasons why the formal design review is of value.
2. List the four basic types of design reviews. Relate each to a program phase.
3. What does CDR mean?
4. What does PDR mean?
5. What are the objectives of the conceptual design review in terms of maintainability?
6. What are the objectives of the critical design review in terms of maintainability?
7. Describe your contribution to a preliminary design review if you were a maintainability engineer assigned to the design-review board.
8. As a maintainability engineer, describe how the formal design review aids you in doing your job.

THIRTEEN

Maintenance Analysis

In discussing the various maintainability-program functions, a certain amount of confusion develops relative to the intent of the maintainability analysis and the maintenance analysis. As covered in Chap. 6, the *maintainability analysis* is defined as the "before-the-fact" process of translating operational and maintenance concepts, personnel factors, support-equipment criteria, etc., into specific qualitative and quantitative design requirements. The maintainability analysis is accomplished during the proposal and/or definition phases of a program prior to the start of actual equipment design. On the other hand, the *maintenance analysis* is the "after-the-fact" process of identifying detailed maintenance requirements through analysis of a design, and of determining the most effective means of accomplishing the requirements. The maintenance-analysis development is based on design data (e.g., engineering layouts, drawings, lists of material), and commences as early as practicable in the equipment-development cycle once the basic design configuration is established.

Reference to "after-the-fact" in this instance means only that a data basis for analysis has been established. It does not mean that equipment design has been fixed so that it cannot be influenced as regards maintainability-design characteristics.

PURPOSE AND SCOPE

The maintenance analysis serves three basic purposes:

1. It provides an engineering-data package covering all maintenance (scheduled and unscheduled) functions, tasks, and support requirements for the equipment being analyzed. In this respect, the maintenance analysis can fulfill the major objectives of the planned-maintenance aspect of integrated logistics support (refer to Chap. 1).

2. It provides a systematic check on equipment design for maintainability prior to "design freeze." Areas of poor design for maintainability which result in lengthy maintenance times or extensive support requirements readily become evident.

3. It provides criteria for the design of support equipment and training equipment. In this respect, the maintenance-analysis data package supplements the subsequent design efforts associated with supporting hardware.

The maintenance analysis constitutes a systematic evaluation of a proposed design configuration. The evaluation considers size, weight, packaging, accessibility, test provisions, and other related maintainability characteristics (refer to Chap. 8) as incorporated in the proposed design configuration. The intent of the evaluation, accomplished through a thorough review of design data, is to ascertain whether the proposed design incorporates the necessary features for effective support and whether the anticipated method of support reflects the best overall approach. In fulfillment of this intent, maintenance-analysis data covers:

1. All repairable maintenance significant items (i.e., system, subsystem, unit, assembly)

2. All maintenance requirements (i.e., troubleshooting, disassembly, remove and replace, repair, alignment and adjustment, functional test, inspection, servicing, calibration, overhaul)

The maintenance analysis should be developed on a concurrent basis with the accomplishment of detailed design commencing during the early part of the acquisition phase (refer to Chap. 3). As design progresses, the analysis effort continues. Thus, development of the analysis becomes an iterative process extending to system/equipment integration and test. The relative timing of the maintenance analysis in the overall system-development cycle is illustrated in Fig. 13-1 (block 14.0). Timing is critical in view of the fact that the analysis serves as a necessary prerequisite in the provisioning and acquisition of support requirements (block 15.0). Support items should be available for delivery to the customer on a concurrent basis with the prime equipment. A late start in the development of the maintenance analysis could cause

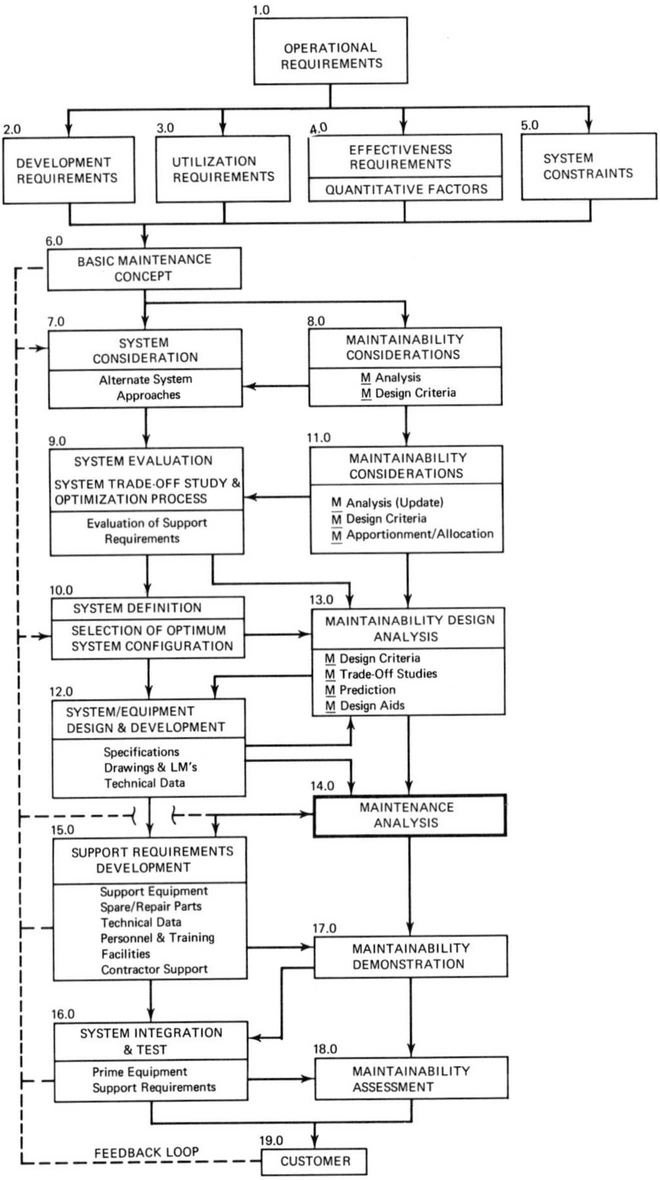

FIG. 13-1 System Development Cycle.

schedule slippages and subsequent contractual delinquencies relative to providing the necessary support items at the time of prime equipment delivery.

The required format and content of maintenance-analysis data can vary considerably depending on individual preferences and on company organizational responsibilities and interests. Such variances not only involve the specific forms utilized, but also cover the methods of presenting quantitative factors. The analysis may include engineering information, spares/support-equipment provisioning data, or a combination of both. This text deals with the specific data elements applicable to the *engineering* aspects of system/equipment design (those factors necessary to support the maintainability engineer's role). It is not intended to fulfill the needs for provisioning (e.g., priced parts lists, nomenclature/FSN requests, spare-part lead times and delivery schedules, method of packaging). The requirements for the provisioning of support constitutes a significant subject by itself. The intent herein is to limit coverage to the engineering-analysis aspect associated with support-requirements development. The paragraphs below highlight the steps involved in completing a maintenance analysis. A format for data presentation is also outlined.

MAINTENANCE-ANALYSIS DEVELOPMENT

The maintenance-analysis development cycle is illustrated through the data flow presented in Fig. 13-2. Referring to the figure, the analysis stems from the maintenance concept and is based on system/equipment-design information, maintainability-design data, reliability reports, and human-factors data. The data content of the analysis is conveyed through the qualitative and quantitative factors listed in block 14.0. Maintenance-analysis data is reflected in the maintenance policy (refer to Chap. 11) and serves as an input to the development of support requirements (block 15.0).

Maintenance Concept (Fig. 13-2, block 6.0)

As conveyed in Chap. 5, the maintenance concept provides the basis for maintainability criteria design and for the establishment of requirements for total support. Development of the maintenance concept is one of the initial steps in the overall system-development cycle, and covers criteria pertaining to applicable maintenance echelons, repair policies, facilities, maintenance environments, etc. Such criteria form the basis for the development of detailed maintenance functions and tasks, task frequencies, and support requirements identified through the main-

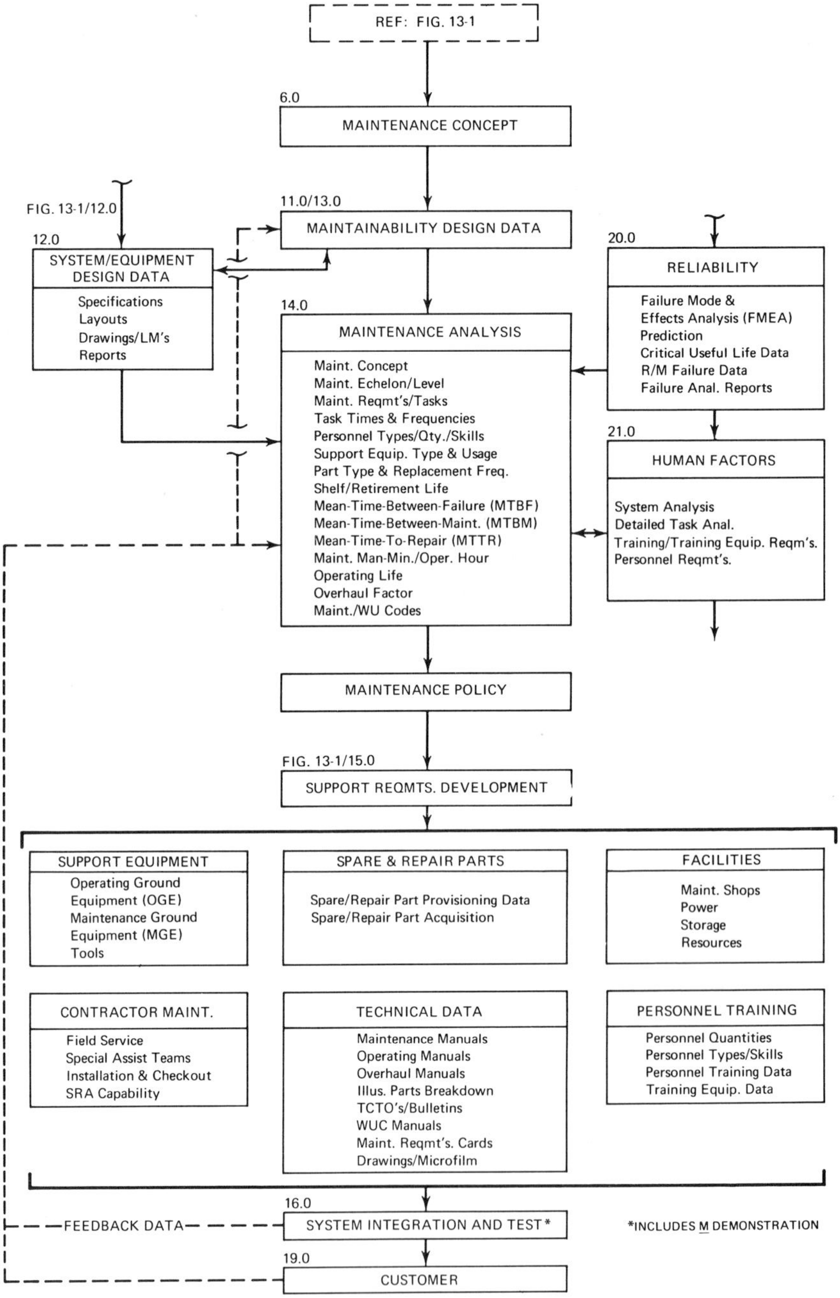

FIG. 13-2 Maintenance Analysis Development and Application.

tenance analysis. Thus, the maintenance analysis is based on, and supplements, the maintenance concept.

System/Equipment-design Data (Fig. 13-2, block 12.0)

Since the maintenance analysis is based on an established design configuration, various design data are necessary input requirements. The maintainability engineer, in completing an analysis, uses design specifications, equipment-packaging layouts, detailed drawings, schematics, lists of material, and pertinent engineering reports to review and evaluate the applicable system or equipment from the standpoint of maintainability features in design. The analysis results in the development of test approaches, disassembly-repair-reassembly sequences, verification procedures, and corresponding support needs.

Maintainability-design Data (Fig. 13-2, blocks 11.0 and 13.0)

Maintainability-design data refers to maintainability-analysis data (Chap. 6), trade-off–study reports (Chap. 7), design recommendations (Chap. 8), and prediction reports (Chap. 10). All such data provide necessary input and backup information (in one way or another) for the maintenance analysis. For instance, trade-off-study reports covering repair versus discard-at-failure decisions, or specific approaches to maintenance support, have direct bearing on the type and level of maintenance to be performed on the equipment. Prediction reports convey maintenance times in conjunction with recommended repair policies. Maintainability-design recommendations (specifically those *not* incorporated in design data) should be reviewed relative to possible support-related problems. The maintainability engineer, through the accomplishment of design services and liaison functions, provides the necessary support required for completion of the maintenance analysis.

Reliability Data (Fig. 13-2, block 20.0)

Reliability data necessary for completion of the maintenance analysis comprise the following:

- Failure-mode-effects-analyses (FMEA) are used in the identification of symptoms of equipment failure and the development of logic troubleshooting flow diagrams, which convey fault isolation and diagnostic sequences.
- Reliability-prediction reports serve as the reference source for equipment *MTBF*, part-failure rates, maintenance-task frequencies, etc.
- Critical useful-life data specify the requirements for forced removals, periodic item replacements, scheduled overhaul frequencies, etc.
- Reliability/maintainability failure data (covering in-plant proto-

type equipment or similiar systems already in the user's operational inventory) serve as a valuable source in the specification of maintenance tasks, task times, and task frequencies in the maintenance analysis.

- Failure-analysis reports. The systematic analysis of outstanding system-design problem areas sometimes results in equipment design change or in definition of special maintenance requirements associated with the existing equipment. Resulting maintenance requirements are reflected in the maintenance analysis.

Human-factors Data (Fig. 13-2, block 21.0)

Human-factors program requirements often result in the generation of a system analysis, operator- and maintenance-task analyses, personnel-requirements data, and training/training-equipment planning information. Certain aspects of such data parallel the maintenance analysis, specifically from the standpoint of facilities, maintenance tasks, personnel quantities and skill levels, and training requirements. In some instances, depending on organizational interests and available manpower, it may be feasible to cover these areas as a joint venture. In any event, the maintainability engineer must maintain full cognizance of available human-factors data when completing the maintenance analysis.

Additional Data

"Additional data" in this instance refers to test-point data reports, support-equipment plans, calibration-requirement summaries, and other similar documentation applicable to maintenance support. The maintainability engineer should review and utilize pertinent information to prepare and update the maintenance analysis.

FUNCTIONAL/TASK ANALYSIS

Development of the maintenance analysis commences with the identification of gross-level maintenance functions and evolves through the definition of detailed tasks combined with an analysis of those tasks relative to support requirements. This approach is conveyed in the paragraphs below.

Functional Analysis

The operational functions derived early in the program identify the various modes of system operation. In addition, these functions are related to proposed system-packaging schemes (refer to Chap. 6). The operational functions and subfunctions or operational modes (terms which are sometimes used synonymously) are then analyzed from the standpoint of failure and the resulting consequences of failure. This is basically accomplished through the reliability failure-mode-effects-analysis (*FMEA*). Failures associated with the various modes of system operation necessitate corrective maintenance and the identification of maintenance functions. The identification of gross-level maintenance functions is accomplished through the functional analysis discussed in Chap. 6. These functions are analyzed relative to the determination of subfunctions and tasks through the maintainability analysis and subsequent maintenance-analysis efforts covered in this chapter.

The identification of functions, through the development of functional-flow diagrams, serves the following basic purposes:

- Ensures that all elements of the system are fully recognized and defined
- Identifies the detailed operational and maintenance tasks, and the support requirements for these tasks
- Necessitates the establishment of task sequences (e.g., diagnostic routines, troubleshooting sequences, preventive-maintenance steps)

Functional-flow diagrams are developed to cover both preventive and corrective maintenance.

Relative to corrective maintenance, the gross-level maintenance functional flows generated through the maintainability analysis (Chap. 6) evolve into the logic troubleshooting-flow diagrams. A typical logic troubleshooting flow is presented in Fig. 13-3. Referring to Fig. 13-3, the logic troubleshooting flow stems from the symptom of failure, the first indication to the system operator that a *NO-GO* condition exists. The failure symptom may occur during normal system operation on a given mission or during system checkout between missions. In any event, the failure symptom is the starting point in the corrective-maintenance cycle. Once a failure is verified, the diagnostic process commences through localization and fault isolation.

Figure 13-3 reflects the recommended sequence of steps in accomplishing troubleshooting. The specific area of coverage is oriented to troubleshooting at the second level of maintenance (e.g., troubleshooting of the computer group). Actually, logic troubleshooting flows are generated to cover all levels of troubleshooting from the system level down to the item level at which repair is to be accomplished.

In generating logic troubleshooting-flow diagrams, the maintainability engineer must determine all possible symptoms of failure at the system level. This is facilitated through review of the reliability failure-mode-effect-analysis (*FMEA*), which simulates failure at the subassembly/part level and indicates failure effects (visual and otherwise) at the system level. The maintainability engineer projects himself into the role of the maintenance technician and looks at the system in terms

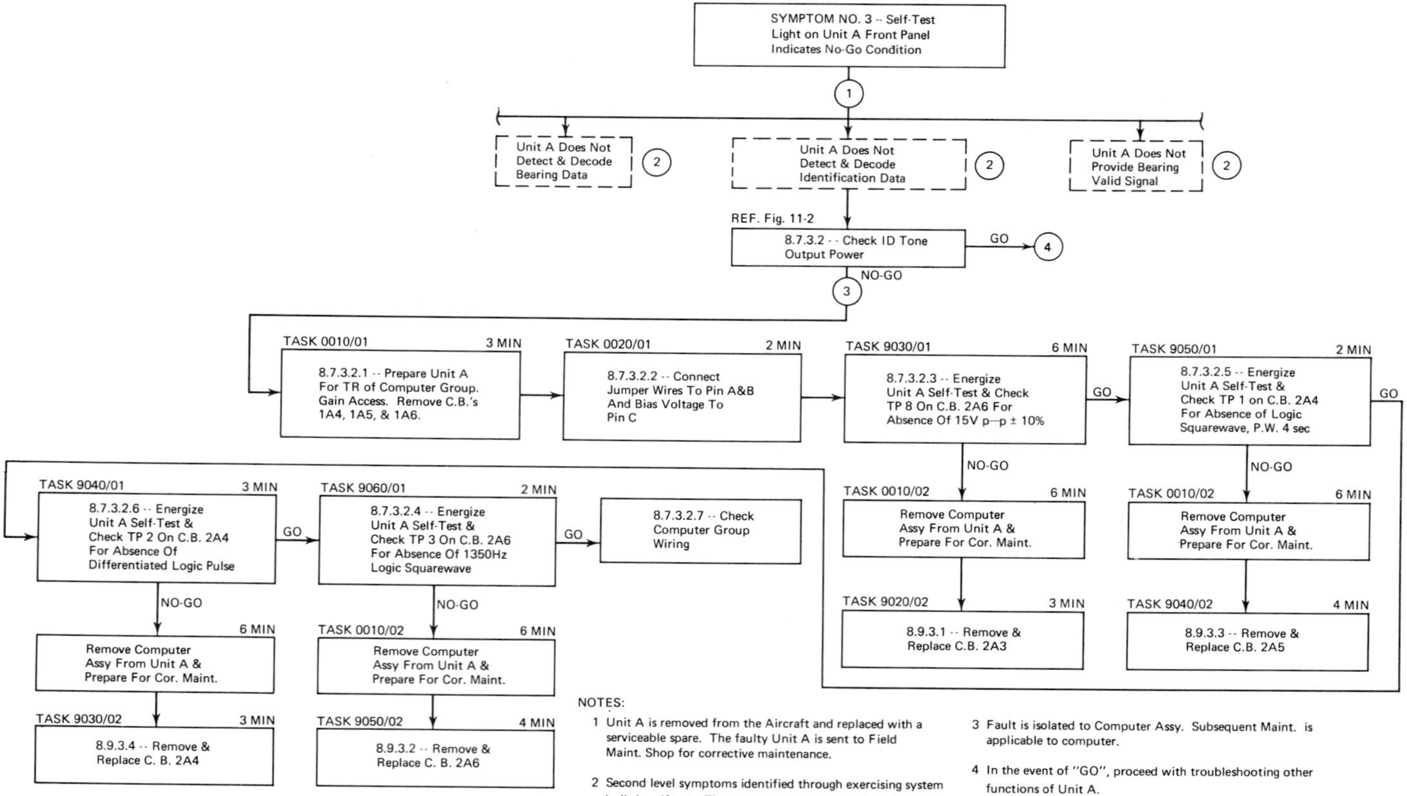

NOTES:

1 Unit A is removed from the Aircraft and replaced with a serviceable spare. The faulty Unit A is sent to Field Maint. Shop for corrective maintenance.

2 Second level symptoms identified through exercising system built-in self-test. These symptoms are also reflected in the Reliability FMEA.

3 Fault is isolated to Computer Assy. Subsequent Maint. is applicable to computer.

4 In the event of "GO", proceed with troubleshooting other functions of Unit A.

FIG. 13-3 Logic Troubleshooting-flow Diagram.

of visual/audio provisions (meter, *GO/NO-GO* light, buzzer, etc.) and proceeds by troubleshooting to the subassembly/part level employing self-test and test-point provisions. In doing such, he may reach the part through several different routes, each route being different relative to maintenance times and support requirements. The best route (minimum maintenance time and support required) is covered by the logic troubleshooting-flow diagram as indicated through the task sequences and maintenance times (see Fig. 13-3). The specific maintenance tasks are further supplemented by the data presented on the maintainability-task-analysis form discussed below and in Appendix C. Once the diagnostic aspect of corrective maintenance is covered, the remaining elements of the corrective-maintenance cycle (i.e., repair, alignment and adjustment, checkout or functional test, etc.) are analyzed. Specific tasks are entered directly on the maintainability-task-analysis form.

Task Analysis

The task analysis, as defined herein, covers those maintenance actions which are significant in defining support requirements. Such actions or tasks evolve directly from the functional analysis.

When defining a maintenance task, one could specify a maintenance action such as: "accomplish troubleshooting of computer assembly." This is quite general and by description is not adequate for the purposes of realistically estimating maintenance times, for example. A more detailed breakdown is required in this instance. Another approach may be to indicate a series of steps to be accomplished by the maintenance technician, such as: "place voltmeter probe at test point 2, rotate switch S1 on unit *A* front panel to test position, measure presence of 28 V dc." If this level of detail were followed, the maintainability engineer would basically be involved in writing the system technical manual procedures or in accomplishing the detailed maintenance-task analysis often conducted as a human-factors program requirement. This is not the intent of the task-analysis portion of the maintenance analysis. The task analysis should cover only enough detail to identify adequately maintenance times, support equipment, replacement parts, personnel types and skill levels, facilities, etc.

Recording of Tasks. Maintenance tasks are identified on the maintainability-task-analysis form, Fig. 13-4. Referring to the figure, a specific maintenance requirement (i.e., troubleshooting, repair, etc.) is identified in blocks 3 and 4. For each requirement, significant maintenance tasks are indicated in block 13. Support requirements (maintenance times, personnel types and skill levels, replacement parts, and support equipment/tools) are identified in blocks 14 through 49. Major maintenance-data elements indicated on the form are summarized in Table

MAINTAINABILITY TASK ANALYSIS

GENERAL DYNAMICS | ELECTRONICS

SYSTEM XYZ SYSTEM

1. PART NUMBER/NOMENCLATURE A58180/COMPUTER ASSY — NEXT ASSEMBLY A37701/UNIT A — 2. MA CONTROL NUMBER A84220

3. REQ. NO.	4. REQUIREMENT	5. CODE	6. TECHNICAL JUSTIFICATION	7. REMARKS
01	TROUBLESHOOT	HI	A MALFUNCTION HAS BEEN ISOLATED TO THE COMPUTER ASSEMBLY WITHIN UNIT A. TROUBLESHOOTING OF THE COMPUTER ASSEMBLY IS REQUIRED FOR CORRECTIVE MAINT.	△ INCLUDES WEIGHTED AVERAGE FOR "9" TASKS. REFER TO TROUBLESHOOTING LOGIC FLOW DIAGRAM.
8. MTBF	**9. OPERATING LIFE**	**10. REPL. INT.**		
1156	4000 HR.	578		
11. REFERENCES				
A84220, REQ. NO. 02				

12. MAINT LEVEL AND TASK NO.	13. TASK/DESCRIPTION	14. ELAPSED TIME – MINUTES	15. ELAPSED TIME TOTAL/TASK	16. JOB SPECIALTY OR RATING	17. BASIC SKILL	18. INTERMED SKILL	19. SUPER SKILL	20. TOTAL	21. QTY/ASSY	22. PART NO. / 27. PART NAME	23. MAINT QTY FACTOR	24. SHELF LIFE	25. TOOL NO. / 28. TOOL NAME	26. TOOL USE
0010	GAIN ACCESS	(1)	003	31350		003		003						
	REMOVE CIRCUIT BOARDS 1A4 1A5 1A6													
	TO GAIN ACCESS TO COMPUTER ASSY													
0020	PREPARE FOR TROUBLESHOOTING	(1)	002	31350		002		002					PS2001	014
	CONNECT JUMPER WIRES & APPLY	(2)		31330	002			002					POWER SUPPLY	
	BIAS VOLTAGE												GX1001	014
													VOLTMETER	
9030	ISOLATE TO CIRCUIT	(1)	006	31350		006		006					XY1002A	006
	BOARD 2A3 (0.00050)	(2)		31330	006			006					OSCILLOSCOPE	
9040	ISOLATE TO CIRCUIT	(1)	011	31350		011		011					XY1002A	011
	BOARD 2A4 (0.00060)	(2)		31330	011								OSCILLOSCOPE	
9050	ISOLATE TO CIRCUIT	(1)	008	31350		008		008					XY1002A	008
	BOARD 2A5 (0.00030)	(2)		31330	008			008					OSCILLOSCOPE	
9060	ISOLATE TO CIRCUIT	(1)	013	31350		013		013					XY1002A	013
	BOARD 2A6 (0.00033)	(2)		31330	013			013					OSCILLOSCOPE	
	REF: A84220 REQ. NO. 02													
	REPAIR													

CHANGE NO. 2

JOB SPEC.-NO. 29. 31350001 31. 31330001 32. 33. 34. 40. 41. 42. 43. 44. 45.

35. 014 △ TOTAL ELAPSED TIME/REQ'MT

36. 040 37. 043 38. 39. 083 TOTAL MAN-MINUTES

46. 001 47. 001 48. 49. 002 TOTAL NO. MEN

SHEET 1 OF 1

PREPARED BY J. JONES

DATE 27 APRIL 1966

FORM 1366E 12/63

11:4.0

FIG. 13-4 Maintainability-task Analysis (Example).

13-1 and discussed in detail in Appendix C (Instructions for Maintainability-task Analysis).

TABLE 13-1 Task-analysis Data Items

	DATA ELEMENT DESCRIPTION	FIG. 13-4 BLOCK NO.
1.	Maintenance Requirement	3, 4, 5, 6
2.	Maintenance Echelon (level)	5
3.	Requirement Frequency	10
4.	Requirement Maintenance Time	35
5.	Maintenance Tasks	12, 13
6.	Maintenance Task Frequency	10, 13
7.	Maintenance Task Times	14, 15
8.	Personnel Type (Career Field)	16
9.	Personnel Skill Level	17, 18, 19
10.	Personnel Quantities	20, 46 - 49
11.	Maintenance Man-Minutes	36 - 39
12.	Replacement Parts (Spares)	22, 23, 27
13.	Shelf Life/Retirement Life	24
14.	Support Equipment/Tools	25, 26, 28

Summary of Tasks. For each repairable unit, assembly, and/or subassembly, a maintenance-analysis data-summary form (Fig. 13-5) is prepared. Referring to Fig. 13-5, the summary form briefly describes the function of the item covered, the proposed maintenance policy, and the maintenance quantitative factors and support requirements associated with individual requirements. The summary provides a quick overview of total maintenance needs and can be readily employed in concept-review meetings, spares/support-equipment provisioning meetings, or other similar events. In addition, the maintainability engineer is provided a ready reference to quantitative time factors (i.e., *MTBM*, *MTTR*, *MMM/OH*, *TBO*). Instructions for completion of the maintenance-analysis data summary are contained in Appendix D.

Determination of Support Requirements

As conveyed in the paragraphs above, each individual maintenance task is analyzed from the standpoint of determining support requirements to include spare/repair parts; personnel types, skill levels, and quantities; support equipment; facilities and resources. These areas of concern are amplified in the following paragraphs.

For troubleshooting and checkout tasks, the prime emphasis is on support equipment. The type of support equipment employed depends on the operational modes being tested, depth of testing, prime-equipment design characteristics (e.g., test points), frequency of testing, duration of testing, and maintenance environment. Testing may be accomplished using built-in, external, manual, semiautomatic, or automatic equipment-

GENERAL DYNAMICS
Electronics Division

MAINTENANCE ANALYSIS DATA SUMMARY

SYSTEM XYZ SYSTEM

1. IDENT PART NUMBER/NOMENCLATURE: A58180/COMPUTER ASSY.
2. IDENT NEXT ASSEMBLY: A37701/UNIT A
3. MA CONTROL NO: A84220
4. DESIGN SPEC: ZA88446 (9/4/65)
5. SUPPORTING DOC: RELIABILITY REPORT RA100-2; TRADE-OFF STUDY REPORT M-106
6. FUNCTIONAL DESCRIPTION: SYSTEM XYZ IS THE AIRBORNE ELECTRONICS PORTION OF THE SHORT RANGE TACTICAL-AIR-NAVIGATION SYSTEM INSTALLED IN THE F-XXX AIRCRAFT. THE COMPUTER ASSY (A58180), INCORPORATED IN UNIT A OF SYSTEM XYZ, PROVIDES FINITE RANGE AND BEARING INFORMATION FOR THE OVERALL SYSTEM. SUCH INFORMATION IS REQUIRED FOR ACCURATE RANGE AND BEARING MEASUREMENTS. THE COMPUTER ASSY CONTAINS FOUR MAJOR CIRCUIT BOARD SUBASSEMBLIES (2A3, 2A4, 2A5, AND 2A6).
7. MAINTENANCE CONCEPT DESCRIPTION: UNSCHEDULED MAINTENANCE -- UPON DETECTION OF A FAULT AT THE SYSTEM LEVEL, THE APPLICABLE FAULT ISOLATED TO UNIT A (A37701), UNIT B (A37702), OR UNIT C (A37703) THROUGH BUILT-IN SELF TEST. THE APPLICABLE UNIT IS SENT TO THE ELECTRONICS FIELD SHOP FOR CORRECTIVE MAINTENANCE. FAULT ISOLATION OF UNIT A TO THE COMPUTER ASSEMBLY LEVEL IS ACCOMPLISHED THROUGH USE OF THE SYSTEM TEST STATION (I.E. -- OSCILLOSCOPE, POWER SUPPLY, VOLTMETER) IN THE FIELD ELECTRONICS SHOP. FAULT ISOLATION IS ACCOMPLISHED TO THE CIRCUIT BOARD LEVEL; THE APPLICABLE CIRCUIT BOARD (I.E. -- 2A3, 2A4, 2A5, OR 2A6) IS REMOVED AND REPLACED; AND THE COMPUTER ASSEMBLY IS CHECKED OUT THROUGH A FUNCTIONAL TEST OF UNIT A. FAULTY CIRCUIT BOARDS ARE DISCARDED AS "THROW-AWAY" ITEMS. MAINTENANCE IS ACCOMPLISHED BY AFSC 31330 & 31350 PERSONNEL IN A NORMAL SHOP ENVIRONMENT.
SCHEDULED MAINTENANCE -- NONE REQUIRED FOR THE COMPUTER ASSEMBLY.

8. REQ. NO.	9. REQ. NOMEN.	10. MAINT. LEVEL			11. MAINT. TYPE		12. FREQUENCY (HOURS)	13. MTBM (HOURS)	14. MTBF (HOURS)	15. MMM/OH	16. T.B.O	17. ELAPSED TIME (HOURS)	18. PERSONNEL DATA				19. SUPPORT EQUIP ⚠1	20. SPARES TYPE ⚠2
		O	I	D	SCH.	UNSCH.							JOB SPEC	B	I	S		
01	TROUBLESHOOT		X			X	0.00173	578	1163	0.1436	N/A	0.233	31330 31350	1 –	– 1		OSCILLOSCOPE POWER SUPPLY VOLTMETER	
02	REPAIR		X			X				0.0657		0.283	31330	1	–		SCREWDRIVER	C.B.2A3 C.B.2A4 C.B.2A5 C.B.2A6
03	FUNCT. TEST		X			X				0.0208		0.100	31330 31350	1 –	– 1		SPECIAL TEST SET	
TOTAL							0.00173	578	1163	0.2301		0.616		1	1			

21. NOTES:
⚠1 REFER TO ADDENDUM A (SHEET 1) FOR SPECIFIC QUANTITIES AND TYPE OF SUPPORT EQUIPMENT.
⚠2 REFER TO ADDENDUM A (SHEET 2) FOR SPECIFIC QUANTITIES AND TYPE OF SPARES.

FORM 22978 8/67

PREPARED BY: J. JONES 27 APRIL 1966

FIG. 13-5 Maintenance-analysis Data Summary (Example).

test features. However, the recommended approach must comply with the required system-maintainability time constraints and must reflect the optimum approach derived through trade-off studies (see Chap. 7).

Relative to repair tasks, primary emphasis is on replacement parts. Part types and frequency of replacement are indicated as part of the repair requirement. These data lead to the determination of spare-part requirements. Part type is identified through the adopted repair policy, and replacement frequency is based on primary failures, dependent or secondary failures, operator/maintenance-induced failures, manufacturing defects, condemnation factor, etc. Determination of the part-replacement frequency is critical when considering the cost of spares in the overall support picture.

Another primary area of support involves personnel. The assignment of personnel (career field, skill level, and quantity) is based on the analysis of each task in terms of complexity. Highly complex tasks may require additional personnel quantities and high skill levels (necessitating additional training). The phasing of various maintenance tasks may in one instance specify the need for a single technician while in another instance indicate the need for two technicians. Thus, the complexity of tasks and proper phasing, as conveyed through the maintenance analysis, is critical relative to optimizing personnel quantities and resultant costs.

When considering various approaches to providing support for a specific task or group of tasks, such consideration should be oriented to the basic principles of maintainability in terms of minimum time, least quantities, minimum cost, etc., as presented in Chap. 1. The method for determining the most effective approach is facilitated through the accomplishment of system trade-off studies by comparing the various alternate approaches.

MAINTENANCE-ANALYSIS UPDATE

The maintenance-analysis development effort is an iterative process commencing from the availability of initial design data to the system integration and test. Throughout this period, the design configuration initially established is subjected to many changes, some resulting from performance, interference, and other such problems, which are the responsibility of the contractor to resolve, and others resulting from customer redirection. In any event, changes of a formal nature are covered by engineering-change proposals (ECPs), and all such changes should be tightly controlled through configuration management procedures. The maintenance-analysis effort must be directly tied in with all design-configuration changes, as should be any other formalized data product.

TABLE 13-2 Data Application

AREA OF SUPPORT	DATA SOURCE		DIRECT APPLICATION †
	M TASK ANALYSIS FORM BLOCK NO.*	MAINT. ANAL. DATA SUMMARY FORM BLOCK NO. †	
1. Support Equipment/Tools	10,13,25,26,28	7,9,12,13,17,19	Level of Test (1366E/13; 2297B/7,9) Duration of Test (1366E/26; 2297B/17) Frequency of Test (1366E/10,13; 2297B/12,13) Type of Support Equip. (1366E/25,28; 2297B/19) Test Environment (2297B/7)
2. Spare/Repair Parts	13,21,22,23,24,27	7,9,12,13,20	Quantity of Spares (1366E/21,23,24; 2297B/12,13) Type of Spares (1366E/13,22,27; 2297B/9,20) Location of Spares (2297B/7) Stockage Level & Turnaround Times (Maintenance Concept/Policy)
3. Personnel	16,17,18,19,20,29-34,40-45, 46-49	7,18	Personnel Types (1366E/16,29,31,33,40,42,44; 2297B/7,18) Personnel Skill Levels (1366E/17,18,19; 2297B/7) Personnel Quantities (1366E/20,30,32,34,41,43,45,46-49; 2297B/18) Maintenance Environment (2297B/7)
4. Training	4,13,16,22,27,25,28	7,9,10,11,18,19,20	Training Data (All Blocks As Indicated) Technical Manuals (Refer to Item 5)
5. Technical Manuals	2,4,5,13	7,9,10,11	Maintenance Concept/Echelons (1366E/5; 2297B/7,10) Maintenance Requirements/Tasks (1366E/2,4,5,13; 2297B/7,9,11) Personnel (Refer to Item 3) Spare/Repair Parts (Refer to Item 2) Support Equipment (Refer to Item 1) Maintenance Environment (2297B/7)
6. Facilities	2,4,5,13	7,9,11	Maintenance Requirements/Tasks (1366E/2,4,5,13; 2297B/7,9,11) Spare/Repair Parts Storage (Refer to item 2) Support Equipment (Refer to Item 1)
7. Maintenance Quantitative Factors	8,10,9,13,14,15,17,18,19,20,21, 36-39	12,13,14,15,16,17	Maintenance Task Times/Clock Time & MMM/OH (1366E/14,15,17,18,19,20,21,26-39; 2297B/15,17) Maintenance Requirement-Task Frequencies/MTBF & MTBM & TBO (1366E/8,10,13; 2297B/12,13,14,16) Operating Life (1366E/9)

* Figure 13-4 (Form 1366E)
† Figure 13-5 (Form 2297B)

This is necessary since support requirements obviously may be affected through prime-equipment design changes.

MAINTENANCE-ANALYSIS APPLICATION

The maintenance analysis provides an evaluation of system design for maintainability through a direct comparison of quantitative maintenance factors against initial customer requirements and against the prediction (refer to Chap. 10). In addition, the analysis serves as a primary source of data necessary in the definition of detailed maintenance-support requirements. The maintenance analysis, in conjunction with the maintenance concept/policy, covers all of the major aspects of support and basically fulfills the objectives of the planned-maintenance aspects of integrated logistics support. Thus, the analysis ideally represents a single engineering-documentation source from which all support requirements are derived. Utilization of the analysis as such tends to ensure a common maintenance-policy approach for the development of support equipment, spares, technical manuals, etc. Basic applications of the analysis are indicated in Fig. 13-2.

Referring to Fig. 13-2, the respective areas of support are listed. These areas are more directly related to the data elements on the maintenance-analysis forms (maintainability-task analysis and maintenance-analysis data summary) in Table 13-2. Further application in these areas is better illustrated through review of Appendices C and D.

In summary, the maintenance analysis contains many useful elements of data. The analysis itself serves primarily as a "working engineering document" leading to the development of more formalized product outputs such as spares-provisioning documentation and technical manuals. The analysis provides the maintainability engineer with the necessary data to facilitate his basic objectives relative to influencing system design for optimum support.

QUESTIONS AND PROBLEMS

1. Describe the difference between *maintainability* analysis and *maintenance* analysis.
2. What are the three main reasons for performing a maintenance analysis of a system or equipment?
3. What is the scope of a maintenance analysis?
4. The maintenance-analysis forms (formats) described in this chapter are the only recognized means of recording data. True or false? Why?
5. The maintenance concept should not influence the maintenance-analysis coverage of the system or equipment. True or false? Why?
6. List three reliability-data items that are used in the development of the maintenance analysis.

7. What is the purpose of a *functional analysis* as it relates to the maintenance analysis?
8. What is a *logic troubleshooting-flow diagram?* What is its purpose?
9. Using Fig. 13-4 as a reference, calculate the weighted downtime (total elapsed time) for the requirement (show your calculations).
10. What does the difference indicate between task numbers such as 0015 and 9055?
11. A maintenance-task description should be of such detail that it can be inserted directly into the equipment manual. True or false?
12. What is the purpose of a *maintenance-analysis data summary?*
13. What does the abbreviation ECP mean?
14. What is the significance of ECPs to the maintenance-analysis process?
15. List five typical maintenance requirements.
16. The equipment failure rate (reliability factor) and the maintenance frequency are usually identical. True or false? Why?
17. Given the following data, calculate the total requirement time:

Task no.	Task time (min)	Frequency
0010	6	0.010
0015	3	0.010
0020	4	0.010
9030	10	0.003
9040	8	0.002
9050	6	0.005
0060	4	0.010

18. In the above problem, what is the frequency of the requirement? What is the *MTBM* or replacement interval?

FOURTEEN

Demonstration

Up to this point, text material has included the specification of maintainability requirements, the establishment of design criteria, the accomplishment of maintainability prediction, the design review, and the analysis of a specific design configuration in terms of the requirements for maintenance support. Through the accomplishment of these functions, the maintainability engineer has been able to specify design objectives and later evaluate a given design configuration relative to compliance with these objectives. The evaluation process thus far has been analytical in nature, providing confidence that both qualitative and quantitative maintainability requirements will be met. Although the analytical process fulfills a particular need, it is limited because it does not reflect practical experience with the applicable hardware. A more realistic evaluation of whether or not a system complies with the initially specified customer requirements is accomplished through actual maintainability demonstrations involving the prime equipment and its associated supporting resources (i.e., support equipment, personnel, technical data, etc.).

TYPES OF DEMONSTRATIONS

A true maintainability demonstration constitutes the evaluation of a system deployed in an operational environment and subjected to actual-use conditions. For instance, an equipment item designed for shipboard use should preferably be demonstrated on-board ship, at sea, and employing shipboard personnel, support equipment, technical data, etc. In such a situation, actual maintenance experience (e.g., equipment downtime) in a realistic environment can be recorded and subsequently analyzed to reflect a true representation of the system design for maintainability. A demonstration of this type can best be accomplished by the customer during standard operations and supported through the employment of normal resources.

Relative to the fulfillment of maintainability-demonstration requirements for a typical program, the contractor is obligated to demonstrate to the customer that his equipment meets the initially specified maintainability requirements. The accomplishment of such a demonstration in a realistic operational environment is often impractical. In certain instances, demonstrations involving large ground electronic systems in fixed installations (e.g., radar systems at satellite-tracking stations) can be accomplished in an environment which closely approximates a true operational situation. In other words, contractor demonstration may be conducted at the customer's facility, employing customer personnel, on equipment installed and ready for operational use. However, such demonstrations are generally scheduled at specific times, and faults are simulated in the equipment in order to simulate maintenance requirements. Although this type of situation does not *completely* reflect normal customer/user operations (since equipment failures are induced, and the accomplishment of subsequent demonstrations are planned, thus eliminating some of the randomness normally involved), it can provide a close simulation. In other instances, the ultimate end item (i.e., test vehicle) may be an airplane, a ship, or a missile. The availability of such as a test vehicle at the specific time of need is not likely. In addition, even if it were available, the costs of conducting such a demonstration would in all probability be prohibitive. Therefore, the method of demonstration must take an alternate approach.

When considering various approaches to accomplishing a formal maintainability demonstration, there are a number of possible alternatives:

1. Formal maintainability demonstration may be accomplished at the customer's facility or in a user's environment utilizing customer personnel and resources. In this instance, the specific equipment item involved may be demonstrated as part of an overall system (e.g., large ground electronic system—satellite-tracking station). The contractor assumes the role of planning, monitoring, data collection and analysis, and reporting.

2. Formal maintainability demonstration may be accomplished at the customer's facility utilizing contractor personnel and resources. The type of equipment involved may be a large ground electronic system,

or a small system. The contractor assumes the role of planning, accomplishment of the demonstration, data collection and analysis, and reporting. The customer monitors all activities.

3. Formal maintainability demonstration may be accomplished at the contractor's facility utilizing customer personnel and resources. In this instance, the customer provides the personnel and resources which are not available to the contractor. The contractor assumes the role of planning, monitoring, data collection and analysis, and reporting. Customer personnel accomplish the demonstration and monitor all other activities.

4. Formal maintainability demonstration may be accomplished at the contractor's facility utilizing contractor personnel and resources. The contractor assumes the role of planning, accomplishment of the demonstration, data collection and analysis, and reporting. The customer monitors all activities.

5. Formal maintainability demonstration may be accomplished in two phases.

Phase I comprises an equipment demonstration at the contractor's facility utilizing contractor personnel and resources. In this instance, the contractor accomplishes all demonstration activities, and the customer assumes the monitoring role.

Phase II comprises an equipment demonstration at the customer's facility or in a user's environment utilizing customer personnel and resources. In this instance, the customer may accomplish all demonstration tasks, and the contractor assumes the monitoring role.

The two-phased effort is normally applied to large programs, where phase I constitutes a direct evaluation of a contractor's equipment (usually a small system) by itself, and phase II covers the evaluation of the contractor's equipment installed as part of the end-item system and deployed in an environment more closely approximating user operations. A phase II demonstration covers the evaluation of the overall system and its various individual equipment items. Since the scope of a phase II demonstration is extensive, and since considerable cost is involved, it is often desirable to verify equipment-design characteristics at an earlier point in the program cycle. Thus, a phase I demonstration may be feasible.

As inferred above, selection of the most appropriate demonstration approach depends on customer requirements and on the type of equipment involved. In the absence of specific direction, the contractor will indicate a recommended approach in the maintainability-program plan (refer to Chap. 3) submitted as part of the contractor's proposal to the customer. This approach will be amplified at a later time during the acquisition phase, prior to the start of the formal demonstration.

When considering a particular demonstration approach, the contractor must consider program schedules and costs, contractor versus customer risks, the availability of facilities and resources at the time of need, etc. For instance, the demonstration may be scheduled at a critical point in the contractor's qualification program (the availability of a prototype system as a demonstration model may be limited due to delivery requirements). In the event that customer personnel and resources are specified, the applicable personnel must be adequately trained and available at the start of demonstration. In addition, customer-furnished support equipment, data, etc., must be available. These conditions are sometimes impossible to meet without completely upsetting normal customer operations. Thus, the alternative involves the utilization of contractor personnel and resources. In any event, the contractor will recommend the most feasible approach consistent with cost and schedule plus the availability of supporting resources. In addition, the approach selected shall simulate, *to the maximum degree possible,* a realistic operational demonstration.

TIME PHASING

The accomplishment of the maintainability-demonstration function as part of the contractor's effort in the overall system-development cycle is illustrated in Fig. 14-1 (blocks 17.0 and 18.0). Referring to the figure, block 17.0 represents the formal demonstration activity, while block 18.0 covers the actual assessment of maintainability characteristics in design. The assessment constitutes the evaluation of the system through the analysis of demonstration-test data. Thus, for the first time in the overall program cycle, system evaluation results from actual equipment exercising.

The procedural approach to accomplishing a maintainability demonstration includes several major steps. Initially, the proposed demonstration concept is outlined in the program plan submitted as part of the contractor's proposal. During the early part of the acquisition phase, the organization and planning commences, resulting in the preparation and submittal of a formal demonstration plan. Once the demonstration plan is approved by the customer, actual test preparations commence, followed by the formal demonstration itself. Throughout the formal demonstration, test data covering all maintenance actions are obtained. Subsequently, data reduction and analysis are accomplished to verify whether the contractor has or has not met the initially specified maintainability requirements. The concluding effort in the contractor's maintainability program constitutes the preparation of a final report covering test results and conclusions. The specific sequence of events in con-

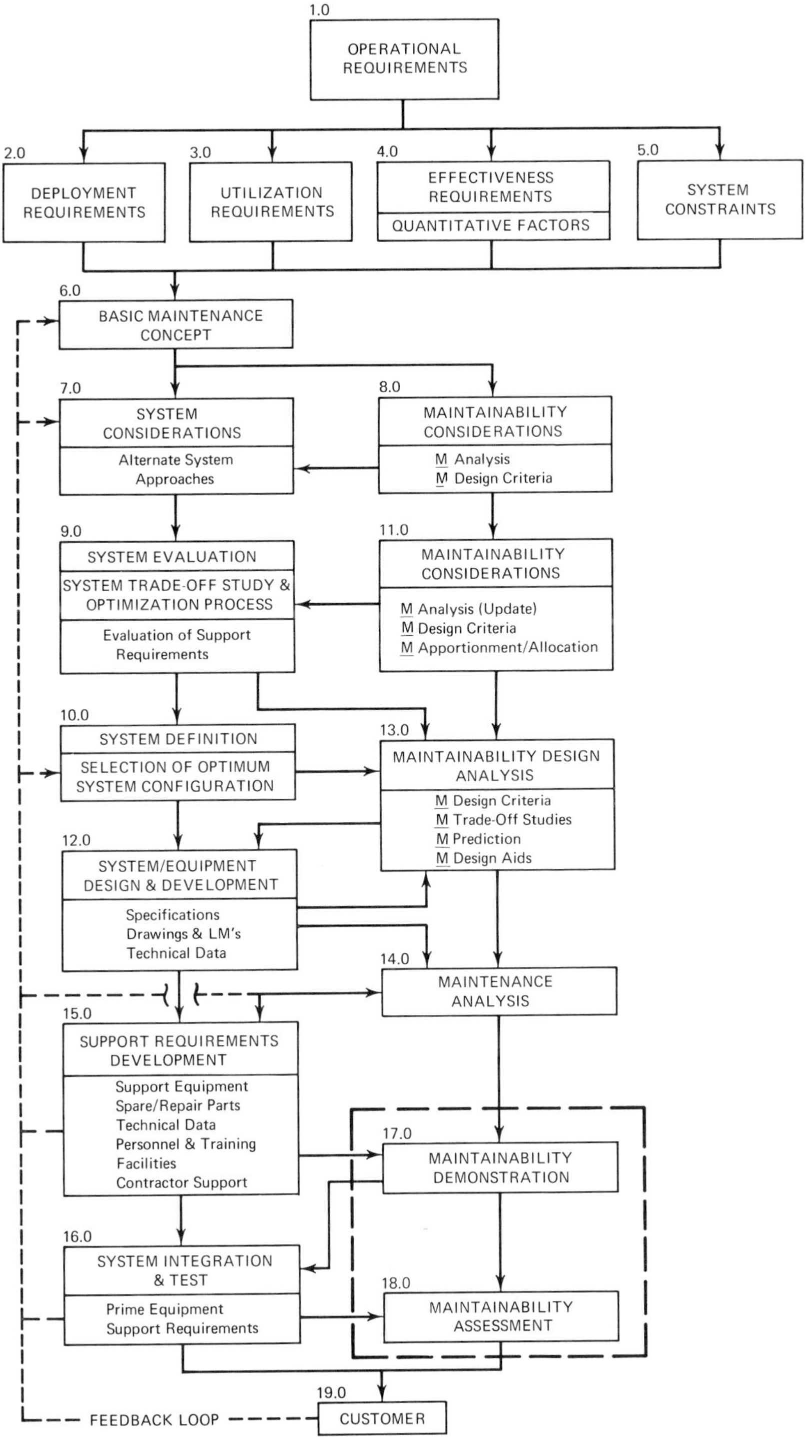

FIG. 14-1 System Development Cycle.

ducting a maintainability demonstration is illustrated in Fig. 14-2 and discussed in the following paragraphs.

TEST CONDITIONS

Test conditions, as referred to herein, encompass the necessary prerequisites to accomplishing a formal maintainability demonstration. These prerequisites include the initial definition of maintainability requirements, the establishment of the maintenance policy, the selection of a demonstration test model, the accomplishment of a maintenance analysis, and the definition, provisioning, and acquisition of support material and data.

Maintainability Requirements

As the maintainability-demonstration process is basically a verification of compliance with a given measurable parameter(s), the proper specification of that parameter(s) is a necessary prerequisite. For instance, maintainability requirements may be specified in a *qualitative* manner such as: "the contractor shall implement a formal program to ensure that equipment design incorporates good maintainability principles to the maximum degree feasible," or, "equipment design shall be optimized relative to accessibility, modularization, handling, and safety." Requirements specified in a qualitative manner are difficult (if not impossible) to measure, and the conductance of a formalized maintainability demonstration is usually meaningless. On the other hand, maintainability requirements may be specified in a *quantitative* manner such as: "the equipment shall be designed to meet an *MTTR* of 30 min at organizational maintenance and 2 hr at intermediate maintenance, and shall be demonstrated to a confidence level of 80 percent with a 75 percent accuracy." In this instance, a measurable parameter has been established, and maintainability demonstration is deemed feasible.

The measurable parameters of maintainability are summarized in Chap. 1. Thus far, the accepted measures are primarily limited to elasped maintenance time ($\bar{M}_{ct}$, $\bar{M}_{pt}$, *MTTR*, *ERT*, etc.) and man-time (*MMH/OH*). Industry and government have not as yet produced meaningful methods of quantitatively expressing figures of merit for personnel, support equipment, spares, and other elements of support which are significant areas of concern in meeting overall maintainability objectives. Until such time as these other aspects of maintainability can be adequately specified and measured, maintainability demonstration will be oriented to the maintenance-time factors. The ultimate determination of demonstration requirements (i.e., whether or not a maintainability demonstration is required, and if so, what demonstration ap-

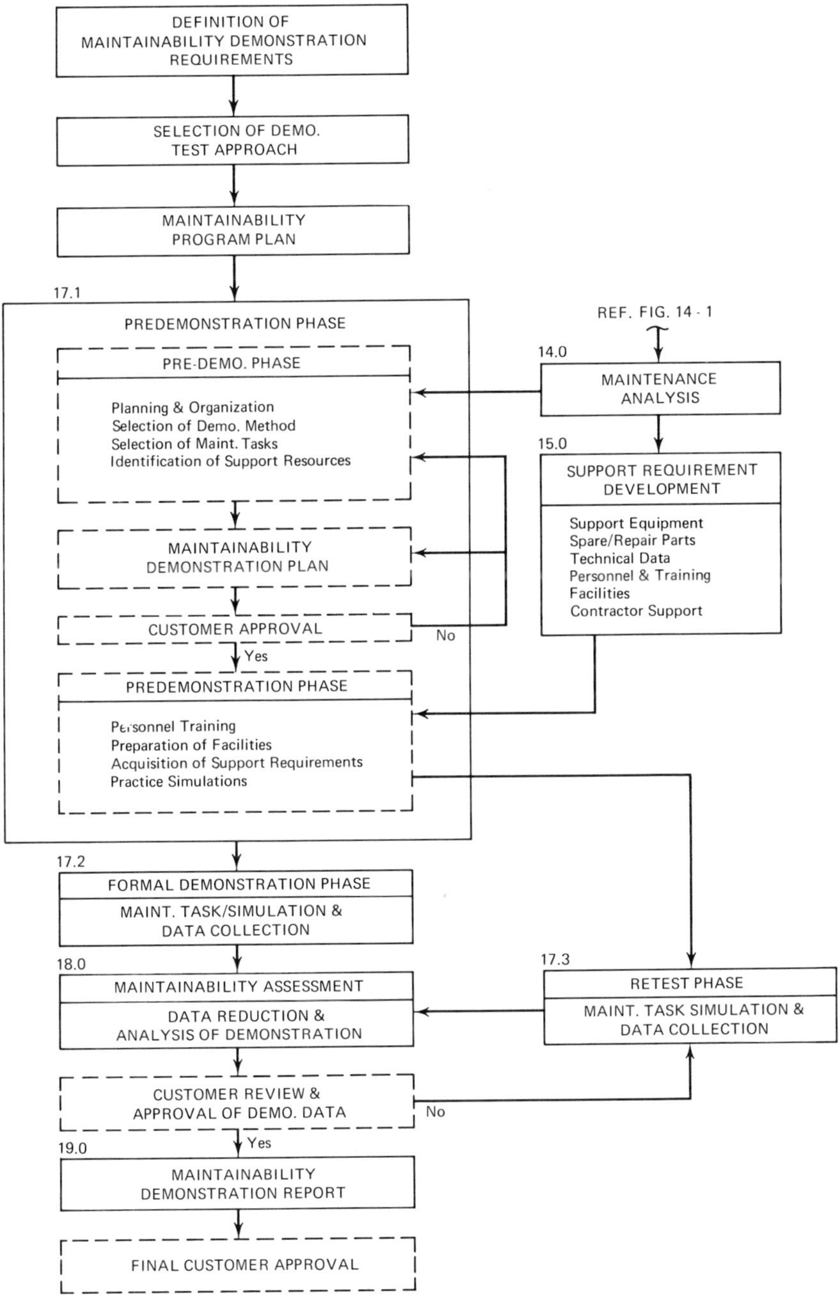

FIG. 14-2 Maintainability Demonstration Procedure.

proach is to be employed) is based on data specified in the customer's RFP and in the contractor's proposal response. As stated above, the feasibility of demonstration depends on whether meaningful and measurable quantitative requirements have been imposed and adequately defined. The approach to demonstration (i.e., type of test, location of test, etc.) depends on the type of equipment, schedule and cost, and the specific quantitative measures involved.

Maintenance Policy

Chapters 5 and 6 discuss the maintenance concept and the analysis where initially specified maintainability requirements are translated into established design criteria. Such criteria are then employed to guide equipment design in areas pertaining to diagnostic provisions, packaging concepts, accessibility, and other related maintainability features. The resulting design configuration is then analyzed relative to support requirements (through the maintenance analysis covered in Chap. 13), and the maintenance policy discussed in Chap. 11 presents a summary of this analysis. The maintenance policy illustrates the extent of maintenance accomplished at each echelon, time constraints, personnel skill levels required, support-equipment needs, etc. Verification of the proposed maintenance policy, which reflects the most effective approach to meeting maintainability requirements, is realized through maintainability demonstration. Demonstration testing is designed to simulate maintenance tasks at various echelons, and such simulation should employ the recommended support equipment, personnel, technical data, facilities, and resources. When demonstrating compliance with a specific quantitative time requirement, the maintenance tasks involved (along with support requirements) must be oriented to the applicable echelon specified by the maintenance policy. Hence, the maintenance policy specifies the framework relative to the accomplishment of maintainability demonstration.

Demonstration-model Configuration

Quite often in the development of a new system or equipment, there may be several hardware configurations developed at different stages in the program. The first of these configurations is a *breadboard model*, which may functionally operate in a manner similar to the production model, but which does not at all reflect the size or packaging configuration of the production model. Next in line may be the *service-test model*, which is far superior to the breadboard, but still is not fully representative of the production model. Functional operation and packaging are similar, but substitute (alternate) subassemblies and parts may be employed. Finally, a *prototype model* is developed to represent fully the proposed forthcoming production equipment. Since maintainability demonstration must simulate (to the maximum degree possible) actual field operations, the most realistic equipment must be employed in its accomplishment. Thus, the prototype model (not the breadboard model or service-test model) should be utilized. If there are several prototype models, each representative of a different production lot (the variance resulting from the incorporation of different engineering-change proposals), the latest or most up-to-date configuration should be used for demonstration purposes. For systems or equipment where different configurations are developed, there may be some confusion as to the actual model to be demonstrated. To avoid such confusion, the contractor should specifically identify, by part number and serial number, the configuration to be utilized for maintainability demonstration. Such identification should be included in both the program plan submitted as part of the contractor's proposal and the demonstration plan submitted to the customer during the acquisition phase. Where applicable, the contractor (in the demonstration plan) should list major engineering-change proposals which are to be incorporated in the equipment prior to the start of demonstration.

Test Environment (Facilities/Resources)

As inferred earlier, a true maintainability demonstration should, if possible, be accomplished in a realistic operational environment. However, fulfillment of this objective often fails to materialize due to schedule, cost, and/or unavailability of associated equipment. Thus, the contractor's facility is generally used as the test site. When selecting a test site, regardless of general location, maximum effort should be made to simulate the operational environment. Test-site planning and preparation must consider size of facility (outside area or laboratory space), temperature, noise, illumination, power requirements, maintenance benches, special clothing (if required), and related resources necessary for demonstration accomplishment. The majority of facility requirements may be derived through the maintenance analysis discussed in Chap. 13 and from supporting human-factors data. A detailed description of facilities proposed for the maintainability demonstration is included in the contractor's demonstration plan (submitted to the customer during the acquisition phase).

Test Personnel

Test personnel in this instance include the technicians who will actually be accomplishing the maintenance tasks during the demonstration. These personnel should possess background and skill levels similar to the ultimate customer/user personnel who will be operating and maintaining the system or equipment when in operational use. Specific backgrounds, skill levels, and personnel-quantity requirements are derived from maintenance-analysis and human-factors data.

In regard to personnel selection and acquisition, there are three basic possibilities to include:

1. The selection of customer/user personnel with the appropriate background and skill levels. Such personnel will be provided by the customer and assigned to the test site (contractor or customer's facility) on a loan basis. Formalized training covering the equipment being demonstrated will be conducted during the predemonstration phase (refer to Fig. 14-2, 17.1) by the contractor. Although this is perhaps the best approach for accomplishing a relatively realistic demonstration, it is often difficult (if not impossible) for the customer to provide the right type of personnel at the time required without completely upsetting normal operations. In addition, the assignment of such personnel to a test site is sometimes quite costly, particularly if the demonstration is conducted at the contractor's facility. In spite of the apparent difficulties encountered in the majority of cases, this approach should be considered during the initial planning stages of demonstration.

2. The selection of contractor personnel (engineering aides or factory technicians) regularly assigned to programs not associated to the equipment being demonstrated. Such personnel will be assigned by the contractor to the test site on a loan basis. Although these personnel are employed by the contractor, instilling some degree of bias into the demonstration, they are not overly familiar with the equipment. Thus, whatever bias is instilled is minimized when considering alternate possibilities involving contractor personnel. Formalized training covering the applicable equipment will be conducted during the predemonstration phase (refer to Fig. 14-2, 17.1) by the contractor.

3. The selection of *contractor* personnel (engineering aides or factory technicians) currently assigned to the same program involving the equipment being demonstrated. In this instance, the personnel available will generally have the advantage of a considerable amount of formal and/or on-the-job training concerning the equipment being demonstrated. Therefore, the maintenance tasks demonstrated may be accomplished in a more efficient manner (e.g., shorter maintenance times) than would normally be experienced by customer/user personnel, and the resulting test data would be distorted. To compensate for the added efficiency, individual maintenance-task-time elements recorded from the demonstration may be adjusted (upward—longer maintenance times) through the use of multiple K factors; however, the use of such factors should be the result of a previous agreement between the customer and the contractor.

From the standpoint of approaching a realistic operational situation relative to demonstration, the preference for personnel selection involves items 1, 2, and 3, in that order. If the demonstration is to be accomplished at the customer's facility, the above order of preference is likely to materialize. If the demonstration is to be accomplished at the contractor's facility, the order of realization is likely to be reversed.

When following either of the approaches involving the use of contractor personnel (items 2 and 3 above), the contractor will generally specify by name the individuals being considered for participation in the demonstration. A personnel data sheet with background and experience information, as illustrated in Fig. 14-3, is completed on each individual concerned and is included in the demonstration plan. The listing of personnel specified in the demonstration plan should provide enough names (e.g., 8 to 10 names to support requirements for 2 maintenance technicians) to allow for a selection by the customer. This tends to maintain a certain degree of randomness from the customer's standpoint. In addition, if a specific personnel selection (by name) was made 60 days or more prior to the start of demonstration, the personnel involved would intuitively tend to prepare themselves for the forthcoming event, which would add increased bias to the demonstration.

Technical Data

Fulfillment of demonstration objectives involves the accomplishment of both corrective- and preventive-maintenance tasks. Completion of the steps involved in these tasks (e.g., troubleshooting, repair, system checkout, servicing, and related actions) should follow formal approved maintenance procedures. Such procedures, generally in the form of technical manuals or special maintenance instructions, are developed on a concurrent basis with equipment design and development (refer to Fig. 14-1, 15.0) and should be available in time to support maintainability demonstration. The contractor, in the demonstration plan, should specify the procedures (formal technical manual or appropriate equivalent) to be employed for demonstration purposes. The selection of approved procedures is necessary to ensure that the demonstration is conducted properly and in a manner which simulates a true operational condition. Deviation from approved procedures may cause the introduction of maintenance-induced failures, and will distort the maintenance times as recorded in the test data.

Support Equipment

The accomplishment of certain maintenance tasks (particularly troubleshooting and system-checkout sequences) in the times allotted, based on maintainability quantitative requirements, is often dependent

PERSONNEL DATA SHEET

NAME: Robert F. Smith AGE: 27 DATE: 9-10-64

JOB CLASSIFICATION: Sr. Lab Technician TYPE OF WORK: Test & Evaluation - - Installation, Checkout, and Repair

EDUCATIONAL RECORD (Indicate Number of Years):

COLLEGE: 1/2 Year HIGH SCHOOL: 4 Years GRADE SCHOOL: 8 Years

MAJOR SUBJECTS: General College Preparation

TECHNICAL SCHOOLS/COURSE NAME	Duration
Brown Technical School - - Electricity I & II, Transistor	2 Years
Theory I, Waveguide Theory, Radar Principles	
Ulcer University - - Electrical Engineering	1/2 Year

MILITARY TRAINING

Branch of Service	Name of School	Duration
U. S. Air Force	Electronics School (Keesler AFB)	6 Months
	Armament Systems School (Lowry AFB)	5 Months
U. S. A. F. I.	Electronics I (Correspondence)	3 Months

EXPERIENCE (Civilian and/or Military)

Company/Branch of Service	Job Title/Rank	Type of Work	Duration
U. S. Air Force	AFSC 30151/A1C	Maintenance Tech	2 Years
Merkins Electronics	T. V. Repairman	T. V. Repair	1 Year
General Dynamics	Lab Technician	Electronic System	3 Years
		Test & Repair	

Date Assigned To Project: 16 August 1964

FIG. 14-3 Personnel Data Sheet (Example).

on the type of support equipment employed. For example, the use of automatic support equipment will generally reduce maintenance times, whereas manually operated support equipment may extend maintenance times. The selection of support equipment stems from the maintenance analysis and supplemental provisioning data. Support-equipment acquisition (whether make or buy) is accomplished on a concurrent basis with equipment design and development (refer to Fig. 14-1, 15.0), and the support equipment should be available in time for the maintainability demonstration. The contractor, in the demonstration plan, should indicate the specific items of support equipment (by part number and serial number) to be employed for demonstration purposes. In the event that alternate items of support equipment are required (due to the unavailability of operational support equipment), such items shall also be identified in the plan. The use of alternate items generally results in distorted maintenance-task times and causes a change in maintenance procedures. In such instances, the necessary maintenance procedural deviations, plus adjustment time factors to be applied to demonstration-test data, must be identified and approved by both the customer and contractor prior to the start of actual demonstration.

Spare/Repair Parts

When considering normal customer/user field operations, spare parts are required to supplement remove-and-replace actions as part of the normal corrective-maintenance cycle. The type of spare parts required at each echelon of maintenance depends on the maintenance policy; and the quantity of spares is based on availability requirements, *MTBR*, stockage level, maintenance turnaround times, etc., as conveyed in the maintenance analysis and supplemental spares-provisioning data. When conducting full-scale maintainability demonstrations on large systems, spare parts are needed to support normal removal and replacement actions. It is important that the complete maintenance cycle (i.e., troubleshooting, remove and replace, repair, checkout) be simulated to reflect actual conditions. In addition, from the customer's standpoint, it is often desirable to perform an evaluation of the logistics support system when conducting demonstrations of this magnitude. On smaller demonstrations, the primary objective is to measure only active elements of the maintenance cycle—those chargeable to contractor design. In such instances, the requirements for spares may not be as critical! For example, maintenance tasks involving troubleshooting, disassembly, remove and replace, and system-checkout verification are evaluated. If the maintenance sequence involves the replacement of a major repairable assembly, that particular assembly may be the same one that was removed. Maintenance times other than actual removal and replacement times are discounted. In other words, the quantity of spares required may be negligible with careful planning. Recorded data is oriented only to active maintenance-task elements. The contractor, in the demonstration plan, should indicate the specific types and quantities of spare parts anticipated. Such items should be acquired, set aside, and stored in a bonded store prior to the start of maintainability demonstration.

TEST PLANNING, ADMINISTRATION, AND CONTROL

The planning aspects of maintainability demonstration actually commence during the proposal phase. The contractor, in the maintainability-program plan (refer to Chap. 3), summarizes his intended approach to demonstration. At that time intended location, method of demonstration, anticipated schedule, and organizational responsibilities are designated. As the program progresses through the definition phase and early part of the acquisition phase, demonstration-test planning continues relative to updating demonstration-test schedules, identification of facility/resource needs, personnel selection, support-equipment coordination, etc. The results of this planning are incorporated into the maintainability-demonstration plan, which is submitted to the customer for approval approximately 60 days prior to the start of formal demonstration.

The demonstration plan should cover the following:

1. Test conditions. Describe major requirements relative to test environments, maintenance concept and policy, facilities, test personnel, technical data, spare/repair parts, and support equipment.
2. Test planning, administration, and control. Describe organizational approach (i.e., test team), responsibilities, organizational interfaces, test schedules, monitoring of test activities, cost control, and reporting.
3. Predemonstration phase. Describe the recommended demonstration method, corrective- and preventive-task selection, personnel training, preparation of facilities, and acquisition of support requirements.
4. Formal demonstration phase. Describe methods of task simulation, type and format of data requirements, data-collection methods, and data reduction and analysis.
5. Retest phase. Discuss the methods for conducting required additional demonstrations.
6. Test documentation. Discuss test-reporting requirements.

Normally, the maintainability-demonstration plan is approved by the customer within 30 days after submittal. Subsequent to approval, the contractor continues activity relative to the predemonstration phase. Required personnel training is accomplished along with facilities preparation and equipment acquisition (refer to Fig. 14-2, 17.1).

Throughout the entire formal demonstration phase, all test exercises are covered by data collection and monitoring. Depending on the location and type of maintainability demonstration accomplished, either contractor or customer personnel will collect data on each maintenance-task element, and customer personnel will monitor, review, and concur with the applicable test sequence prior to commencing to the next step. Following the completion of all maintainability-test sequences, the test data is collated, reviewed, analyzed, and compared against the initially specified customer requirements. The results are summarized and included in the final maintainability-demonstration report (refer to Chap.

3) submitted to the customer approximately 30 days after completion of formal demonstration. Approval of this report concludes all formal maintainability-program effort.

PREDEMONSTRATION PHASE

Demonstration Method

Demonstration approach, as discussed earlier, refers to the proposed location of test and the agency (i.e., contractor or customer) conducting the test. Demonstration method, which is discussed below, refers to the particular test scope and sequence utilized for demonstrating the achievement of specified quantitative maintainability requirements.

For a given maintainability demonstration, any one of several different methods may be employed depending on the particular maintainability quantitative factor(s) given. For example, if the maintainability indices $\bar{M}_{ct}$ and M_{max} are specified in the initial customer RFP, one method may be employed. If $\tilde{M}_{ct}$ and $\tilde{M}_{pt}$ are specified, another method may be employed. The test method used should be the one best suited to a given situation having considered risk, cost, time, and validity of assumptions that have been made in planning the demonstration test. The selected method and reasons for selection are included in the maintainability-demonstration plan.

For the purpose of this text, two different methods are covered. Method 1 is oriented to verify $\tilde{M}_{ct}$, $\tilde{M}_{pt}$, and M_{max}. Method 2 is designed to consider $\bar{M}_{ct}$, $\bar{M}_{pt}$, $\bar{M}$, and M_{max}. The differences in these methods will be discussed in detail later in this chapter.

Maintenance-task Selection

The assurance that the proposed maintainability demonstration reflects total system design is facilitated through the task-selection process. This process involves the identification of a representative sample (based on expected percent contribution toward total maintenance requirements) of maintenance tasks to be demonstrated. The process does not include actual random (unplanned) failures occurring during the test, but it does include a variety of induced failures to ensure adequate coverage.

A sample or demonstration task in this instance refers to all preventive or corrective sequences necessary to complete a single maintenance action. For example, if an equipment failure is induced, all subsequent maintenance steps (i.e., troubleshooting, disassembly, remove and replace, repair, verification checkout) constitute one sample or demonstration task.

For the purpose of demonstration, a sample stems from a single failure (not a multiple or series of failures) or a specific preventive-maintenance requirement.

In completing a demonstration task, the steps involved may cover several echelons of maintenance as defined in the maintenance policy. In such instances, test data must be recorded accurately to ensure the proper analysis and comparison of data results against the maintainability requirements specified for each echelon.

Selection of sample size. The quantity or size of samples to be demonstrated should be based on the variances of the tasks to be represented, on equipment complexity, and on the probability of expected error (i.e., contractor risk and customer risk). The sample size should be determined independently of the availability of prime equipment and supporting resources and of alloted program-test time. It is desirable to select a sample size large enough to be representative, and small enough to be compatible with total program-cost and schedule requirements.

The sample size often preferred and selected for the two demonstration methods discussed in this text is 50. In other words, at least 50 samples or demonstration tasks should be accomplished for each primary test requirement. A larger sample size will, of course, provide more definitive test results, but time and cost of testing increase considerably as testing continues. A smaller number of samples involves increasingly larger risks and may produce inconclusive test results. The quantity 50 is recommended based on industry experience.

Corrective-maintenance task selection. Corrective (unscheduled) maintenance usually results from a component failure within the prime equipment. For demonstration purposes, corrective-maintenance tasks are simulated by inducing single failures into the equipment by opening, shorting, or grounding circuits, or by inserting known faulty parts. In response to the requirement for a sample size of 50, a single failure is induced in the equipment for each of the 50 demonstration tasks. The distribution by selection of corrective-maintenance tasks to be demonstrated is illustrated in Table 14-1. Referring to Table 14-1, the illustration covers system *XYZ* as reflected by the maintenance-concept flow diagram presented in Chap. 5. System *XYZ* in this instance consists of two major repairable units, unit *A* and unit *B*, each of which contains four major assemblies. The task-selection process is applicable to both demonstration methods, and the sample size employed for both methods constitutes 50 tasks as mentioned previously. Through the task-selection process, the 50 tasks are proportionately distributed among the eight major assemblies of units *A* and *B*. (Refer to Table 14-1).

BLOCK 1—List and categorize all significant assemblies, modules, or

TABLE 14-1 Corrective-maintenance-task Sample Selection (Example)

(1) ITEM	(2) QUANTITY OF EACH ITEM (Q)	(3) FAILURES/ITEM %/1,000 HR (λ)	(4) TOTAL FAILURES/ ITEM/1,000 HR (Q) (λ)	(5) ITEM % CONTRIBUTION TOTAL MAINTENANCE TASK	(6) GROUPING OF BLOCK 5 TASKS (%)	(7) ALLOCATION OF MAINTENANCE TASKS
UNIT A						
Power Amplifier	1	0.55	0.55	25.2	25.2	13
Power Supply	1	0.24	0.24	11.0	11.0	6
IF Amplifier	1	0.05	0.05	2.4	2.4	1
Computer Assembly	1	0.86	0.86	39.4	39.4	19
UNIT B						
Bearing Converter	1	0.06	0.06	2.8	2.8	1
Range Converter	1	0.11	0.11	5.0	5.0	3
Digital Output	1	0.10	0.10	4.6	4.6	2
Power Supply	1	0.21	0.21	9.6	9.6	5
TOTAL			2.18	100%		50

parts that make up the item being tested. The example illustrated covers the eight major assemblies of units *A* and *B*.

BLOCK 2—Indicate the quantity Q of each item listed in block 1.

BLOCK 3—Determine the failure rate (percent per 1,000 hours) of each item listed in block 1. Failure rates for each of these items are derived from reliability-prediction data.

BLOCK 4—Determine the total failures per 1,000 hours for each item listed. This value is determined from the product of blocks 2 and 3 ($Q\lambda$).

BLOCK 5—Determine the percent contribution of each item to the total corrective-maintenance tasks. This is computed by dividing the total failures/item/1,000 hours for each item (block 4) by the sum of all item failures/1,000 hours.

$$\text{Item percent contribution} = \frac{Q\lambda}{\Sigma Q\lambda}$$

BLOCK 6—Where applicable, group together items within the same category for which the percent contribution (block 5) is less than 2 percent (or as otherwise specified), and assign a percent contribution equal to the sum of the individual percent contribution for these items. (In the illustration presented in Table 14-2, some grouping was accomplished.)

BLOCK 7—Apportion the number of corrective-maintenance tasks to be demonstrated in proportion to the percent-contribution grouping in block 6. This is derived from the product of the value in block 6 and the total sample size of 50 tasks as originally specified. In instances where the resulting quantitative value is less than 1, at least one task shall be demonstrated, and the value 1 is entered in block 7. Other fractional values should be rounded off to nearest whole number. The sum of block 7 should equal 50. If, when totaled, the sum exceeds 50, the items which contribute least (lowest percent) to the total maintenance requirement are adjusted downward. If the sum is less than 50, the items which contribute most (highest percent) to the total maintenance requirements are adjusted upward.

Table 14-1 apportions the 50 maintenance tasks among the eight major assemblies of units *A* and *B*. In order to ascertain specifically (within the eight assemblies) what maintenance tasks will be demonstrated below the assembly level, the same apportionment procedure is accomplished to the next level, as illustrated in Table 14-2. Each lower maintenance indenture level is analyzed until all of the 50 demonstrated tasks are completely identified.

In instances where the lowest repairable item (constituting a number of parts) is covered by one or two sample demonstrations, the question

TABLE 14-2 Corrective-maintenance-task Sample Selection (Example)

(1) ITEM	(2) QUANTITY OF EACH ITEM (Q)	(3) FAILURES/ITEM %/1000 HRS (λ)	(4) TOTAL FAILURES/ ITEM/1000 HRS (Q) (λ)	(5) ITEM % CONTRIBUTION TOTAL MAINTENANCE TASKS	(6) GROUPING OF BLOCK 5 TASKS (%)	(7) ALLOCATION OF MAINTENANCE TASKS
POWER AMPLIFIER*						
RF Power Amplifier	1	0.3800	0.380	69.1	69.1	9
Duplexer	2	0.0025	0.005	0.9		
Coupler	2	0.0060	0.012	2.2	3.8	1
Detector	4	0.0010	0.004	0.7		
Modulator Control	1	0.0650	0.065	11.8	11.8	1
Modulator Tuner	1	0.0510	0.051	9.3	9.3	1
Self-Test Oscillator	1	0.0300	0.015	2.7	6.0	1
Self-Test Amplifier	1	0.0030	0.018	3.3		
TOTAL			0.550	100%		13

*Table 14 - 2 Presents an Allocation of Maintenance Tasks for the Power Amplifier Listed in Table 14 - 1.

arises as to what specific maintenance sequences are to be selected, and what particular part(s) will be chosen for failure-inducement purposes. When a variety of possibilities exist, a table of random numbers can be employed to facilitate the selection process.

All maintenance tasks to be demonstrated, and the specific items of hardware where failures are to be induced, are identified at this point in time.

Preventive-maintenance task selection. There are two basic categories of preventive-maintenance tasks:

1. Preventive-maintenance tasks that can be performed while the equipment is operating (No maintenance downtime involved)
2. Preventive-maintenance tasks which require the equipment to be shut down for the duration of task accomplishment (Maintenance downtime involved)

When analyzing test requirements, one must determine whether the test occurs during a period when the equipment is not needed, or whether the test occurs during an equipment operational use period. Only those maintenance tasks which result in an interruption of normal equipment operation (i.e., downtime which occurs during a normal equipment use period) will be demonstrated. Preventive-maintenance tasks that can be performed while the equipment is operating are not demonstrated, since these tasks do not affect the availability of the system or equipment.

In regard to preventive-maintenance tasks subject to demonstration, if the quantity is small (less than 50), all identified tasks will be demonstrated. If the quantity is large (greater than 50), a representative sampling will be employed. In this instance, the following steps are applicable:

a. Identify all applicable preventive-maintenance tasks.

b. Determine the frequency of occurrence for each preventive-maintenance task listed.

c. Determine the number of preventive-maintenance tasks with the same frequency of occurrence.

d. Determine the total frequency of occurrence for each preventive-maintenance task (product of *b* and *c* values, above).

e. Determine the percent contribution of each preventive-maintenance task to the total by dividing the frequency of occurrence for each by the sum of all preventive-maintenance frequencies of occurrence.

f. Group together in the same category preventive-maintenance tasks for which the percent contribution is less than 2 percent, and assign a percent contribution equal to the sum of the individual percent contributions for these tasks.

g. Apportion the number of preventive-maintenance tasks to be demonstrated in proportion to the percent contribution. A minimum of one task should be apportioned to each category.

The sample selection procedure for both corrective- and preventive-maintenance tasks is basically the same.

The maintenance-task selection process is completed by the contractor during the early portion of the predemonstration phase (refer to Fig. 14-2, 17.1), and resulting data are included in the demonstration plan for customer review and approval. Once approved, the selected tasks serve as the basis for formal maintainability demonstration.

Demonstration-test Preparations

The remaining effort in the predemonstration phase comprises final preparations at the test site.

Equipment. The prime-equipment demonstration model and associated support-equipment requirements are deployed, installed, and operationally checked. An equipment operational checkout at the test site ensures that the nontest abnormalities (i.e., manufacturing defects, transportation induced problems, etc.) have been eliminated prior to the start of demonstration. On a concurrent basis, spare/repair parts and technical data are dispatched to the test site for direct-support purposes.

Training. The maintenance technicians scheduled to participate in the demonstration who are not familiar with the equipment to be demonstrated must receive adequate formal training on the equipment. Such training, accomplished either at the test site or at the contractor's facility, should be oriented to the depth needed to be compatible with the personnel skill-level requirements as identified in the maintenance-analysis and human-factors data.

Test-team Briefing. A formal briefing of test-team personnel is required to ensure complete continuity relative to demonstration-test accomplishment. The test team includes the test director, the maintenance technician(s) who induces the failures into the demonstration model, the maintenance technicians who perform the corrective- and preventive-maintenance tasks being demonstrated, the maintainability personnel monitoring the demonstration, the data recorder, and the customer's representative monitoring the demonstration. Depending on the type of demonstration and the specific equipment involved, the test team may consist of additional personnel. The test-team briefing should cover maintainability requirements, the purpose and intent of the demonstration, the demonstration plan, specific organizational and personnel responsibilities, approach and procedures to be employed during the demonstration, data-recording methods, data analysis, and report requirements.

The briefing should be held at the test site prior to the start of formal demonstration.

FORMAL DEMONSTRATION PHASE

The formal demonstration includes the complete simulation of all preventive- and corrective-maintenance steps associated with the representative sample of maintenance actions specified in the demonstration plan. The purpose of such simulation is to:

1. Verify that the contractor-produced system or equipment meets the measurable parameter requirements initially specified in the RFP and contract. As indicated previously, currently accepted measurable parameters are primarily limited to maintenance downtime elements (i.e., $\bar{M}_{ct}$, $\bar{M}_{pt}$, $\bar{M}$, $\tilde{M}_{ct}$, $\tilde{M}_{pt}$, M_{max}, etc.). Thus, the methods covered herein are oriented to an evaluation of maintenance time.
2. Verify that the support requirements (i.e., support equipment, personnel, spare/repair parts, technical data) specified in the maintenance policy and the supplemental maintenances analysis are adequate to sustain the system or equipment.

Demonstration Methods

For the purposes of this text, two demonstration methods are discussed:

Method 1. This method employs a test of proportion to demonstrate achievement of $\tilde{M}_{ct}$, $\tilde{M}_{pt}$, $M_{max(ct)}$, and $M_{max(pt)}$ when the distribution of corrective- and preventive-maintenance repair times is unknown. The method assumes a confidence level at 75 percent or 90 percent as may be desired and requires a minimum sample size of 50.

This constitutes method 4 as currently specified in MIL-STD-471.

Method 2. This method is oriented toward demonstrating the achievement of $\bar{M}_{ct}$, $\bar{M}_{pt}$, $\bar{M}$ and $M_{max(ct)}$. The procedures for demonstrating $\bar{M}_{ct}$, $\bar{M}_{pt}$, and $\bar{M}$ are based on the central-limit theorem and are applied to any distribution. The procedures for demonstrating $M_{max(ct)}$ are valid for those instances where the underlying distribution of corrective-maintenance tasks is log-normal. A minimum sample size of 50 is required for this method.

This constitutes method 2 as currently specified in MIL-STD-471.

Both methods employ the same demonstration sample size, the same test procedures, and the same mechanisms for recording and gathering data. The difference in the two methods is highlighted in the data analysis and in the criteria for system acceptance or rejection. Both methods are equally applicable to the various demonstration approaches. However, accept/reject criteria, which depend on the initially specified customer requirements, may vary from one demonstration approach to the next.

In an attempt to maintain continuity, discussion in subsequent paragraphs is oriented to system *XYZ*. It is assumed that contractual requirements specify that formal demonstration be accomplished at the contractor's facility utilizing contractor personnel and resources.

Test Procedure (Corrective Maintenance)

The following steps, applicable to each of the 50 simulations, constitute the general test procedural approach.

1. The maintainability test monitor provides data covering the techniques governing failure simulation. Such data may be in the form of a single or a series of data sheets. One approach employs the use of demonstration data sheets as illustrated in Fig. 14-4. One data sheet covers each of the 50 simulations. Referring to Fig. 14-4, the top part (through item *A*3) of each data sheet will be completed prior to the start of demonstration. The 50 simulation requirements stem from the sample-selection process illustrated in Table 14-1, and the method of failure simulation is determined during the predemonstration test phase subsequent to approval of the demonstration plan. Simulation can be accomplished through opening leads, shorting leads, installation of known faulty parts, application of tape to connector pins, etc.

Simulations which could result in damage to the equipment should be avoided.

2. The maintenance technicians assigned to complete the tasks being demonstrated are requested to leave the immediate premises where the test is being conducted. During their absence, the maintainability-test monitor and/or the customer's representative may select at random any one of the 50 demonstration data sheets (Fig. 14-4). The selected data sheet indicates the type of failure to be simulated and the method of simulation. A technician (assigned to assist the test director and maintainability-test monitor) then induces the failure into the equipment. The test director, maintainability-test monitor, and customer's representative verify the induced failure. The "bug" is checked to ensure that it provides the desired malfunction in the equipment. The maintenance technicians (assigned to complete the tasks to be demonstrated) are then recalled to the test premises.

3. The technicians then perform an operational check of the demonstration model utilizing technical-manual or approved alternate procedures. At some point during the operational checkout, a *NO-GO* condition will exist. When this occurs, the data recorder notes the symptom of failure and the mode of system operation, at the time of symptom detection, on the demonstration-data sheet (Fig. 14-4, items *A*4 and *A*5).

DEMONSTRATION DATA SHEET

DEMO NO: 29 CM: X PM: ___ RECORDER: J. SMITH DATE: 7/1/65

EQUIPMENT	NOMENCLATURE	MFGR PART NO.
SYSTEM	System	A37700
UNIT	Unit A	A37701
ASSEMBLY	Computer	A58180
SUBASSEMBLY	C. B. No. 2A5	A58183

CORRECTIVE MAINTENANCE (CM):

A. FAULT SIMULATION:

1. Type of Failure Simulated: Open emitter lead on transistor
2. Method of Simulation: Replaced Q5 with known faulty transistor
3. Designation of Part Simulating Failure: F10
4. Symptom of Failure: Unit A front panel light indicates NO-GO condition. No bearing output data. Absence of signal at T. P. 1 of C. B. 2A4
5. Mode of operation at Detection: T/R Mode and A/A Mode
6. Additional Information: ___

B. TIME DATA:

MAINT. REQMT.	TIME	MAINT. REQMT.	TIME
LOCALIZATION	1.0	REASSEMBLY	17.0
ISOLATION	20.0	ALIGN/ADJUST	–
DISASSEMBLY	10.0	CHECKOUT	6.0
REMOVE/REPLACE	2.0	OTHER	–
REPAIR	5.0	TOTAL MAINT. (MIN)	61.0

PREVENTIVE MAINTENANCE (PM):

INSP: ___ FUNCT. TEST: ___ CAL: ___ SERV: ___ OH: ___ FORCE REM: ___

A. MAINT. PROCEDURE: ___

B. TOTAL MAINT. TIME (ACTIVE): ___

FIG. 14-4 Demonstration Data Sheet (Example).

4. Once a symptom of failure has been detected, the maintenance technician(s) proceeds to accomplish the necessary corrective-maintenance tasks (i.e., localization, fault isolation, disassembly, repair, checkout, etc.). Maintenance tasks are completed utilizing technical manuals or approved alternate maintenance procedures. Such procedures should specify required support equipment. The data recorder monitors every detailed step in the maintenance process and records data as illustrated in the demonstration worksheet presented in Fig. 14-5.

DEMONSTRATION WORKSHEET SHEET 41 OF 83

DEMO NO: 29 CM: X PM: ____ RECORDER: J. SMITH DATE: 7/1/65

EQUIPMENT	NOMENCLATURE	MFGR. PART NO.
SYSTEM	System XYZ	A37700
UNIT	Unit A	A37701
ASSEMBLY	Computer	A58180
SUBASSEMBLY	C. B. No. 2A5	A58183

DETAILED MAINTENANCE TASKS

TASK NO.	TASK DESCRIPTION	QTY MEN	ACTIVE MAINT. TIME (HR:MIN)		ADMIN/LOG TIME (HR:MIN)	
			START	STOP	START	STOP
1	Verify System XYZ NO-GO Through Self-Test	1	9:15	9:16		
2	Remove Unit A From System XYZ	1	9:16	9:19	9:19	9:22
3	Remove Unit A Access Cover (six quick-disconnect fasteners)	1	9:22	9:26		
4	Connect Jumper Wires To Pins A & B	1	9:26	9:28		
5	Obtain Power Source	1	-	-	9:28	9:30
6	Apply Bias Voltage to Pin C	1	9:30	9:34		
7	Check for 15V p-p Signal at TP8 on C. B. 2A6	2	9:34	9:41		
8	Check For Logic Square-wave Pulse (4 μsec) at TP1, C. B. 2A4	2	9:41	9:48		
9	Remove Computer Assembly From Unit A	1	9:48	9:51		
10	Remove C. B. 2A5	1	9:51	9:52		
11	Repair C. B. 2A5	1	9:52	9:57		
12	Install Spare C. B. 2A5	1	9:57	9:58		
13	Break For Coffee	2	-	-	9:58	10:28
14	Replace Computer Assembly	1	10:28	10:34		
15	Accomplish Checkout of Unit A	2	10:34	10:39		
16	Install Unit A Access Cover	1	10:39	10:44		
17	Install Unit A In System *XYZ*	1	10:44	10:50		
18	Accomplish Final Check Utilizing Self-Test	1	10:50	10:51		

FIG. 14-5 Demonstration Worksheet (Example).

The technician must notify the test director when the cause of failure is detected. The test director ensures that the detected failure and the initially induced failure are one and the same before permitting the test to continue.

In the event that the maintenance technician is unable to diagnose and find the induced failure (through some abnormality or equipment-design deficiency), the point at which the test should be discontinued must be determined through mutual agreement by the maintainability-test monitor and the customer's representative. A provision covering this possibility (e.g., "test should be discontinued after a period of time equal to two times the predicted *MTTR*") should be included in the demonstration plan.

5. Data elements recorded on the demonstration worksheets (Fig. 14-5) are summarized, and total times are entered in the applicable blocks in the demonstration-data sheet (Fig. 14-4).

Illustrative photographs of a demonstration involving system *XYZ* are contained in Figs. 14-6 through 14-9. Figure 14-6 shows a technician inducing a fault into the equipment. Figure 14-7 shows the same tech-

FIG. 14-6 The Maintenance Technician Induces a Fault into Assembly of System *XYZ*. (*General Dynamics Corporation.*)

nician verifying that the bug provides the desired malfunction in the equipment. Figure 14-8 covers one of the steps in the troubleshooting process accomplished by a different maintenance technician. (The step is similar to tasks 7 and 8 as listed on the demonstration worksheet in Fig. 14-5.) The data recorder is monitoring the detailed maintenance actions involved. Figure 14-9 covers the repair action (correction of the simulated failure), which is similar to task 11 in Fig. 14-5.

Test Procedure (Preventive Maintenance)

1. The maintainability-test monitor provides data covering the preventive-maintenance tasks to be demonstrated. As in the case of corrective-maintenance demonstrations, each preventive-maintenance requirement may be covered by a demonstration-data sheet (illustrated in Fig. 14-4). Referring to Fig. 14-4, the top part of the form, covering equipment, and the bottom part, covering the maintenance requirement to be demonstrated (i.e., inspection, functional test, calibration, etc.), are completed prior to the start of demonstration.
2. The maintainability-test monitor and the customer's representative may randomly select any one of the preventive-maintenance requirements to be demonstrated. The selected requirement will then be demonstrated by the assigned maintenance technician(s) utilizing the technical manual or approved procedures. The data recorder monitors every detailed step in the maintenance process and records such data on the demonstration worksheet.

FIG. 14-7 The Maintenance Technician Verifies that the Induced Fault Produces a Specific Symptom of Assembly Malfunction. (*General Dynamics Corporation.*)

Data Recording

An important activity throughout the accomplishment of formal demonstration is the data-recording function. Every step in the active maintenance process must be monitored and covered by accurate data. An individual, designated as the data recorder, is assigned the function of completing demonstration worksheets (Fig. 14-5). If several maintenance steps are accomplished in parallel, the data-recording function becomes complex. In this instance, either two or more data recorders are required, or a single data recorder is supported by visual and/or audio aids. As regards support aids for data-recording purposes, there are a number of possibilities:

1. Time-lapse photography (movies). Slow motion time-oriented photography may be employed to monitor demonstration tasks. This, of course, is one of the best methods for recording and retaining data, and the data are easy to analyze.

2. Still photography. In the event that the employment of time-lapse movie photography is too costly, the use of still pictures may facilitate data coverage for specific demonstration steps (instances where photographs will enhance the data recorded on the demonstration worksheets).

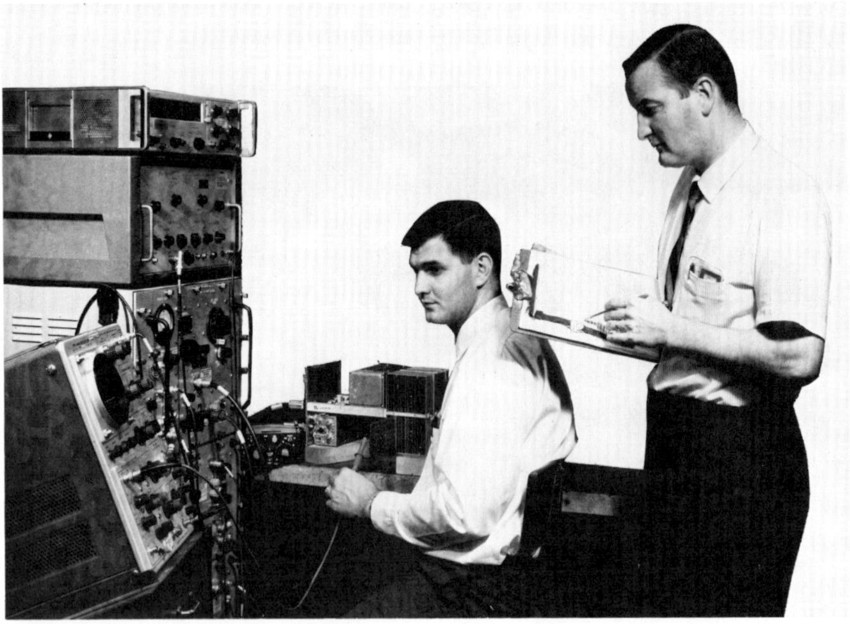

FIG. 14-8 A Maintenance Technician Troubleshoots System XYZ to Isolate the Faulty Assembly. The Data Recorder Monitors the Procedure. (*General Dynamics Corporation.*)

FIG. 14-9 A Maintenance Technician Repairs the Faulty Assembly. A Data Recorder Monitors the Procedure. (*General Dynamics Corporation.*)

3. Tape recorder. A tape recorder may be employed to record verbal communication between maintenance technicians accomplishing the demonstration tasks. In addition, a tape recorder may be used by the data recorder for describing the maintenance steps being demonstrated in relationship to a time reference.

4. Chart recorder (brush or equivalent). A chart recorder may be used to facilitate the recording of time data. This may prove beneficial when there are many elements of time to be measured on a simultaneous basis.

5. Stopwatch. A stopwatch is normally used by each data recorder to measure task times.

Data Analysis

Corrective Maintenance. The formal demonstration of corrective-maintenance tasks on system *XYZ* (50 independent simulations) is recorded on the demonstration-data sheet (Fig. 14-4) and on a set of demonstration worksheets (Fig. 14-5). The results, in terms of active

TABLE 14-3 Demonstration-test Data (Example)

(1) Demonstration Number	(2) Observed Repair Time (RP) (Minutes)	(3) Log Rp	(4) (Log Rp)^2	(5) $(Rp)^2$
1	52	1.716	2.944	2704
2	47	1.672	2.795	2209
3	69	1.839	3.381	4761
4	72	1.857	3.448	5184
5	86	1.935	3.744	7396
6	67	1.826	3.334	4489
7	93	1.968	3.873	8649
8	38	1.580	2.496	1444
9	81	1.908	3.640	6561
10	64	1.806	3.261	4096
11	75	1.875	3.515	5625
12	101	2.004	4.016	10201
13	71	1.851	3.426	5041
14	48	1.681	2.825	2304
15	65	1.813	3.286	4225
16	83	1.919	3.682	6889
17	97	1.987	3.948	9409
18	103	2.013	4.052	10609
19	52	1.716	2.944	2704
20	114	2.057	4.231	12996
21	132	2,120	4.494	17424
22	96	1.982	3.928	9216
23	119	2.075	4.305	14161
24	118	2.071	4.289	13924
25	91	1.959	3.837	8281
26	44	1.643	2.699	1936

(1) Demonstration Number	(2) Observed Repair Time (Rp) (Minutes)	(3) Log Rp	(4) (Log Rp)^2	(5) $(Rp)^2$
27	80	1.903	3.621	6400
28	86	1.934	3.740	7396
29	61	1.785	3.186	3721
30	45	1.653	2.732	2025
31	40	1.602	2.566	1600
32	77	1.886	3.556	5929
33	71	1.851	3.426	5041
34	109	2.037	4.149	11881
35	67	1.826	3.334	4489
36	76	1.880	3.534	5776
37	84	1.924	3.701	7056
38	99	1.995	3.980	9801
39	55	1.740	3.027	3025
40	70	1.845	3.404	4900
41	63	1.799	3.236	3969
42	49	1.690	2.856	2401
43	72	1.857	3.448	5184
44	78	1.892	3.579	6084
45	36	1.567	2.441	1296
46	32	1.505	2.265	1024
47	35	1.544	2.383	1225
48	46	1.662	2.762	2116
49	43	1.633	2.666	1849
50	49	1.690	2.856	2401
TOTALS	3601	91.573	168.841	290536

maintenance times, are summarized in Table 14-3. Note that the data are presented in a similar format to that specified in Chap. 4.

Referring to Table 14-3, the totals presented are as follows:

$$(1)\ \sum_1^N R_p = \sum_1^N M_{ct} = 3601 \text{ (block 2)}$$

$$(2)\ \sum_1^N \log R_p = \sum_1^N \log M_{ct} = 91.573 \text{ (block 3)}$$

$$(3)\ \sum_1^N (\log R_p)^2 = \sum_1^N (\log M_{ct})^2 = 168.841 \text{ (block 4)}$$

$$(4)\ \sum_1^N R_p{}^2 = \sum_1^N M_{ct}{}^2 = 290{,}536 \text{ (block 5)}$$

Using the above quantities, the observed $\tilde{M}_{ct}$, $\bar{M}_{ct}$, $M_{max(ct)}$, and standard deviation can be calculated as follows:

1. Calculation of $\tilde{M}_{ct}$ ($MTTR_G$)

$$\tilde{M}_{ct} = \text{antilog}\left[\frac{\sum_1^N \log M_{ct_i}}{N}\right]$$

where $N = 50$

$$\tilde{M}_{ct} = \text{antilog}\left[\frac{91.573}{50}\right]$$

$$\tilde{M}_{ct} = \text{antilog } 1.831 = 67.8 \text{ min}$$

2. Calculation of $\bar{M}_{ct}$ ($MTTR$)

$$\bar{M}_{ct} = \frac{\sum_1^N M_{ct}}{N} = \frac{3601}{50} = 72 \text{ min}$$

3. Calculation of $M_{max(ct)}$

$$M_{max} = \text{antilog}\,(\log \tilde{M}_{ct} + 1.645\,\sigma_{\log M_{ct_i}})$$

where

$$\sigma_{\log M_{ct}} = \sqrt{\frac{\sum_1^N (\log M_{cti})^2 - \left(\sum_1^N \log M_{cti}\right)^2/N}{N-1}}$$

$$M_{max} = \text{antilog}\left(1.831 + 1.645\sqrt{\frac{168.841 - (91.573)^2/50}{49}}\right)$$

$$M_{max(ct)} = 121 \text{ min}$$

4. Calculation of standard deviation σ

$$\sigma = \sqrt{\frac{N \sum_{1}^{N} (M_{ct})^2 - \left(\sum_{1}^{N} M_{ct}\right)^2}{N(N-1)}}$$

$$\sigma = \sqrt{\frac{50(290536) - (3601)^2}{50(49)}}$$

$$\sigma \cong \sqrt{636}$$

$$\sigma \cong 25.2$$

Employing the statistical principles presented in Chap. 4, a histogram can be constructed using observed repair-time data.

$$\text{Class interval width} = \frac{\text{range of observations}}{K}$$

$$\text{Range of observations} = 132 - 32 = 100$$

(See Table 14-3.)

$$K = 1 + 3.3 \log N \qquad \text{where } N = 50$$

$$K = 1 + 3.3(1.7) = 6.61$$

$$\text{Class interval width} \cong \frac{100}{6.6} = 15.15$$

Using the class interval of 15.15, the frequency distribution of repair times is calculated and presented in Table 14-4.

Based on the frequency distribution indicated in Table 14-4, the data can be plotted in a histogram as shown in Fig. 14-10.

Preventive Maintenance. Normally, the demonstration of preventive-maintenance tasks is accomplished in a manner similar to the demonstration of corrective-maintenance tasks, and the calculation of $\bar{M}_{pt}$ follows the groundrules specified in Chap. 4. However, in the case of system *XYZ*, design is such that no preventive-maintenance requirements are anticipated. Therefore, the preventive-maintenance aspect of demonstration is not required.

TABLE 14-4 Frequency Distribution

Group	Class Interval	Frequency	Cumulative Frequency
1	32.0 - 47.15	10	10
2	47.15 - 62.30	7	17
3	62.30 - 77.45	14	31
4	77.45 - 92.60	8	39
5	92.60 - 107.75	6	45
6	107.75 - 122.90	4	49
7	122.90 - 138.05	1	50

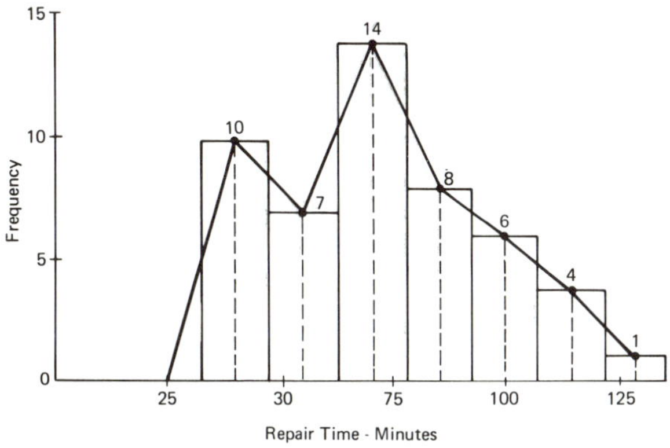

FIG 14-10 Repair-time Histogram.

Accept/Reject Decision

Demonstration method 1. Application of this method requires the initial specification of $\widetilde{M}_{ct}$, $\widetilde{M}_{pt}$, $M_{max(ct)}$, or $M_{max(pt)}$, and the selection of a 75 percent or 90 percent confidence level. For the purpose of this text, it is assumed that the requirements (specified by the customer at contract award) for system *XYZ* are as follows:

1. $\widetilde{M}_{ct} = 75$ min
2. $M_{max(ct)} = 120$ min
3. $\widetilde{M}_{pt} = 120$ min
4. $M_{max(pt)}$: not specified
5. Confidence level: 75%

Thus, the overall demonstrated value of $\widetilde{M}_{ct}$ for system *XYZ* should not exceed 75 min. Considering a 75 percent confidence level, no more than 22 of the 50 individual sample repair times can exceed 75 min. Checking the repair time values presented in Table 14-3, a total of 21 of the 50 samples fall above 75 min. Therefore, system *XYZ* successfully passes the demonstration requirements relative to $\widetilde{M}_{ct}$.

Suppose in the case of system *XYZ* that a 90 percent confidence level had been specified. According to the criteria of method 1, no more than 20 of the 50 individual samples are allowed to exceed the 75 min. In this instance, system *XYZ* would not pass demonstration and would require some form of corrective action and retest.

For $M_{max(ct)}$, the criteria of method 1 state that no more than one of the fifty individual samples is allowed to exceed the given $M_{max(ct)}$ of 120 min for a confidence level of 75 percent. Referring to Table 14-3, only one of the fifty samples exceeds 120 min; therefore, system *XYZ* successfully meets the requirements.

In the event that a 90 percent confidence level had been specified, the criteria of method 1 would indicate that none of the 50 samples could exceed the given 120-min value. In this instance, system *XYZ* would not pass demonstration, since one sample does exceed this value.

As regards preventive maintenance, system *XYZ* obviously passes the $\tilde{M}_{pt}$ requirement of 120 min since there are no preventive-maintenance requirements anticipated.

Demonstration method 2. Application of this method requires the initial specification of $\bar{M}_{ct}$, $\bar{M}_{pt}$, $\bar{M}$, and $M_{max(ct)}$. For the purpose of this text, it is assumed that the requirements (specified by the customer at contract award) for system *XYZ* are as follows:

1. $\bar{M} = 65$ min
2. $\bar{M}_{ct} = 75$ min
3. $\bar{M}_{pt} = 120$ min
4. $M_{max(ct)} = 120$ min
5. Contractor's risk = 20% (by mutual agreement)

The determination of $\bar{M}_{ct}$ from the observed repair-time data in Table 14-3 is based on the following expression (represents upper limit):

$$\bar{M}_{ct} + \frac{\phi \sigma_{M_{ct}}}{\sqrt{N}}$$

where ϕ = numerical value corresponding to the contractor's risk α, which is obtained from a table of normal distribution factors.

Numerical Value (ϕ)	Contractor's Risk (α)
0.84	20%
1.04	15%
1.28	10%
1.65	5%

$\sigma_{M_{ct}}$ = standard deviation observed from test data,
N = sample size (quantity of demonstrations)

In regard to the accept/reject decision, the following criteria shall govern:

1. *Accept* if $\bar{M}_{ct}$ (specified) $\geqq \bar{M}_{ct} + \phi\sigma_{M_{ct}}/\sqrt{N}$
2. *Reject* if $\bar{M}_{ct}$ (specified) $< \bar{M}_{ct} + \phi\sigma_{M_{ct}}/\sqrt{N}$

Applying demonstration-test data with an assumed 20 percent contractor's risk, the following results are obtained:

$$\bar{M}_{ct}\text{ (specified)} \geqq \bar{M}_{ct} + \frac{\phi\sigma_{M_{ct}}}{\sqrt{N}}$$

$$75 \geqq 72 + \frac{0.84(25.2)}{\sqrt{50}}$$

$$75 \geqq 72 + 2.99 \text{ or } 74.99$$

In this instance, system *XYZ* has successfully passed the $\bar{M}_{ct}$ requirement of 75 min at a contractor's risk of 20 percent.

Suppose in the case of system *XYZ* that a 10 percent contractor's risk had been assumed. According to this assumption, the following results are obtained:

$$75 \geqq 72 + \frac{1.28(25.2)}{\sqrt{50}}$$

$$75 \geqq 72 + 4.56 \text{ or } 76.56$$

In this instance, the demonstrated value (with standard deviation applied) exceeds the specified value. Therefore, system *XYZ* would not pass demonstration and would require some form of corrective action and retest.

In the area of preventive maintenance, the criteria for the accept/reject decision are the same as for corrective maintenance. That is:

1. Accept if $\bar{M}_{pt}$ (specified) $\geqq \bar{M}_{pt} + \dfrac{\phi\sigma_{M_{pt}}}{\sqrt{N}}$
2. Reject if $\bar{M}_{pt}$ (specified) $< \bar{M}_{pt} + \dfrac{\phi\sigma_{M_{pt}}}{\sqrt{N}}$

For the purpose of this method (to illustrate the use of various formulas), it is now assumed that system *XYZ* requires some preventive maintenance and that the observed value of $\bar{M}_{pt}$ (through actual demonstration) is 45 min. The sample size N is assumed to be 10, and the standard deviation $\sigma_{M_{pt}}$ is assumed as 20. Based on these assumptions, the following results are obtained (20 percent contractor risk):

$$\bar{M}_{pt}\text{ (specified)} \geqq \bar{M}_{pt} + \frac{\phi\sigma_{M_{pt}}}{\sqrt{N}}$$

$$120 \geqq 45 + \frac{0.84(20)}{\sqrt{10}}$$

$$120 \geqq 45 + 5.32 \text{ or } 50.32$$

Thus, system *XYZ* successfully passes demonstration by a wide margin.

The next step in method 2 is to evaluate the observed mean downtime

$\bar{M}$. Determination of $\bar{M}$ from observed demonstration repair-time data is based on the following expression:

$$\bar{M} + \phi \sqrt{\frac{N_p(f_c \sigma_{M_{ct}})^2 + N_c(f_p \sigma_{M_{pt}})^2}{N_c N_p (f_c + f_p)^2}}$$

where N_p = preventive-maintenance demonstrated sample size (assumed to be 10).

N_c = corrective-maintenance demonstrated sample size (50).

f_p = quantity of expected preventive-maintenance tasks occurring during a representative operating-time period T. f_p can be calculated by dividing T by the expected mean-time-between-preventive-maintenance-actions (normally obtained from maintenance-analysis data). With T assumed as 5,000 hr and the mean-time-between-preventive-maintenance-actions as 833 hr, f_p is 6.

f_c = quantity of expected corrective-maintenance tasks occurring during a representative operating-time period T. f_c can be calculated by dividing T by the expected mean-time-between-corrective-maintenance-actions (i.e., $MTBF$). With T assumed as 5,000 hr and the $MTBF$ as 450 hr, f_c is 11.

$\sigma_{M_{ct}}$ = standard deviation for corrective maintenance (25.2).

$\sigma_{M_{pt}}$ = standard deviation for preventive maintenance (20).

In regard to the accept/reject decision, the following criteria shall govern:

1. *Accept* if $\bar{M}$ (specified) $\geqq \bar{M} + \phi \sqrt{\frac{N_p(f_c \sigma_{M_{ct}})^2 + N_c(f_p \sigma_{M_{pt}})^2}{N_c N_p (f_c + f_p)^2}}$
2. *Reject* if $\bar{M}$ (specified) $< \bar{M} + \phi \sqrt{\frac{N_p(f_c \sigma_{M_{ct}})^2 + N_c(f_p \sigma_{M_{pt}})^2}{N_c N_p (f_c + f_p)^2}}$

Applying demonstration-test data, the following results are obtained:

$$\bar{M} = \frac{\bar{M}_{ct} f_c + \bar{M}_{pt} f_p}{f_c + f_p}$$

$$\bar{M} = \frac{72 \times 11 + 45 \times 6}{11 + 6}$$

$$\bar{M} = 62.5 \text{ min}$$

The next step involves the accept/reject formula:

$$65 \geqq 62.5 + 0.84 \sqrt{\frac{10[(11)(25.2)]^2 + 50[(6)(20)]^2}{(10)(50)(17)^2}}$$

$$65 \geqq 62.5 + 0.84(3.21) \text{ or } 65.2 \text{ min}$$

In this case, test results indicate that system *XYZ* has failed to pass demonstration at a contractor's risk of 20 percent.

The final proof of method 2 involves $M_{\max(ct)}$, which is based on the following expression (at 95th percentile point):

$$M_{\max(ct)} = \text{antilog}\,(\log \tilde{M}_{ct} + \phi\sigma_{\log M_{ct}})$$
$$M_{\max(ct)} = \text{antilog}\,(1.831 + 1.645\sqrt{0.02304})$$
$$M_{\max(ct)} = 121 \text{ min}$$

Refer to calculation of $M_{\max(ct)}$ in paragraph on data analysis.

In regard to the accent/reject decision, the following criteria shall govern:

1. *Accept* if $M_{\max(ct)}$ (specified) $\geqq M_{\max(ct)}$ (demonstrated).
2. *Reject* if $M_{\max(ct)}$ (specified) $< M_{\max(ct)}$ (demonstrated).

Applying these criteria, the demonstration value of 121 min exceeds the 120-min specified requirement. Therefore, system *XYZ* has failed to pass the demonstration.

Summary of demonstration-method results. A summary of the two methods covering the specified requirements, demonstration results, and accept/reject decision is presented in Table 14-5.

As a normal maintainability program requirement, only one demonstration method will be specified. In the case of system *XYZ*, had method 1 been specified as the requirement, the system would have

TABLE 14-5 Summary of Demonstration-test Data

Demonstration Method	Specified Value	Demonstration Criteria	Demonstration Value	Accept or Reject
1	$\tilde{M}$ct -- 75 min (75% confidence)	No more than 22 of 50 samples should exceed 75 min.	21 samples exceed 75 min	Accept
	$\tilde{M}$ct -- 120 min	————	No preventive maintenance required	Accept
	Mmaxct -- 120 min. (75% confidence)	No more than 1 to 50 samples should exceed 120 min.	Only 1 sample exceeds 120 min	Accept
2	$\bar{M}$ct -- 75 min (20% contractor risk)	$\bar{M}$ct specified must be $\geqslant$ than $\bar{M}$ct demonstrated	74.99 min	Accept
	$\bar{M}$pt -- 120 min (20% contractor risk)	$\bar{M}$pt specified must be $\geqslant$ than $\bar{M}$pt demonstrated	50.32 min	Accept
	$\bar{M}$ -- 65 min (20% contractor risk)	$\bar{M}$ specified must be $\geqslant$ than $\bar{M}$ demonstrated	65.2 min	Reject
	Mmaxct -- 120 min (20% contractor risk)	Mmaxct specified must be $\geqslant$ than Mmaxct demonstrated	121 min	Reject

passed the maintainability demonstration. Had method 2 been specified (considering that test data employed an arbitrary $\bar{M}_{pt}$ value) the system would have failed the maintainability demonstration.

RETEST PHASE

The retest phase is designed to repeat certain specific demonstrations and/or accomplish additional maintainability testing to investigate deficiencies detected during the first demonstration. Generally, a retest may be required due to the following conditions:

1. The system or equipment fails to meet formal demonstration requirements, and the margin of failure is small (refer to the time differences for method 2 in Table 14-5). In this instance, the customer may request that an additional quantity of tasks be demonstrated to increase confidence in the test-data findings. If, after additional testing, the data indicates compliance with the initially specified requirements, the customer may choose to accept the system, and testing is concluded. If, on the other hand, the data still indicate noncompliance, the contractor may be required to accomplish some redesign or to accept a penalty.
2. The system or equipment may fail to meet formal demonstration requirements by a considerable margin. In this instance, the contractor may be required to redesign the system or equipment and/or to accept a penalty.

In the event that system/equipment redesign is required, possible areas of concentration are easily identified through the maintenance-time data recorded on the demonstration-data sheet (Fig. 14-4) and demonstration worksheet (Fig. 14-5). Extensive maintenance times will indicate problem areas. These areas are reviewed and evaluated by the design engineer and maintainability engineer to recommend corrective action.

Subsequent to the approval (by the customer) and incorporation of proposed design modifications, the system will be subjected to additional demonstration testing. Possibilities include either a complete retest (50 new sample simulations) or a partial retest covering only specific equipment areas where problems were identified. In either case, additional testing will be accomplished following the same ground rules as stated for the initial demonstration.

DEMONSTRATION REPORTING

The final effort in fulfilling maintainability-demonstration requirements is the preparation of the final report (refer to Fig. 14-2, 19.0). The final report, covering demonstration-test results (plus supporting data), is generally submitted to the customer approximately 30 days following completion of the demonstration.

QUESTIONS AND PROBLEMS

1. What is the main purpose of maintainability demonstration?
2. What is the ideal condition to demonstrate (test) the system?
3. Demonstration may be accomplished at the contractor's facility utilizing production test equipment. True or false?
4. List three sets of conditions concerning manpower, equipment, and facilities as they relate to demonstration of a system. That is, what are some alternatives in the approach to demonstration?
5. What are three constraints in specifying the approach to demonstration?
6. The planning, preparation, and execution of the demonstration all take place during the latter part of the program acquisition phase. True or false? Why?
7. Test objectives are normally specified by the customer 30 days before the scheduled demonstration. True or false? Why?
8. Test objectives are always expressed in quantitative terms. True or false? Why?
9. The maintenance concept and the maintenance analysis are used to establish the parameters of demonstration. True or false? Why?
10. Describe the criteria you would apply in selecting personnel to perform the demonstration of a system.
11. List the technical data that you would authorize to be used for support of the demonstration.
12. What are the criteria you would apply in specifying the test (support) equipment to be used for a demonstration?
13. What would you consider in selecting the test article to be demonstrated?
14. Allocation of maintenance tasks to subsystems (or items) is based on their contribution to maintenance. In your own words, describe the meaning behind this statement.
15. Preventive-maintenance tasks that can be performed during normal operation (use) of the system are not usually subject to demonstration. Why? Can you suggest one or more reasons why one may wish to demonstrate these tasks?
16. Regardless of the total number of valid preventive-maintenance tasks, all are demonstrated. True or false? Why?
17. In preparing for a demonstration, what would be your primary areas of concern if you had been designated the test director?
18. The test methods described in this chapter are the only ones currently being used by industry and government. True or false?
19. The test is continued in all cases until the task is completed, regardless of the time involved. True or false? Why?
20. List four support aids that may be used by the maintainability engineer to record test data.
21. The definition of a confidence-level factor is necessary before an accept/reject decision can be made. True or false?
22. Under what conditions can a retest requirement be specified?

APPENDIX A

Maintainability Detailed Checklist

In accomplishing an informal maintainability review, specific questions which reflect good maintainability principles serve as the basis in determining whether or not the item being reviewed incorporates optimum maintainability considerations. These questions, in the form of a checklist, are listed below. Not all questions are applicable in all reviews; however, the answer to those questions as applicable should be yes to reflect good maintainability.

GENERAL

1.0 Standardization
- 1.1 Are standard GFE/commercial equipment items incorporated to the maximum degree possible (except for such items not compatible with system time constraints)?
- 1.2 Are the same GFE/commercial equipment items used in similar applications?

- 1.3 Are corresponding rack/console positions used throughout all test stations for the same test-equipment items?
- 1.4 Are corresponding rack/console positions used throughout all test stations for peculiar control panels?
- 1.5 Are preferred standard parts, AN, NAS, or MIL-STD parts utilized to the maximum extent possible?
- 1.6 Are control-panel layout (from panel to panel) the same or similar?
- 1.7 Is panel nomenclature/color coding the same for all panels?
- 1.8 Are like modular units employed throughout rack/console drawers?
- 1.9 Are maintainability checklists, test procedures, and technical manual formats standardized?
- 1.10 Are electrical/electronic circuit types kept to a minimum?

2.0 Component Functional Grouping

Packaging of hardware should be by direct association with primary signal flow. The following are in order of preference:

- 2.1 Are electrical circuit functions packaged in individual plug-in modules?
- 2.2 Are electrical circuit functions, not packaged in individual plug-in modules, packaged in a minimum number of plug-in modules or printed circuit boards?
- 2.3 Are electrical circuit components, not packaged in plug-in modules or printed circuit boards, functionally grouped?

3.0 Console Layout

Related rack/console features are covered under Equipment Racks-General

- 3.1 Is rack/console layout optimized from the human-factors standpoint (consult human-factors personnel)?
- 3.2 Are rack/console drawers optimally placed according to functional relationships, frequency of schedule maintenance, etc?

4.0 Complexity

- 4.1 To the degree ascertainable, is the number of parts minimized?
- 4.2 Does each part have an absolute function or need?
- 4.3 Are difficult maintenance procedures, adjustments, etc., justified?

5.0 Self-test

- 5.1 Have self-test provisions been incorporated?
- 5.2 Is the degree or depth of self-testing in accordance with specification requirements?
- 5.3 Are self-test provisions automatic?
- 5.4 Have direct fault indications been provided? (This means that either a fault light or an audio signal is provided or there is a direct indication of a malfunction through a meter reading or similar means.)

6.0 Maximum-time-to-repair

- 6.1 Has equipment turnaround time been minimized through:
  a. Capability for rapid and positive fault localization?

b. Maximum accessibility to enable rapid removal and replacement of assemblies or subassemblies?
c. Selection of assemblies or subassemblies requiring minimum time to self-test (warm-up time included)?
d. Rapid equipment-performance-verification capability?

6.2 Does equipment maximum-time-to-repair meet the required time constraint?

7.0 Auxiliary Tools and Test Equipment

7.1 Are the number of required auxiliary tools and test-equipment items minimized?
7.2 Are required tools and test-equipment items standard?
7.3 Are required tools and test-equipment items simple to operate? Can applicable maintenance tasks be accomplished in a minimum span of time?
7.4 Have peculiar and nonstandard auxiliary tools and test-equipment items been eliminated?

8.0 Labeling

8.1 Is labeling maximized?
8.2 Is labeling adequate in the following areas?
a. Racks/consoles
b. Panel controls and meters
c. Test and adjustment points
d. Access openings
e. Safety areas
f. Servicing/lubrication areas
g. Cables and connectors

Labeling relative to specific areas is covered in other sections of this checklist. Size and type of labeling should be discussed with human-factors personnel.

9.0 Weight

Equipment should be of the lightest weight possible consistent with sturdiness, safety, and reliability.

9.1 To the degree ascertainable, is equipment weight minimized through the elimination of unnecessary structural members, cases, heavy parts, etc.?
9.2 Are assemblies and subassemblies which are removed for maintenance minimized relative to weight (preferably less than 10 lb)?

10.0 Calibration Requirements

10.1 Are calibration requirements known?
10.2 Do standards exist for calibrating test equipment?
10.3 Are calibration tolerances known?
10.4 Are calibration frequencies known?
10.5 Are calibration procedures prepared?
10.6 Are calibration requirements possible within the existing capabilities of a standard base level Precision Measurement Equipment Laboratory (PMEL)?

11.0 Repair/Replace Philosophy
- 11.1 Is the basic maintenance concept known?
- 11.2 Has the level of maintenance for each repairable item been defined (whether the item will be repaired at field level or at depot level)?
- 11.3 Has the maintenance frequency (*MTBM*) been defined or predicted?
- 11.4 Has the maintenance-downtime requirement (*MTTR*) been specified?
- 11.5 Have the maintenance-facility and resource requirements been specified?
- 11.6 Have the maintenance personnel-quantity and skill-level requirements been specified?
- 11.7 Are test-equipment requirements compatible with the maintenance concept?

12.0 Maintenance Procedures
- 12.1 Have maintenance procedures and task sequences been identified (maintenance-requirements task analysis)?
- 12.2 Have maintenance checklists been prepared?
- 12.3 Have maintenance technical publications been prepared?
- 12.4 Are the maintenance procedures compatible with the maintenance concept?

13.0 Personnel Requirements
- 13.1 Are anticipated maintenance tasks simple enough to enable accomplishment by an average technician?

Average Technician

a. Age	21 years
b. Experience	No regular work experience prior to training
c. Time in service	On first 3- or 4-year enlistment
d. Educational level	High school graduate
e. Service schooling	A total of 19 weeks of formal training
f. General level of interest	Varies, but usually low
g. General level of morale	Varies, but usually low
h. General reading level	About 9th grade

- 13.2 Are anticipated maintenance tasks such that no more than two technicians are required for accomplishment?

14.0 Trade-off Studies

Design decisions are made based on weighing factors such as performance, reliability of equipment, weight, schedule, maintainability considerations, and cost. When design decisions are made which compromise maintainability, the trade-off considerations as applicable should be documented.
- 14.1 Have maintainability trade-off studies been documented?
- 14.2 Have trade-offs been justified/validated?

HANDLING

1.0 Equipment Lifting
- 1.1 Are lifting eyes installed on each rack/console which is handled separately? Is the inside diameter of each lifting eye at least 1 in?
- 1.2 Is the base of each console reinforced to enable lifting with a fork-lift truck?
- 1.3 Are hoisting fittings identified relative to hoisting capacity and fork-lift application points?

2.0 Rack/Console Drawer or Panel Handles
- 2.1 Are rack/console drawers provided with handles?
- 2.2 Are two handles provided on each drawer front panel?
- 2.3 Are drawer handles mounted vertically?
- 2.4 Are drawer handles adequate in size?
- 2.5 Is one handle provided on each rack/console panel?
- 2.6 Are handles optimally located (provide at least 2-in. clearance between handles and obstructions)?

3.0 Assembly Handles
- 3.1 Do assemblies weighing over 10 lb within equipment items incorporate handles?
- 3.2 Are handles optimally located from the weight-distribution aspect (handles should be located over the center of gravity)?

4.0 Console Castors

For heavy racks or consoles which are likely to be moved periodically within the maintenance shop, castors are sometimes desirable. However, when castors are employed, jacks should be installed adjacent to each castor to enable leveling and to preclude undesirable movement.
- 4.1 Are castors provided on racks or consoles which are likely to be moved periodically within the maintenance shop?

5.0 Damage Susceptibility
- 5.1 Are covers or cases provided to protect equipment vulnerable areas from damage?
- 5.2 Are connectors and other parts which normally extend beyond outside covers or cases recessed or adequately protected by some other means?

6.0 Weight Labeling
- 6.1 Are weight labels provided on each rack and console?
- 6.2 Are weight labels provided on each equipment weighing over 50 lb?
- 6.3 Are weight labels provided on three external surfaces? Three surfaces should be front, top, and rear.

EQUIPMENT RACKS-GENERAL

1.0 Drawer Mounting
- 1.1 Are rack/console drawers mounted on roll-out slides?

1.2 Are drawer slides easily and quickly detached for removal from the rack or console without use of hand tools?
1.3 Do slides incorporate provisions for locking in the open position?

2.0 Panel Mounting
2.1 Are panels hinge mounted?
2.2 Are separable hinges used in panel mounting?
2.3 Are provisions provided for bracing the panel when in the open position?

3.0 In-position Maintenance
3.1 Is in-position maintenance possible with the rack/console drawer or panel in the open position?
3.2 Can rack/console drawers or panels be extended in the open position with power cables attached?

4.0 Cable Inlets
4.1 Are test station power or interconnecting cables routed such as to avoid connection on the front of the rack or console?

5.0 Equipment Mounting (Weight)
5.1 Are the heavier items on the bottom of each rack or console? Unit weight should decrease with the increase in installation height above the floor.

6.0 Operator Panels Optimally Positioned. Optimum panel positioning is based on a human-factors system analysis. However, as a guide, the following questions are applicable:
6.1 For maintenance personnel in the standing position, are panels located between 40 and 70 in. above the deck? Are key, critical, or precise controls and displays located between 48 and 64 in. above the deck?
6.2 For maintenance personnel in the sitting position, are panels located 30 in. above the deck?

7.0 Air Intake/Exhaust Provisions
7.1 Are air intake/exhaust provisions adequate? Are hot spots covered?
7.2 Are air intake ports insulated from exhaust ports?
7.3 Are blower motors of the self-lubricating type?
7.4 Are air filters used?
7.5 Are air filters readily accessible?

PACKAGING

1.0 Plug-in Modules and Components
1.1 Are drawer modules and components of the plug-in type?
1.2 Are plug-in modules and components removable without the use of tools? If tools are required, are they of the standard variety?
1.3 Are accesses between plug-in modules and components adequate to allow for hand grasping?

2.0 Module/Component Stacking

- 2.1 Are modules and components (not of the plug-in type) mounted with no more than four fasteners? Are the fasteners standardized?
- 2.2 Are modules and components mounted such that the removal of any single unit for maintenance should not require the removal of other units?

3.0 Accessibility Based on Replacement Frequency

- 3.1 Are modules and components mounted such that access priority has been assigned according to predicted removal/failure rates? (The component with the greatest anticipated removal frequency should be readily accessible.)

4.0 Wrong Installation Prevention

- 4.1 Are modules and components designed to preclude installation in the wrong position?

5.0 Module/Component and Mounting-plate Labeling

- 5.1 Are modules and components labeled?
- 5.2 Are module/component labels located on top of each unit or in plain sight?
- 5.3 Is label information adequate (full identification nomenclature)?
- 5.4 Are labels attached to the module/component mounting plate?
- 5.5 Are mounting-plate labels adjacent to the applicable module or component?
- 5.6 Are labels etched or embossed into the applicable surface rather than merely painted or stamped on the surface?

6.0 Guides for Module/Component Installation

- 6.1 Are guide pins provided on modules and components requiring alignment during installation?
- 6.2 Are guides provided on the chassis to facilitate printed-circuit-board installation and removal?

7.0 Interchangeability

- 7.1 Are modules and components having like functions both electrically and mechanically interchangeable?
- 7.2 Are module/component mounting fasteners interchangeable (i.e., fastener type, size, and length)?
- 7.3 Are cable harness sections interchangeable?

ACCESSIBILITY

1.0 Access Doors Provided

- 1.1 Are access openings provided in areas (not readily accessible through an open drawer or panel in front of the rack or console) where periodic maintenance is anticipated?
- 1.2 Are access openings optimally located for the access required?
- 1.3 Are hinged doors used where physical access is required (instead of a cover plate held in place by screws or other fasteners)?
- 1.4 Is a transparent window or quick-opening cover used for visual inspection accesses?

1.5 Are access doors shaped such as to permit passage of components and implements which must pass through?

2.0 Access-door Support

2.1 On hinged console panels and access doors, is there some means of holding the panel or door in the open position?

2.2 Are small access doors which are not hinge mounted (held with fasteners) attached with a chain or by some other means to preclude loss?

3.0 Access-door Labeling

3.1 Is each access door or opening labeled to indicate the items which are accessible?

3.2 Does each access label indicate what auxiliary equipment is to be used at the access?

3.3 Is each access labeled uniquely so each one can be clearly identified in maintenance instructions?

3.4 Does each access label indicate the recommended time period for performing maintenance operations?

4.0 Access Size

4.1 Are access openings adequate in size?

4.2 Do hinged panels open wide enough to allow access to components?

5.0 Access Fasteners Minimized

5.1 Are access panels and door held in the closed position with a minimum quantity of fasteners? No more than four fasteners should be used.

5.2 Are access panel/door fasteners of the quick-release variety? (Hand operated fasteners are preferred.)

6.0 Special Tools Minimized

6.1 Are the number of tools required to gain access held to a minimum?

6.2 Are required tools of the standard variety? Have special tools been eliminated?

7.0 Component Accessibility

7.1 Are accesses between plug-in modules and components adequate to allow for hand grasping?

7.2 Are accesses between fastener-mounted modules and components adequate to allow for proper tool use?

8.0 Guides for Dangerous Accesses

8.1 Are screwdriver guides provided on adjustment points near high voltages?

8.2 When blind accesses exist, are protective features provided to preclude personnel injury?

Fasteners

1.0 Quick-release Fasteners

1.1 Are quick-release fasteners employed?

2.0 Fasteners Standardized
- 2.1 Are the number of different types of fasteners minimized?
- 2.2 Is the same type and size of fastener used for a given application (i.e., all mounting bolts for a given type of item)? Are bolt/screw lengths the same? Are bolt/screw thread sizes the same?
- 2.3 If fasteners require torque application, are the number of differing torque requirements minimized?
- 2.4 Have fasteners been chosen based on the requirement for standard tools, in lieu of requiring special tools?
- 2.5 Are all external fasteners which are manipulated during normal maintenance of a contrasting color to the surface on which they appear? Are all other fasteners and assembly screws of the same color as the surface on which they appear?

3.0 Fasteners Minimized
- 3.1 Is the quantity of fasteners minimixed? No more than four fasteners per application is preferable.
- 3.2 Are a few large fasteners used in lieu of many small fasteners?

4.0 Hexagonal Socket-head Fasteners
- 4.1 Are hexagonal socket-head screws used?

5.0 Captive Fasteners
- 5.1 Are captive fasteners used wherever lost screws, bolts, or nuts might cause a malfunction or excessive maintenance time?
- 5.2 Can captive fasteners be operated by hand or using a standard tool?
- 5.3 Can captive fasteners be easily replaced in case of damage?

6.0 Fasteners Turns
- 6.1 Are bolt/screw lengths no more than required for the application?
- 6.2 Is the number of turns required to tighten a bolt/screw less than 10?
- 6.3 When tightened, does the bolt or screw extend two threads beyond the nut?

Panel displays and controls

Panel displays and controls are defined as a result of a human-factors system analysis. There are many factors which influence panel arrangements, and the human-factors group should be consulted relative to criteria. The following questions cover some of the areas of maintainability interest (this checklist, covering panel displays and controls, is not intended to represent the entire picture):

1.0 Controls Standardized
- 1.1 Are preferred standard, AN, NAS, or MIL-STD parts utilized to the maximum extent possible?
- 1.2 Do knobs performing the same function have the same shape?
- 1.3 Are pointer-type knobs used to indicate a marking or to indicate relative position from a fixed point?
- 1.4 Are bar-type knobs used where rotation is more than 360°?

1.5 Are control knobs secured by two setscrews?
1.6 Are toggle switches employed when control functions require no more than two discrete positions?
1.7 Are toggle switches mounted vertically (i.e., the switch should move upward for ON, START, or INCREASE, and downward for OFF, STOP, or DECREASE)?
1.8 Is there a guard for those toggle switches susceptible to accidental activation which is considered serious?
1.9 Do values increase with clockwise control rotation?
1.10 Are display scales standardized?
1.11 Are circular display scales used in lieu of straight scales?

2.0 Controls Sequentially Positioned
2.1 Are controls and displays that are used sequentially aligned horizontally from left to right or vertically from top to bottom, or in rows from top to bottom row and from left to right within each row?
2.2 Are controls and displays that are used sequentially arranged in an order analogous to the actual order of events?

3.0 Control Spacing
3.1 Are controls properly spaced?

4.0 Control Labeling
4.1 Are controls properly labeled?
4.2 Are controls and displays identified as to their function?
4.3 Are labels located either on or immediately adjacent to (preferably above) the controls and displays to be identified?
4.4 Is the location of control and display labels consistent on all equipment?
4.5 Is the labeling brief?

5.0 Control/Display Relationship
5.1 Are related controls and displays located on the same panel?
5.2 Are controls adjacent to the applicable display?
5.3 Is the control and display relationship unambiguous and unmistakable?

6.0 Meters
6.1 Are ruggedized meters incorporated?
6.2 Are meters externally removable from the outside of the panel?
6.3 Can meters be calibrated from the front panel?

7.0 Panel Lighting
7.1 Is panel lighting employed?
7.2 Is panel lighting adequate?
7.3 Are panel indicator lights of the "press-to-test" variety?

8.0 Fuse Requirements
8.1 Are all line fuses mounted on the applicable front panel?
8.2 Are fuses replaceable without use of tools and without moving other parts?

8.3 Are fuse holders of the self-indicating type (light indication for blown fuse)?
8.4 Are fuse-holder caps of the quick-disconnect type rather than screw-in type?
8.5 Are spare fuses provided?
8.6 Are spare fuses located adjacent to the applicable line fuse?
8.7 Are fuses labeled both as to function and value? If space is limited, provide fuse value rather than fuse function.

9.0 Warning Lights Employed
9.1 Are warning lights employed when a malfunction has occurred or when equipment marginal operation exists?

10.0 Color of Indicating Lights
10.1 Are the following colors used for the applicable indications as stated?
White—indicates those conditions that are not identified to provide a right or wrong implication (i.e., power-on light).
Green—indicates that a unit is in tolerance or that a condition is satisfactory and all right to proceed (i.e., go-ahead, in tolerance, ready, acceptance, normal, etc.).
Yellow—operation in progress transitory condition, or a condition that exists which is marginal (i.e., adjustment required or tuning in process).
Red—equipment inoperable or dangerous condition exists.

11.0 Controls Placed According to Frequency of Use
11.1 Are emergency controls and displays placed in readily accessible positions with critical emergency controls and displays located in the optimum position?
11.2 Are primary controls aid displays placed in optimal manual and visual areas?

Test points

1.0 Front Panel Location
1.1 Are test-point locations (major, intermediate, minor) compatible with the requirements of MIL-STD-415?
1.2 Are major (primary) test points located on the front panel?

2.0 Functionally Grouped
2.1 Are test points grouped conveniently to allow for sequential testing (signal flow), testing of similar functions (power supply voltages, i-f amplifier stages, etc.), or frequency of use (when access is limited)?

3.0 Test-point Labeling
3.1 Is each test point labeled with the proper signal (with tolerance limits) that should be measured at the test point?
3.2 Is each test point identified with a unique number (i.e., TP-24)?

3.3 Are test points which are located in groups sequentially numbered from left to right, or from top to bottom.

3.4 Are test-point numbers located above each test point, and are signal values located on the right side of each test point?

3.5 Are test points color-coded so that they are easily distinguishable?

4.0 Internal Test Points Accessible

4.1 Are secondary test points readily accessible within each assembly? Are they brought out to a single point?

4.2 Are secondary test points which are not brought out to a single point adjacent to the item being tested?

5.0 Degree of Test

5.1 Are test points provided for direct test of all replaceable parts?

5.2 Are test points provided at the input and output for each major unit or assembly?

6.0 Adequately Protected

6.1 Are test points recessed or otherwise protected from damage by personnel, dust, or moisture?

7.0 Adequately Illuminated

7.1 Is adequate illumination provided to allow the technician to see the test-point number and labeled signal value?

7.2 Are luminescent markings used so test points can be read in very low illumination?

8.0 Adjacent to Applicable Control or Display

8.1 Are test points located adjacent to the controls and displays that are used with these test points?

ADJUSTMENTS

1.0 Adjustment Points Accessible

1.1 Are adjustment-point locations keyed to the maintenance level at which the adjustment will be made?

1.2 Are adjustment points required for overall equipment adjustment located on the front panel?

1.3 Are routine alignment/adjustment points located on the front panel so that individual equipment assemblies do not require removal?

1.4 For hermetically sealed units requiring periodic adjustment, are adjustment points externally located to permit adjustment without breaking the hermetic seal?

2.0 Periodic Adjustment Known

2.1 Are periodic adjustment requirements known? Is the extent of adjustment known?

2.2 Are periodic adjustment frequencies known?

2.3 Have equipment *internal* adjustments (after 10 min of warm-up time) been eliminated?

2.4 Have equipment calibration adjustments during the duty cycle been eliminated?

2.5 Can adjustments compensate for tolerance change?

3.0 Interaction Effects Eliminated

3.1 When maintenance adjustments are required, are single adjustments for a given function possible?

3.2 Have interaction effects been eliminated?

3.3 If interaction adjustments are required, they should be bench-type.

4.0 Adjustment Locking Devices

4.1 Are adjustment locking devices provided to preclude inadvertent changes to adjustment-point settings?

5.0 Factory Adjustments

5.1 Are factory adjustments specified?

5.2 Are critical adjustments which, if improperly made, could result in serious degradation to system performance or reliability identified as factory adjustments?

6.0 Adjustment-point Labeling

6.1 Is each adjustment point labeled with the adjustment-point number and the adjustment tolerances?

7.0 Fine Adjustments through Large Movements

7.1 Are fine adjustments caused by large adjustment increments?

7.2 Are stopping devices employed to prevent overriding the limit of adjustment?

8.0 Built-in Jacks for Meter Calibration

8.1 Are built-in jacks provided to allow for meter calibration with the meter in the installed position?

9.0 Clockwise Adjustments for Increasing Values

9.1 Does clockwise rotation of an adjustment point result in an increased adjustment value?

PARTS AND COMPONENTS

1.0 Grouping

1.1 Are parts and components functionally grouped?

1.2 Are parts such as resistors, capacitors, tube sockets, etc., mounted on subassemblies rather than on the unit chassis?

1.3 Are parts which are arranged in family groups outlined by painted borders or outlined in some other manner?

2.0 Labeling

2.1 Are parts and components adequately labeled (i.e., resistor number and resistor value)?

2.2 Are potted parts labeled with current, voltage, impedance, terminal information, etc.?

2.3 Are part/component labels adjacent to the applicable item, and visible after assembly?

3.0 Adequate Space for Tool Access

3.1 Is direct tool access provided to allow removal of a single part or component? This includes space required to solder a joint with a soldering iron and to remove or install a fastener with a screwdriver or socket wrench.

4.0 Individual Parts Directly Accessible
4.1 Are individual parts directly accessible (do not require the removal of other parts for access)?
4.2 If individual parts are not directly accessible, is the number of parts requiring removal to gain access minimized?

5.0 Delicate Parts
5.1 Are delicate parts located where they will not be damaged during maintenance operations? If not, are they adequately protected?

6.0 Part Vulnerability
6.1 Are parts far enough apart so that adjacent parts will not be damaged by a hot soldering iron or by the use of other tools?
6.2 Are leads and terminals far enough apart so that adjacent items will not be damaged by a hot soldering iron or by the use of other tools?
6.3 Is the clearance between the soldered connection and the part between ¼ and ½ in?

CABLES

1.0 Cables Fabricated in Removable Sections
1.1 Are cable harnesses designed so they can be fabricated in removable sections and installed as a unit?
1.2 Are cables routed so they are accessible and not behind panels that are difficult to remove?
1.3 Can cables routed through walls, cabinet holes, etc., be easily removed?

2.0 Cables Routed to Avoid Sharp Bends
2.1 Are cables routed to avoid sharp bends?
2.2 Are cables routed so they need not be bent sharply when connected or disconnected?

3.0 Cables Routed to Avoid Pinching
3.1 Are cables routed so they cannot be pinched by doors, rack/console drawers, panels, etc.?

4.0 Protection of Cables Routed through Holes
4.1 Are cables routed through holes, bulkheads, etc., adequately protected with grommets, pads, or in some other way?

5.0 Cable Labeling
5.1 Is cable labeling provided?
5.2 Are individual wire numbers (wires within the cable bundle) labeled on the outside of the cable bundle?

5.3 Are cable labels repeated (at 2-1 or 3-ft intervals) such that searching for cable identification is minimized?

5.4 Are different cable sections labeled differently (i.e., two sections of the same cable are labeled 101A and 101B respectively)?

6.0 Cable Clamping

6.1 Are cables adequately supported by cable clamps (not more than 24 in. apart)?

6.2 Are the clamps used of the type that will distribute pressure over a wide area of the cable?

6.3 Are the clamps used of the correct size so that the cable will not be squeezed too tightly and cause possible damage?

6.4 Are the clamps used of the quick-disconnect variety?

7.0 Handhold/Step-prevention Considered

7.1 Are cables routed so they will not be walked on or used for handholds?

CONNECTORS

1.0 Quick-disconnect Variety

1.1 Are quick-disconnect connectors used (no more than one full turn for connection/disconnection)? Quick-disconnect provisions should be incorporated when auxiliary test equipment is connected, when units require frequent disconnection or replacement, and when units require replacement within critical readiness times.

1.2 Are connectors removable by hand?

2.0 Are connectors far enough apart from other obstructions so they can be grasped firmly for connecting and disconnecting? In general, minimum separations are 0.75 in. for bare fingers or 1.25 in. for the bare hand or gloved fingers.

3.0 Connector Labeling

3.1 Are connectors and receptacles labeled?

3.2 Are plugs and their corresponding receptacles coded alike?

3.3 Do plugs and receptacles have painted arrows or other indications to show proper position of keys for aligning pins?

3.4 Are the pins in each plug clearly identified?

4.0 Connectors Keyed

4.1 Are plugs keyed or designed so that it is impossible to insert a plug in the wrong receptacle?

4.2 Do aligning pins or keys extend beyond electrical pins?

5.0 Connectors Standardized

5.1 Is the same style or type of connector used (i.e., BNC)? Standardization should not be carried to the extent where a plug can be inserted into the wrong receptacle.

6.0 Spare Pins Provided

6.1 Are spare pins provided in each plug? The total quantity of spare contact should be two for up to 35 contacts in the connector, four for 26 to 100 contacts in the connector, and six for 101 or more contacts in the connector.

7.0 Male Connectors Capped
- 7.1 Are male connectors capped to preclude damage and unsafe conditions?
- 7.2 Are connector caps chained to the equipment item to preclude possible loss?

8.0 Receptacles "Hot" and Plugs "Cold"
- 8.1 Is the equipment designed such that receptacles are "hot" and plugs are "cold"?

9.0 Moisture Prevention
- 9.1 Where applicable, have cable "drip-loops" been provided to preclude moisture from entering the connector?
- 9.2 Where possible, have moisture-proof connector provisions been considered?

SERVICING AND LUBRICATION

1.0 Servicing Requirements
- 1.1 Have servicing requirements been minimized?
- 1.2 When servicing is needed, are specific requirements identified?
- 1.3 Are servicing resources identified?

2.0 Servicing Points Accessible
- 2.1 Are servicing points accessible?

3.0 Servicing Frequencies
- 3.1 Are servicing frequencies known?

ENVIRONMENT

1.0 Temperature and Humidity Ranges
- 1.1 Have temperature and humidity requirements been considered?

2.0 Illumination
- 2.1 Is illumination adequate to enable maintenance operations? The average electronic shop maintains 20 footcandles, and small detailed-type maintenance work usually requires 50 footcandles. Good illumination should include suitable brightness, uniform lighting, suitable contrast between task and background, freedom from glare, and suitable quality and color for illuminants and surfaces.
- 2.2 Has built-in rack/console lighting been provided?

3.0 Transportability
- 3.1 Have transportation considerations been incorporated?
- 3.2 Is equipment designed for air transportation? Are attachment/tie-down provisions incorporated on the AGE rack or console outside packing cases? Are hoisting fittings provided?

3.3 Is labeling adequate? Are weight and hoist fitting labels provided?
3.4 Will equipment in shipping cases pass through AE Shop doorways or Avionics Shop hatches (45 in. high and 2 in. wide)?

4.0 Mobility
4.1 Have mobility provisions been considered?

5.0 Storage Conditions
5.1 Can the equipment be stored for extended periods of time without degrading affects?
5.2 Have special servicing requirements for stored equipment been minimized?
5.3 Have servicing requirements for stored equipment been defined?
5.4 Have servicing resources been identified?
5.5 Have storage environments been identified?

SAFETY

1.0 Electrical Outlets/Junction Boxes
1.1 Are all electrical outlets or junction boxes labeled as to the voltages contained therein?
1.2 Are electrical outlets or junction boxes designed or covered to preclude inadvertent hazardous contact?

2.0 Interlocks
2.1 Are interlocks provided where potentials exceed 40 volts?
2.2 Where interlocks are provided, is there a means of bypassing for servicing with a proper warning indicator?

3.0 Fuse/Circuit-breaker Protection
3.1 Are circuit breakers provided in lieu of fuses?
3.2 Is fuse/circuit-breaker protection adequate (protection on both sides of a line)?
3.3 Are spare fuses provided?
3.4 Are fuses or circuit breakers provided to safeguard against damage which might result from the wrong switch or jack position?

4.0 Warning Decals
4.1 Are adequate warning labels provided in high-voltage areas and in other areas where hazards exist?
4.2 Are labels placed adjacent to the hazardous area and in plain sight?

5.0 Guards/Safety Covers
5.1 Are guards, safety covers, and warning plates provided for high potentials (in excess of 350 volts) on contacts, terminals, and like devices?
5.2 On terminals that are used as test points, are adequate barriers between adjacent terminals provided to alleviate the possibility of a probe's slipping to the adjacent terminal?

6.0 Protruding Devices
- 6.1 Are unnecessary protruding devices eliminated?
- 6.2 Are delicate protruding parts suitably protected?
- 6.3 When performing bench maintenance, are legs or standoffs provided to protect parts?

7.0 External Metal Parts
- 7.1 Are all external metal parts adequately grounded?

8.0 Drawer/Panel/Structure Edges
- 8.1 Are drawer/panel/structure edges rounded?
- 8.2 Are accesses provided with fillets, rubber, fiber, or plastic protection to preclude personal injury?
- 8.3 Are covers on outside cases adequately grounded?

9.0 Tools
- 9.1 Are tools which must be used near high-voltage areas adequately insulated at the handle or at other parts of the tool which the technician is likely to touch?
- 9.2 Are screwdriver guides used for adjustments which are near high-voltage areas?

Reliability

The following questions concern maintainability but relate directly to reliability. For assistance in this area, refer to reliability-group personnel.

1.0 Allocated *MTBF*
- 1.1 Is the allocated or predicted *MTBF* known?

2.0 Fail-safe Provision
- 2.1 Are fail-safe features designed into the equipment where failures or malfunction will preclude injury to the operator or to the equipment itself?

3.0 Critical/Service Life
- 3.1 Is the equipment operating life known?
- 3.2 Is the equipment retirement life known?
- 3.3 Is the equipment shelf life known?

4.0 Wear-in/Wear-out Cycles
- 4.1 Has the wear-out cycle for components been considered?
- 4.2 Have parts subject to early wear-out been identified?

5.0 Failures Traceable by Test
- 5.1 Can every fault (degrading or catastrophic) which can possibly occur in the unit be detected by the use of the proposed test equipment and standard test procedures?

APPENDIX B

Maintainability-prediction Checklist

These data were extracted directly from MIL-HDBK-472, Maintainability Prediction

DESIGN CHECK LISTS AND SCORING CRITERIA:

The three Design Check Lists A, B, and C and their respective scoring criteria follow:

Check List A—Scoring Physical Design Factors

(1) *Access (External):* Determines if the external access is adequate for visual inspection and manipulative actions. Scoring will apply to external packaging as related to maintainability design concepts for ease of maintenance. This item is concerned with the design for external visual and manipulative actions which would precede internal maintenance actions. The following scores and scoring criteria will apply:

SCORES

(a) Access adequate both for visual and manipulative tasks (electrical and mechanical) 4

(b) Access adequate for visual, but no manipulative, tasks 2
(c) Access adequate for manipulative, but not visual, tasks .. 2
(d) Access not adequate for visual or manipulative tasks 0

SCORING CRITERIA

An explanation of the factors pertaining to the above scores is consecutively shown. This procedure is followed throughout for other scores and scoring criteria.

(a) To be scored when the external access, while visual and manipulative actions are being performed on the exterior of the subassembly, does not present difficulties because of obstructions (cables, panels, supports, etc.).

(b) To be scored when the external access is adequate (no delay) for visual inspection, but not for manipulative actions.
External screws, covers, panels, etc., can be located visually; however, external packaging or obstructions hinder manipulative actions (removal, tightening, replacement, etc.).

(c) To be scored when the external access is adequate (no delay) for manipulative actions, but not for visual inspections. This applies to the removal of external covers, panels, screws, cables, etc., which present no difficulties; however, their location does not easily permit visual inspection.

(d) To be scored when the external access is inadequate for both visual and manipulative tasks. External covers, panels, screws, cables, etc., cannot be easily removed nor visually inspected because of external packaging or location.

(2) *Latches and Fasteners (External):* Determine if the screws, clips, latches, or fasteners outside the assembly require special tools, or if significant time was consumed in the removal of such items. Scoring will relate external equipment packaging and hardware to maintainability design concepts. Time consumed with preliminary external disassembly will be proportional to the type of hardware and tools needed to release them and will be evaluated accordingly.

SCORES

(a) External latches and/or fasteners are captive, need no special tools, and require only a fraction of a turn for release 4
(b) External latches and/or fasteners meet two of the above three criteria 2
(c) External latches and/or fasteners meet one or none of the above three criteria 0

SCORING CRITERIA

(a) To be scored when external screws, latches, and fasteners are:
(1) Captive
(2) Do not require special tools
(3) Can be released with a fraction of a turn

Releasing a "DZUS" fastener which requires a 90 degree turn using a standard screw driver is an example of all three conditions.

(b) To be scored when external screws, latches, and fasteners meet two of the three conditions stated in (a) above. An action requiring an Allen wrench and several full turns for release shall be considered as meeting only one of the above requirements.

(c) To be scored when external screws, latches, and fasteners meet only one or none of the three conditions stated in (a) above.

(3) *Latches and Fasteners (Internal):* Determines if the internal screws, clips, fasteners or latches within the unit require special tools, or if significant time was consumed in the removal of such items. Scoring will relate internal equipment hardware to maintainability design concepts. The types of latches and fasteners in the equipment and standardization of these throughout the equipment shall tend to affect the task by reducing or increasing required time to remove and replace them. Consider "internal" latches and fasteners to be within the interior of the assembly.

SCORES

(a)	Internal latches and/or fasteners are captive, need no special tools, and require only a fraction of a turn for release	4
(b)	Internal latches and/or fasteners meet two of the above three criteria	2
(c)	Internal latches and/or fasteners meet one or none of the above three criteria	0

SCORING CRITERIA

(a) To be scored when internal screws, latches and fasteners are:
   (1) Captive
   (2) Do not require special tools
   (3) Can be released with a fraction of a turn

Releasing a "DZUS" fastener which requires a 90-degree turn using a standard screw driver would be an example of all three conditions.

(b) To be scored when internal screws, latches, and fasteners meet two of the three conditions stated in (a) above. A screw which is captive can be removed with a standard or Phillips screw driver, but requires several full turns for release.

(c) To be scored when internal screws, latches, and fasteners meet one of three conditions stated in (a) above. An action requiring an Allen wrench and several full turns for release shall be considered as meeting only one of the above requirements.

(4) *Access (Internal):* Determines if the internal access is adequate for visual inspection and manipulative actions. This item applies to internal packaging concepts in relation to design for ease of maintenance. Internal is to mean all work accomplished after gaining access to some portion of the equipment.

SCORES

(a) Access adequate both for visual and manipulative tasks (electrical and mechanical) 4
(b) Access adequate for visual, but manipulative, tasks 2
(c) Access adequate for manipulative, but not visual, tasks .. 2
(d) Access not adequate for visual or manipulative tasks 0

SCORING CRITERIA

(a) To be scored when the internal access, while performing manipulative or visual actions in a sub-assembly or unit, does not present difficulties because of the internal construction or part location.
(b) To be scored when the internal access is adequate (no delay) for visual inspection, but not for manipulative actions. Components and parts can readily be located visually during the maintenance task; however, internal construction or part location hampers manipulative actions (testing, removal, etc.).
(c) To be scored when the internal access is adequate for manipulative actions, but not for visual inspections. Components or parts can be easily tested or removed; however, their physical location does not easily permit visual inspection.
(d) To be scored when internal access is adequate for both visual and manipulative tasks. Components or parts cannot be easily tested or identified because of internal construction or location during the maintenance action.

(5) *Packaging:* Determines the access (within the subassembly) to components or parts requiring mechanical disassembly. This question concerns itself with the internal packaging of parts relative to the maintenance action. Current design concepts have been concentrated on module type packaging; however, even these vary in a mechanically held module, while others are plug-in type only. This item deals with the mechanical problems involved in gaining access to failed components or parts.

SCORES

(a) Internal access to components and parts can be made with no mechanical disassembly 4
(b) Little disassembly required (less than 3 min.) 2
(c) Considerable disassembly is required (more than 3 min.) .. 0

SCORING CRITERIA

(a) To be scored when less than one minute is required to gain access to the failed components or part.
(b) To be scored when less than three minutes is expended in gaining access to the failed component or part.
(c) To be scored when more than three minutes is expended in gaining access to the failed component or part.

(6) *Units-Parts (Failed):* Determines the manner in which units or parts are removed or replaced during the maintenance action. Since units and parts are electrically and/or mechanically secured in equipments in many different ways, the time to remove such items varies considerably. Mechanically held items include tubes protected from vibrations by special shields or clamps, printed boards clipped into their sockets, and parts and components held by brackets. Soldered items include resistors, capacitors, etc.

(a) Units or parts of plug-in nature 4
(b) Units or parts of plug-in nature and mechanically held 2
(c) Units of solder-in nature 2
(d) Units of solder-in nature and mechanically held 0

SCORING CRITERIA

(a) To be scored when units or parts are plug-in types requiring only to be pulled out. Plug-in type parts such as tubes, some relays, crystals, etc., would be included in this category.
(b) To be scored when units or parts are plug-in types, but are mechanically held by clips, shields, clamps, etc. Also applies to maintenance requiring the removal of a tube having external grid or plate connections, anti-vibration shields, etc.
(c) To be scored when units or parts are soldered-in types such as resistors, capacitors, etc., when the removal of parts requires the unsoldering of part terminations.
(d) To be scored when units or parts are soldered-in mechanically held types such as transformers, jacks, etc. The removal or replacement of parts requires mechanical disassembly and unsoldering.

(7) *Visual Displays:* Determines if sufficient visual information pertaining to the equipment malfunction is displayed within one area or unit. Circuit indicators and meters provide, to some extent, symptom analysis. Therefore, it is important that these indications be displayed within one area to ensure rapid analysis and action. If several areas must be consulted before a qualified estimation of the difficulty can be made, much time is required.

SCORES

(a) Sufficient visual information on the equipment is given within one display area 4
(b) Two display areas must be consulted to obtain sufficient visual information 2
(c) More than two areas must be consulted to obtain sufficient visual information 0

SCORING CRITERIA

(a) To be scored when visual information associated with the fault or malfunction is displayed with one area. Applicable if diagnosis and repair can be accomplished successfully following symptoms derived from one display area or subassembly of the system.

(b) To be scored when two display areas must be consulted to provide visual information associated with the fault or malfunction. Two separate display areas on the system (meter panel and fault indicators) must be consulted to diagnose malfunctions successfully.

(c) To be scored when more than two areas or subassemblies must be consulted to provide visual information associated with the fault or malfunction. This would be indicative of a least maintainable condition.

(8) *Fault and Operation Indicators (Built-in Test Equipment):* Determines if an equipment malfunction or fault is clearly discernible via audible alarms, indicators, etc., and that such information is clearly presented for rapid maintenance action. The use of indicators is increased as complexity increases, and equipment availability becomes more important. Although visual and audio alarms usually indicate that a problem exists, they do not always determine the exact location of the malfunction. The more precise the indication, the better the maintenance condition.

SCORES

(a) Fault or malfunction information is provided clearly and for rapid action 4

(b) Fault or malfunction information clearly presented, but requires operator interpretation 2

(c) Fault or malfunction information requires no operator interpretation, but is not clearly presented 2

(d) Fault or malfunction information not clearly presented and requires operator interpretation 0

SCORING CRITERIA

(a) To be scored when an equipment fault or malfunction occurs and is evidenced by alarms, indicators, etc., which provide for rapid diagnosis and maintenance action. An example of this would be when a power supply failure occurs because of an open fuse which is pointed out by an indicator or alarm.

(b) To be scored when an equipment fault or malfunction occurs and is evidenced by alarms, indicators, etc., but requires further tests for isolation of the fault. Loss of output power is evidenced by an alarm; however, further diagnosis must be made to determine the exact cause of trouble.

(c) To be scored when an equipment fault or malfunction occurs and is not clearly determined by alarms, indicators, etc., however, provisions for rapid diagnosis and maintenance action are available. Applies when some preliminary testing might be required to determine if a fault or malfunction such as the loss of some voltage (B+, Bias, etc.) exists. Once determined, however, maintenance is expedited, such as in the case of an open fuse.

(d) To be scored when an equipment fault or malfunction occurs and is not clearly discernible, and which requires symptom interpretation. Testing is also necessary to determine the equipment status and cause of failures.

(9) *Test Points (Availability):* Determine if test points are available for needed tests pertaining to maintenance action. A test point shall be considered as any test probe receptacle where specific system operation data can be obtained. This definition eliminates as test points connector pins on printed circuit boards, terminals, tube pins, etc. The number of test points available and the amount of information yielded will affect the time to establish the cause and location of fault.

SCORES

(a) Task did not require use of test points 4
(b) Test points available for all needed tests 3
(c) Test points available for most needed tests 2
(d) Test points not available for most needed tests 0

SCORING CRITERIA

(a) To be scored when the maintenance action did not require the use of test points, but when, instead, the malfunction can be diagnosed and repaired via built-in test equipment.

(b) To be scored when all needed tests were accomplished at test points. Sufficient information to diagnose and repair the trouble was available at test points.

(c) To be scored when at least 51% of the required tests were accomplished at test points. Troubleshooting required that several separate tests, most of which made use of test points, had to be made.

(d) To be scored when the majority of needed tests were not accomplished at test points. Malfunction diagnosis and repair required the making of tests for which few or no test points were available.

(10) *Test Points (Identification):* Determines if all test points required during the maintenance action are properly identified by circuit symbol and pertinent test data. This precise information provides diagnostic data to aid in troubleshooting the malfunction.

SCORES

(a) All test points are identified with required readings given . . 4
(b) Some are suitably marked 2
(c) Points are not marked and test data are not given 0

SCORING CRITERIA

(a) To be scored when all test points needed for task completion are identified (circuit symbol) with required readings given (+6VDC, − 18VDC, 115VAC, etc.). This is indicative of a best maintainable condition.

(b) To be scored when the majority of test points required for task completion are suitably identified.

(c) To be scored when test points required for task completion are not suitably identified. Troubleshooting at test points is a cause for delay because required voltage readings, signal characteristics, etc., are not specified. This would indicate that a least maintainable condition exists.

(11) *Labelling:* Determines if parts associated with the maintenance actions are identified with respect to circuit symbol and part identification. Proper identification of parts can be an important asset to the maintenance task in that, if part circuit number is omitted from the equipment, considerable time could be wasted tracing the circuit to identify it. Similarly, if information is "hidden," requiring removal of other parts to read it, much time will be consumed.

SCORES

(a) All parts labelled with full identifying information, and all identifying information clearly visible 4

(b) All parts labelled with full identifying information, but some information hidden 2

(c) All information visible, but some parts not fully identified.. 2

(d) Some information hidden and some parts not fully identified 0

SCORING CRITERIA

(a) To be scored when all parts associated with the maintenance action are identified and this information is clearly visible. To include testing or removing of parts that are clearly identified (V401-6BE6) or (R-1225-400Ω).

(b) Applies when all parts associated with the maintenance action are identified, but some of this information is not visible. Applies to testing or removing parts that are labelled, but which information is hidden by obstructions.

(c) Applicable when all circuit symbols are visible, but some parts associated with the tasks are not identified. Parts required for testing or removal are not identified with reference to part value, etc.

(d) To be scored when some parts associated with the maintenance task contain hidden circuit symbols and are not fully identified. Parts required in testing or removal are not identified and information is also hidden.

(12) *Adjustments:* Determines if adjustments such as tuning and alignment are required, after a maintenance action, to make the equipment operate according to specifications. An adjustment will be any action which resets or changes variable components such as potentiometers, variable capacitors, slug-tuned coils, etc., whereby the operation of the system assembly or subassembly is affected. These actions, depending upon their criticality and frequency, will affect the overall maintenance time.

SCORES

(a) No adjustments or realignment are necessary to place equipment back in operation 4
(b) A few adjustments, but no major realignments are required 2
(c) Many adjustments or major realignments must be made. . 0

SCORING CRITERIA

(a) To be scored when no adjustments are required to bring the equipment back to normal operating specifications. Applies to repair of the malfunction, if the equipment need only be turned on.
(b) To be scored when a few adjustments of a minor nature are required to place equipment back in operation according to specifications.
(c) To be scored when many adjustments (time-consuming) or a major tuning or alignment is required to place equipment back to normal operating specifications.

(13) *Testing (in Circuit):* Determines if the defective component or part can be tested without removal from the circuit. This question is based on the nature of the equipment and the repair concepts associated with the particular design.

SCORES

(a) Defective part or component can be determined without removal from the circuit 4
(b) Testing requires removal 0

SCORING CRITERIA

(a) Applicable when the component or part can be decisively determined as being defective without removal of any part from the circuit.
(b) To be scored when the component or part must be removed from the circuit to be decisively determined as defective. When testing has isolated the trouble to a particular part or component, however, a definite opinion cannot be made until such part or component is electrically or physically removed from the circuit for further testing.

(14) *Protective Devices:* Encompasses equipment design provisions for self-protection against damage to components or parts after a malfunction has occurred. If a system has protection devices such as fuses, circuit breakers, etc., then the equipment can be protected from further damage as well as aiding in isolating the malfunction. If no provisions have been made, further damage and increased repair time could result.

SCORES

(a) Equipment was automatically kept from operating after malfunction occurred to prevent further damage. (This refers to malfunction of such areas as bias supplies, keep-alive voltages, etc.) 4
(b) Indicators warned that malfunction has occurred 2
(c) No provisions have been made 0

SCORING CRITERIA

(a) To be scored when automatic shut-off devices protected parts or components from further damage after a malfunction occurred in a critical area. A typical example of such a malfunction would be if the Bias supply fails and B+ voltage is automatically cut off by circuit breakers, fuses, or relay action.
(b) To be scored when automatic shut-off-devices do not protect parts or components from further damage, but when visual indicators or audible alarms warn personnel of the situation.
(c) To be scored when a critical malfunction occurs and parts or components are not protected by automatic shut-off devices, indicators, or alarms. Involves malfunction which damages parts or components because automatic shut-off devices or alarms were not provided.

(15) *Safety (Personnel)*: Determines if the maintenance action requires personnel to work under hazardous conditions such as close proximity to high voltage, radiation, moving parts, high-temperature components, or on elevated structures, etc.

SCORES

(a) Task did not require work to be performed in close proximity to hazardous conditions (high voltage, radiation, moving parts and/or high temperature parts) 4
(b) Some delay encountered because of precautions taken 2
(c) Considerable time consumed because of hazardous conditions 0

SCORING CRITERIA

(a) To be scored when the maintenance action did not require personnel to work under hazardous conditions. The maintenance action did not require precautions to be taken, in that the task was not associated with high voltage, moving parts, etc.

(b) To be scored when precautions were taken because of hazardous conditions causing slight delays in the maintenance action. A typical example would be when a shorting probe must be used to discharge high-voltage capacitors.

(c) To be scored when precautions taken because of hazardous conditions caused a considerable delay to the maintenance action. Maintenance required that testing be done in close proximity to high voltage where extreme caution was necessary, or the closeness of moving parts (gears, motors, etc.), caused delay because of precautions taken.

Check List B—Scoring Design Dictates-Facilities

The intent of this questionnaire is to determine the need for external facilities. Facilities, as used here, include material such as test equipment, connectors, etc., and technical assistance from other maintenance personnel, supervisor, etc.

(1) *External Test Equipment:* Determines if external test equipment is required to complete the maintenance action. The type of repair considered maintainably ideal would be one which did not require the use of external test equipment. It follows, then, that a maintenance task requiring test equipment would involve more task time for set-up and adjustment and should receive a lower maintenance evaluation score.

SCORES

(a)	Task accomplishment does not require the use of external test equipment	4
(b)	One piece of test equipment is needed	2
(c)	Several pieces (2 or 3) of test equipment are needed	1
(d)	Four or more items are required	0

SCORING CRITERIA

(a) To be scored when the maintenance action does not require the use of external test equipment. Applicable when the cause of malfunction is easily detected by inspection or built-in test equipment.

(b) To be scored when one piece of test equipment was required to complete the maintenance action. Sufficient information was available through the use of one piece of external test equipment for adequate repair of the malfunction.

(c) To be scored when 2 or 3 pieces of external test equipment are required to complete the maintenance action. This type of malfunction would be complex enough to require testing in a number of areas with different test equipments.

(d) To be scored when four or more pieces of test equipment are required to complete the maintenance action. Involves an extensive testing requirement to locate the malfunction. This would indicate that a least maintainable condition exists.

(2) *Connectors:* Determines if supplementary test equipment requires special fittings, special tools, or adapters to adequately perform tests on the electronic system or sub-system. During troubleshooting of electronic systems, the minimum need for test equipment adapters or connectors indicates that a better maintainable condition exists.

SCORES

(a) Connectors to test equipment require no special tools, fittings, or adapters 4
(b) Connectors to test equipment require some special tools, fittings, or adapters (less than two) 2
(c) Connectors to test equipment require special tools, fittings, and adapters (more than two) 0

SCORING CRITERIA

(a) To be scored when special fittings or adapters and special tools are not required for testing. This would apply to tests requiring regular test leads (probes or alligator clips) which can be plugged into or otherwise secured to the test equipment binding post.
(b) Applies when one special fitting, adapter or tool is required for testing. An example would be if testing had to be accomplished using a 10 DB attenuator pad in series with the test set.
(c) To be scored when more than one special fitting, adapter, or tools is required for testing. An example would be when testing requires the use of an adapter and an RF attenuator.

(3) *Jigs or Fixtures:* Determines if supplementary materials such as block and tackle, braces, dollies, ladder, etc., are required to complete the maintenance action. The use of such items during maintenance would indicate the expenditure of a major maintenance time and pinpoint specific deficiencies in the design for maintainability.

SCORES

(a) No supplementary materials are needed to perform task 4
(b) No more than one piece of supplementary material is needed to perform task 2
(c) Two or more pieces of supplementary material are needed . . 0

SCORING CRITERIA

(a) To be scored when no supplementary materials (block and tackle, braces, dollies, ladder, etc.) are required to complete maintenance. Applies when the maintenance action consists of normal testings and the removal or replacement of parts or components can be accomplished by hand, using standard tools.

(b) To be scored when one supplementary material is required to complete maintenance. Applies when testing or when the removal and replacement of parts requires a step ladder for access or a dolly for transportation.

(c) To be scored when more than one supplementary material is required to complete maintenance. Concerns the maintenance action requiring a step ladder and dolly adequately to test and remove the replaced parts.

(4) *Visual Contact:* Determines if the nature of the equipment, location, or maintenance action causes the members of a team to be hidden from the view of each other at times during the task.

SCORES

(a) The activities of each member are always visible to the other member 4

(b) On at least one occasion, one member can see the second, but the reverse is not the case 2

(c) The activities of one member are hidden from the view of the other on more than one occasion 0

SCORING CRITERIA

(a) Applies when the team members are visible to each other during the entire maintenance action.

(b) To be scored if one member of the team becomes hidden from view of the other member or members during the maintenance action.

(c) Applicable if team members are hidden from view on more than one occasion.

(5) *Assistance (Operations Personnel):* Determines whether or not information or assistance from operations personnel is required, and if required, to what extent.

SCORES

(a) Task did not require consultation with operations personnel 4

(b) Some contact was required 2

(c) Considerable coordination required 0

SCORING CRITERIA

(a) To be scored when the maintenance action does not require the assistance of operations personnel. This would apply if physical or verbal aid to the technical personnel was not required. (Less than one minute.)

(b) To be scored when the maintenance action requires a small amount of assistance from operations personnel. (One to five minutes.)

(c) To be scored when the maintenance action requires considerable assistance from operation personnel in the operation or repair of the malfunctioning equipment. (Over five minutes.)

(6) *Assistance (Technical Personnel):* Determine the number of technical personnel required to complete the maintenance action, not including administrative or operations type personnel.

SCORES

(a)	Task required only one technician for completion	4
(b)	Two technicians were required	2
(c)	Over two were used	0

SCORING CRITERIA

(a) To be scored when only one technician was required to complete the maintenance action.
(b) To be scored when two technicians were required to complete the maintenance action.
(c) To be scored when more than two technicians were required to complete the maintenance action.

(7) *Assistance (Supervisors or Contractor Personnel):* Determines whether or not the services of supervisor or contractor personnel (TECH. REPS.) were required to complete the maintenance action and the extent of their participation in the task.

SCORES

(a)	Task completion did not require consultation with supervisor or contract personnel	4
(b)	Some help needed	2
(c)	Considerable assistance needed	0

SCORING CRITERIA

(a) To be scored when no supervisor or contractor personnel are consulted during the maintenance action.
(b) To be scored when a small amount of assistance from supervisor or contractor personnel is required to complete the maintenance action.
(c) To be scored when considerable assistance from supervisor or contractor personnel is required to complete the maintenance action.

Check List C—Scoring Design Dictates-Maintenance Skills

This check list evaluates the personnel requirements relating to physical, mental, and attitude characteristics, as imposed by the maintenance task.

Evaluation procedure for this check list can best be explained by way of several examples. Consider first question which deals with arm, leg and back strength. Should a particular task require the removal of an equipment drawer weighing 100 pounds, this would impose a severe requirement on this characteristic. Hence, in this case the question would be given a low score (0 to 1). Assume another task which, due to small size and delicate construction, required extremely careful handling. Here question 1 would be given a high score (4), but the question dealing with eye-hand coordination and dexterity would be given a low score. Other questions in the check list relate to various personnel characteristics important to maintenance task accomplishment. In completing the check list, the task requirements for each of these characteristics should be viewed with respect to average technician capabilities.

SCORES

	Score
1. Arm, Leg, and Back Strength	______
2. Endurance and Energy	______
3. Eye-Hand Coordination, Manual Dexterity, and Neatness	______
4. Visual Acuity	______
5. Logical Analysis	______
6. Memory—Things and Ideas	______
7. Planfulness and Resourcefulness	______
8. Alertness, Cautiousness, and Accuracy	______
9. Concentration, Persistence and Patience	______
10. Initiative and Incisiveness	______

SCORING CRITERIA

Quantitative evaluation of these items range from 0 to 4 and are defined in the following manner:

4. The maintenance action requires a minimum effort on the part of the technician.
3. The maintenance action requires a *below average* effort on the part of the technician.
2. The maintenance action requires an *average* effort on the part of the technician.
1. The maintenance action requires an *above average* effort on the part of the technician.
0. The maintenance action requires a maximum effort on the part of the technician.

These criteria will be used in scoring the following specific divisions of physical, mental, and motor requirements.

(1) *Arm, leg and back strength:* Determines the degree of arm, leg, and back strength required to complete the maintenance action. Refers to any effort, no matter how minimal. Varying degrees of strength are required for various maintenance actions are related to equipment design.

(2) *Endurance and energy:* Determines the degree of endurance and energy required to complete the maintenance action. Endurance might be referred to as the physical counterpart of patience, where a sustained physical effort is required. Energy required to complete

the maintenance action when the task requires vigorous activity or exertion by the technician is also assessed. This applies to the necessity of lifting and carrying heavy assemblies, tools, or parts.

(3) *Eye-hand coordination, manual dexterity, and neatness:* Determines the degree of eye-hand coordination required to complete the maintenance action. Refers to any act involving the use of the eyes while manipulating the hands to accomplish the same action. This type of action would be applicable mostly in testing and measuring activities; however, it is not inconceivable that this item would also be applicable in other areas of the maintenance action. Scoring shall be proportional to the degree or the intensity of the requirements of the task.

Determines the degree of manual dexterity required to complete the maintenance action. When the skilled use of the hands is required to accomplish the task, appropriate degrees of necessity shall be established. Those types of actions involving manual dexterity would more naturally apply to the repair, assembly, or disassembly of equipments rather than the troubleshooting processes.

Also determines the degree of neatness required by the maintenance action. Applies specifically to the requirement of the actual repair where tidiness is of prime importance to accomplish the task adequately. Since equipment is designed and constructed in accordance with quality control specifications, it is important to consider the care which has to be exercised during a particular repair.

(4) *Visual Acuity:* Determines the degree of visual acuity required to complete the maintenance task. When the maintenance action is such that the visual accuracy of the technician is required to accomplish the task, a degree of requirement shall be established. Such actions shall include the need for accurate and precise visual activity in finding indications of trouble, faulty components, or the visual sensitivity sometimes necessary in reading certain oscilloscope presentations.

(5) *Logical Analysis:* Determines the degree of logical analysis required to complete the maintenance action. Refers to the need for involved logical analysis or for extensive mental reasoning to determine the origin of the fault or malfunction. If the problem is such that it requires orientation on the logical signal sequence, then this shall also be considered as part of this question.

(6) *Memory: Things and ideas:* Determines the degree to which the maintenance action requires a knowledge of the equipment past history with reference to component or part failure, tools to be used, and sequences to be followed (assembly, disassembly, etc.).

Also determines the degree to which the maintenance action requires a previous knowledge of the equipment. Refers to the degree that the task requires recall of concepts or principles of operation, function and operation of circuits and parts, or electronic theory and maintenance procedures.

(7) *Planfulness and resourcefulness:* Determines the degree of planning

required to complete the maintenance action successfully. Refers to the extent to which the task requires a planned and methodical approach to assure rapid diagnosis and repair of the equipment fault or malfunction.

Also determines the degree of resourcefulness required to complete the maintenance action. Refers to the capabilities necessary in dealing with a situation or in meeting difficulties pertaining to the diagnosis and repair of the equipment. Conditions sometimes exist where certain needed materials such as tools, test equipment, or technical publications are not available, although substitution is possible, by some improvised method, to accomplish the task adequately.

(8) *Alertness, cautiousness, and accuracy:* Alertness is a readiness or promptness in comprehending and a keen awareness and knowledge of all events or factors affecting the maintenance action. Cautiousness is the exercise of forethought so that risks may be avoided or minimized during the maintenance action. (A surveyance of all possible consequences before making a decision.) Accuracy is attained by the exercise of care by showing close attention to the details of the maintenance task and cautiousness in avoiding errors. The design requirements for these characteristics are to be assessed.

(9) *Concentration, persistence, and patience:* Concentration is the close mental application or exclusive attention to the maintenance task and the direct focusing of the mind upon one thing to the exclusion of everything else. Persistence refers to maintenance tasks with the implication of being able to carry performance to a successful conclusion. Patience is the quiet perseverance, calmness in working, and being undisturbed by obstacles, delays, or failures which might occur during the maintenance task.

(10) *Initiative and incisiveness:* Initiative is the energy or aptitude displayed in the initiation of action and the ability or power to introduce a new measure or course of action. Incisiveness is the keenness of mind and acuteness of understanding the task at hand.

APPENDIX C

Instructions for Maintainability-task Analysis

1.0 GENERAL

The maintainability-task-analysis form (Fig. 13-4) commands a great deal of emphasis as it defines maintenance tasks, task times, task frequencies, personnel types and skill levels for task completion, replacement parts, and support-equipment/tool data. This appendix is designed to cover instructions for completion of the form.

2.0 DATA COVERAGE

A maintainability-task analysis shall be completed to cover:

1. All repairable maintenance significant items (i.e., end item, assembly, subassembly, etc.)
2. All corrective- and preventive-maintenance requirements (i.e., troubleshooting and fault isolation, disassembly, remove and replace, repair, reassembly, alignment and adjustment, functional test, inspection, servicing, calibration, and overhaul)

3.0 INSTRUCTIONS FOR FORM COMPLETION

The following instructions should be employed when completing the maintainability-task-analysis form. (Completed forms are illustrated in Fig. 13-4 (Chap. 13) and Figs. C-1, C-2, C-3, and C-5 of this appendix.)

System. Enter the proper nomenclature for the end item covered by the overall analysis.

Block 1—Part Number/Nomenclature. Enter the manufacturer's part number and nomenclature for the item being covered by the specific task-analysis sheet. The item may be a subsystem, unit, assembly, subassembly, etc.

Next Assembly. Enter the next higher assembly by part number and nomenclature. If the task-analysis form covers the *subsystem* level, the system nomenclature is entered.

Block 2—MA Control Number. Enter the maintainability-analysis (MA) control number for the item being covered by the task-analysis form. The maintainability-analysis control number is a numerical identification of maintenance-significant items. It may follow the work-unit code system (Air Force), the MEAR control-number system (Navy), or a contractor's designated numbering system. MA control numbers are assigned by indenture level of maintenance. The following listing is presented as an example:

Equipment Item	MA Control Number	Functional Level
Aircraft	A000 000	System
XYZ Subsystem	A840 000	Subsystem
Unit *A*	A841 000	Unit
IF Amplifier	A841 210	Assembly
Circuit Board No. 1A4	A841 211	Subassembly

MA control numbers are assigned by function and not necessarily by part number. If a modification of unit *A* is accomplished or an alternate configuration of unit *A* is employed, the manufacturer's part number changes, but the MA control number remains the same as long as the basic function of the applicable item is unchanged.

The intent of the MA control number is to serve as a *reference number* for maintenance-requirements identification. With a number assigned to each maintenance-significant item (e.g., i-f amplifier) and all of the maintenance requirements on the task-analysis form being identified against that item, a computer program can readily select that item plus its support requirements from a large volume of data. In addition, when a system is operationally deployed, maintenance time charged to a particular item can be easily identified through the MA control number. Another example involves the requirements for supporting specific aircraft subsystems (deemed critical to the mission) when the aircraft is deployed to a remote operating base. Anticipated support requirements can readily be identified through the MA control-number system. All maintenance actions can be identified by a particular MA control number.

Block 3—Requirement Number. The requirement number is a method to identify for electronic data-processing purposes each requirement that is accomplished on an item. The requirements, identified in block 4, may be sequentially numbered from 01 to 99, as necessary, for each MA control-number item.

MAINTAINABILITY TASK ANALYSIS

GENERAL DYNAMICS | ELECTRONICS

SYSTEM XYZ SYSTEM

1. PART NUMBER/NOMENCLATURE A58180/COMPUTER ASSY — NEXT ASSEMBLY A37701/UNIT A — 2. MA CONTROL NUMBER A84220

3. REQ. NO.	4. REQUIREMENT	5. CODE	6. TECHNICAL JUSTIFICATION	7. REMARKS
01	TROUBLESHOOT	HI	A MALFUNCTION HAS BEEN ISOLATED TO THE COMPUTER ASSEMBLY WITHIN UNIT A. TROUBLESHOOTING OF THE COMPUTER ASSEMBLY IS REQUIRED FOR CORRECTIVE MAINT.	△ INCLUDES WEIGHTED AVERAGE FOR "9" TASKS. REFER TO TROUBLESHOOTING LOGIC FLOW DIAGRAM.
8. MTBF	**9. OPERATING LIFE**	**10. REPL. INT.**		
1156	4000 HR.	578		
11. REFERENCES				
A84220, REQ. NO. 02				

12. MAINT. LEVEL AND TASK NO.	13. TASK/DESCRIPTION	14. ELAPSED TIME - MINUTES	15. ELAPSED TIME TOTAL/TASK	16. JOB SPECIALTY OR RATING	17. BASIC SKILL	18. INTERMED SKILL	19. SUPER SKILL	20. TOTAL	21. QTY/ASSY	22. PART NO. / 27. PART NAME	23. MAINT QTY FACTOR	24. SHELF LIFE	25. TOOL NO. / 28. TOOL NAME	26. TOOL USE
0010	GAIN ACCESS	①	003	31350		003		003						
	REMOVE CIRCUIT BOARDS 1A4 1A5 1A6													
	TO GAIN ACCESS TO COMPUTER ASSY													
0020	PREPARE FOR TROUBLESHOOTING	①	002	31350		002		002					PS2001	014
	CONNECT JUMPER WIRES & APPLY	②		31330	002			002					POWER SUPPLY	
	BIAS VOLTAGE												GX1001	014
													VOLTMETER	
9030	ISOLATE TO CIRCUIT	①	006	31350		006		006					XY1002A	006
	BOARD 2A3 (0.00050)	②		31330	006			006					OSCILLOSCOPE	
9040	ISOLATE TO CIRCUIT	①	011	31350		011		011					XY1002A	011
	BOARD 2A4 (0.00060)	②		31330	011								OSCILLOSCOPE	
9050	ISOLATE TO CIRCUIT	①	008	31350		008		008					XY1002A	008
	BOARD 2A5 (0.00030)	②		31330	008			008					OSCILLOSCOPE	
9060	ISOLATE TO CIRCUIT	①	013	31350		013		013					XY1002A	013
	BOARD 2A6 (0.00033)	②		31330	013			013					OSCILLOSCOPE	
	REF: A84220 REQ. NO. 02													
	REPAIR													

CHANGE NO. 2

JOB SPEC.-NO.: 29. 31350 30. 001 31. 31330 32. 001 33. 34. 40. 41. 42. 43. 44. 45.

35. 014 △ TOTAL ELAPSED TIME/REQ'MT

36. 040 37. 043 38. 39. 083 TOTAL MAN-MINUTES

46. 001 47. 001 48. 49. 002 TOTAL NO. MEN

SHEET 1 OF 1

PREPARED BY J. JONES

DATE 27 APRIL 1966

FORM 1366F 12/63

11:4.0

FIG. C-1 Maintainability-test Analysis (Example). *(General Dynamics Corporation.)*

Block 4—Requirement. The requirement name should be entered in this block. Generally, maintenance requirements are expressed as one of the following types of maintenance actions:

1. SERVICING—The maintenance operation of servicing may require the application of a lubricant, oxygen, fuel, etc., by means of special equipment. Servicing may require removal, disassembly, reassembly, adjustment, installation, or a lesser number of these tasks. Servicing is not a requirement when included as a task under a requirement to repair, replace, or install an item.
2. INSPECTION—Inspection is a maintenance requirement when the basic objective is to ensure that a requisite condition or quality exists. In order to inspect for the desired condition, it may be necessary to remove the item, to gain access by removing other items, or partially to dissassemble the item for inspection purposes. In such cases, these associated actions necessary to accomplish the required inspection would be specific tasks.
3. ADJUSTMENT/ALIGNMENT—Adjustment/alignment is a maintenance requirement when the primary cause of the maintenance action is to adjust or align, or to verify adjustment/alignment of specific equipment. Adjustment/alignment accomplished subsequent to repair of a given item is not considered as a separate requirement (it is included as a task in the repair requirement).
4. FUNCTIONAL TEST—Functional test constitutes a system or subsystem operational checkout either as a condition verification after the accomplishment of corrective-maintenance action or as a scheduled requirement on a periodic basis. The purpose for accomplishment (whether scheduled or unscheduled) is differentiated by the code in block 5.
5. REMOVE AND REINSTALL—Remove and reinstall is a maintenance requirement when an item is removed for any reason, and the same item is reinstalled.
6. REMOVE—Remove is a maintenance requirement when the basic objective is to remove the item from the next higher assembly. This requirement is usually applied for a configuration change (e.g., change in weapon load for an aircraft).
7. REMOVE AND REPLACE—This requirement constitutes the removal of one item and replacement with another like item. Such action can result from a failure or from a scheduled action (e.g., forced removal). The reason for removal and replacement is identified through the code in block 5.
8. TROUBLESHOOTING—Troubleshooting is a maintenance operation which involves the logical process (series of tasks) which leads to the positive identification of the cause of a malfunction. Troubleshooting is best illustrated through the logic troubleshooting-flow diagrams illustrated in the text (see Fig. 13-3 for an example).
9. REPAIR—When the maintenance operation involves a series of

specific tasks required to return the item to a serviceable condition, the basic maintenance requirement is repair. Repair normally involves the replacement of a part, module, or subassembly.

10. OVERHAUL—Overhaul is a maintenance requirement whenever an item is completely disassembled, refurbished, tested, and returned to a serviceable condition meeting all requirements set forth in applicable specifications. Overhaul may result from either a scheduled or unscheduled requirement and is generally accomplished at the depot level of maintenance or at the contractor's factory.
11. CALIBRATION—Calibration is a maintenance requirement whenever an item is checked against a "working standard," a "secondary standard," or a "primary standard." Calibration generally applies to precision measurement equipment (PME) and can be accomplished either on a scheduled basis or on an unscheduled basis subsequent to the accomplishment of corrective maintenance on a PME item.

Whenever one of the above requirements is the basic underlying cause, purpose, or objective to be satisfied, the operation becomes a maintenance requirement.

Block 5—Code. The requirement identified in Block 4 is coded as follows:

SERVICING	AX	REMOVE & REPLACE	GX
INSPECTION	BX	TROUBLESHOOTING	HX
ADJUSTMENT/ALIGNMENT	CX	REPAIR	IX
FUNCTIONAL TEST	DX	OVERHAUL	JX
REMOVE & REINSTALL	EX	CALIBRATION	KX
REMOVE.	FX		

The "X" factor in the above code is replaced by "O," "I," or "D" depending on the level of maintenance where the requirement is accomplished. These levels are:

Organizational maintenance . . . . O
Intermediate (field) maintenance I
Depot maintenance . . . . D

For example, if a repair requirement is to be accomplished at depot, the code is ID. (The levels of maintenance are discussed in Chap. 5 of this text.)

Block 6—Technical Justification. The maintainability engineer shall state in this space the technical considerations which justify the maintenance requirement. The need for performing the requirement must be clearly established. In the event that such a requirement is based on a special study, reliability data, or equivalent, the applicable documentation source shall be referenced.

Block 7—Remarks. Include notes, symbols (△2), or other abbre-

MAINTAINABILITY TASK ANALYSIS

GENERAL DYNAMICS | ELECTRONICS

SYSTEM XYZ SYSTEM

1. PART NUMBER/NOMENCLATURE A58180/COMPUTER ASSY — NEXT ASSEMBLY A37701/UNIT A — 2. MA CONTROL NUMBER A84220

3. REQ. NO.	4. REQUIREMENT	5. CODE	6. TECHNICAL JUSTIFICATION	7. REMARKS
02	REPAIR	II	A FAULTY COMPUTER ASSEMBLY MUST BE REPAIRED TO RESTORE UNIT A TO FULL OPERATIONAL STATUS	△1 WEIGHTED AVERAGE OF "9" TASKS IS 3.36 OR 3 MINUTES.
8. MTBF	**9. OPERATING LIFE**	**10. REPL. INT.**		△2 SAME C.B.'s AS REMOVED FOR ACCESS. REF: A84220, REQ. NO. 01, TASK 0010
1156	4000 HR	578		

11. REFERENCES: A84220, REQ NOS. 01 & 03

12. MAINT LEVEL AND TASK NO.	13. TASK/DESCRIPTION	14. ELAPSED TIME - MINUTES	15. ELAPSED TIME TOTAL/TASK	16. JOB SPECIALTY OR RATING	17. BASIC SKILL	18. INTERMED SKILL	19. SUPER SKILL	20. TOTAL	21. QTY/ASSY	22. PART NO. / 27. PART NAME	23. MAINT QTY FACTOR	24. SHELF LIFE	25. TOOL NO. / 28. TOOL NAME	26. TOOL USE
0010	REMOVE COMPUTER ASSY FROM UNIT A & PREPARE FOR CORRECTIVE MAINTENANCE	(1) (2)	006	31330	006			006						
9020	REMOVE & REPLACE CIRCUIT BOARD 2A3 (0.00050)	(1) △1	003	31330	003			003	01	A58181 C.B. 2A3	0.0500	NA		
9030	REMOVE & REPLACE CIRCUIT BOARD 2A4 (0.00060)	(1) △1	003	31330	003			003	01	A58182 C.B. 2A4	0.0600	NA		
9040	REMOVE & REPLACE CIRCUIT BOARD 2A5 (0.00030)	(1) △1	004	31330	004			004	01	A58183 C.B. 2A4	0.0300	NA		
9050	REMOVE & REPLACE CIRCUIT BOARD 2A6 (0.00033)	(1) △1	004	31330	004			004	01	A58184 C.B. 2A6	0.0330	NA		
0060	INSTALL COMPUTER ASSY IN UNIT A	(1) (2)	004	31330 31330	004 004			004 004					58148-001 SCREWDRIVER	004
0070	INSTALL CIRCUIT BOARDS △2 1A4, 1A5, 1A6 & CLOSE UNIT A.	(1)	004	31330	004			004						
	REF: A84220 REQ. NO. 03 FUNCTIONAL TEST													

CHANGE NO. 2

JOB SPEC.-NO. 29. 31330 30. 002 31. 32. 33. 34. 40. 41. 42. 43. 44. 45.

35. 017 △1 TOTAL ELAPSED TIME/REQ'MT

36. 038 37. 38. 39. 038 TOTAL MAN-MINUTES

46. 002 47. 48. 49. 002 TOTAL NO. MEN

SHEET 1 OF 1

PREPARED BY J. JONES

DATE 27 APRIL 1966

FORM 1366F 12/63 11:4.0

FIG. C-2 **Maintainability-task Analysis (Example).** (*General Dynamics Corporation.*)

viated information which supports data on the form. Reference reports as applicable. Use an addendum sheet for additional space as required.

Block 8—Mean-time-between-failures (MTBF). *MTBF* information is obtained from reliability data. The value entered in this block is the predicted or assessed *MTBF* for the nomenclature item as identified in block 1. State whether the value entered is predicted or assessed and reference the applicable report title, number, and date in block 7.

Block 9—Operating Life. Operating life (or service life) is the assigned life expectancy that an item is expected or intended to undergo from the time of its initial installation until it must be scrapped. It should be assumed that the item will operate within the bounds of its performance specification and will be subject to normal care in handling associated with similar types of equipment. When the end of this period is reached, the item is removed from inventory and disposed of. Operating life may be specified in terms of calendar time, equipment operating hours, or cycles.

Block 10—Replacement Interval/Frequency. The value entered in this block represents the composite frequency of maintenance (resulting in item replacements) anticipated for the nomenclatured item identified in block 1. The cause for item replacements may be based on unscheduled maintenance, on periodic scheduled maintenance, or on forced removals at the end of the anticipated item operating-life period (refer to block 9).

The need for establishing a frequency factor for a requirement is to provide a basis for determining associated resource demands. In other words:

- How often do we expect to need spares?
- How often do we need tools and test equipment to support maintenance actions?
- How often will maintenance personnel be required?
- How often will equipment item or dependent equipment be down for maintenance actions?

Once these questions are answered, it becomes possible to project total downtime and to respond to other system and equipment demands.

In determining the intervals or frequencies representing unscheduled maintenance actions, the inherent reliability characteristics of an item may be significant, but they are not necessarily dominant considerations. For some equipment, field data has indicated that less than one-third of all unscheduled maintenance actions are attributed to random catastrophic failures; therefore, it is obvious that concern must be given to other contributing factors as well. On the other hand, there are instances where reliability failure rates dominate all considerations. In this event, proportional attention must be given to the significance of random failures. In either case, the maintenance/interval frequency-factor determination must consider:

1. INHERENT RELIABILITY CHARACTERISTICS—Predicted catastrophic-failure rates derived through reliability-prediction data are used directly in maintenance interval/frequency determination.
2. WEAROUT OF EQUIPMENT (RATE OF)—Unscheduled requirements resulting from wearout of electronic equipment are quite remote and are *not* significant to establishing a maintenance frequency. This is a significant consideration, however, if mechanical linkage constitutes a great percentage of the assembly. Whether the item is 100 percent mechanical or amounts to a high percentage of the electrical unit assembly, it is necessary to arrive at a frequency for this consideration. This is done by review of the item usage to determine frequency and period of exercising. Based on this review, a factor is selected which best represents the period of useful life before wearout causes replacement of the item. This factor is retained to be considered for use with other factors discussed in this section.
3. PERFORMANCE-DETERIORATION OF EQUIPMENT (RATE OF)—The effects of performance-deterioration will account for the majority of maintenance requirements. This is to say that it can be expected that during a stated period of time an electrically complex unit will drift beyond tolerance limits caused by changes in electrical or mechanical characteristics. This fact is supported by the established need of periodic calibration. From this we can conclude that performance deterioration will likely occur prior to the calibration date and most certainly will occur at a point beyond that date. Since the latter represents idealized conditions, the former must be accepted from the standpoint of maintenance and logistics interests. Therefore, the frequency factor selected for deterioration effects will be one representing a period of time less than the calibration cycle. The time difference between the two factors will depend directly on the degree of complexity of the item under review and will be decided based upon a review of the equipment article. The factor arrived at will be retained to be considered with others obtained during the review (analysis) process.
4. PROBABILITY OF DEPENDENT FAILURES (CHAIN EFFECT)—The dependent failures causing maintenance requirements in electrical equipment are quite remote due to circuit protection provisions. However, this possibility does exist, and an investigation is needed to establish whether or not it is of concern. This is done through circuit analysis of the item and all interfacing equipment articles. If failure of one unit can be caused by failure of an adjacent item, then a maintenance-frequency factor for this possibility must be generated and applied to the article in question. This factor is the same as the composite factor for the adjacent item. If

MAINTAINABILITY TASK ANALYSIS — GENERAL DYNAMICS | ELECTRONICS

SYSTEM ______

1. PART NUMBER / NOMENCLATURE A11002/ HYDRAULIC SUBSYSTEM — NEXT ASSEMBLY N6200000/SHIPS STEERING SYST. — 2. MA CONTROL NUMBER N6200100

3. REQ. NO.	4. REQUIREMENT	5. CODE	6. TECHNICAL JUSTIFICATION	7. REMARKS
01	TROUBLESHOOT	HI	A MALFUNCTION HAS BEEN LOCALIZED TO THE SHIP'S STEERING. TROUBLESHOOTING IS NECESSARY BEFORE CORRECTIVE ACTION CAN BE INITIATED.	△ REFER TO LOGIC TROUBLESHOOT. FLOW DIAGRAM FOR SYMPTOM AND SUBSEQUENT SEQUENCE OF EVENTS.
8. MTBF	**9. OPERATING LIFE**	**10. REPL. INT.**		
4424	20000	226x10⁻⁶		

11. REFERENCES: N6200100 REQ. NOS 02 & 03

12. MAINT LEVEL AND TASK NO.	13. TASK/DESCRIPTION	14. ELAPSED TIME - MINUTES (1–29)	15. ELAPSED TIME TOTAL/TASK	16. JOB SPECIALTY OR RATING	17. BASIC SKILL	18. INTERMED SKILL	19. SUPER SKILL	20. TOTAL	21. QTY/ASSY	22. PART NO. / 27. PART NAME	23. MAINT QTY FACTOR	24. SHELF LIFE	25. TOOL NO. / 28. TOOL NAME	26. TOOL USE
	SYMPTOM: RUDDER ANGLE INDICATOR FAILS TO RESPOND WHEN HELM IS MOVED △													
0010	SHIFT STEERING CONTROL TO STBD TRICK WHEEL	①	003	MM2		003		003						
0020	CHECK PRESSURE GAUGE FOR STBD HYDRAULIC PUMP		001	MM2		001		001						
0030	CHECK OPERATION OF STBD HYDRAULIC PUMP	①	005	MM2		005		005						
0040	CHECK THAT SELECTOR SWITCH IS AT PROPER SETTING	①	001	MM2		001		001						
0050	CHECK PLUNGER FOR FREEDOM OF MOTION	①	004	MM2		004		004					5B144-001 SCREWDRIVER	004
0060	START EMERGENCY PUMP & CHECK OPERATION	①	006	MM2		006		006						
	NOTE: FAULT IS ISOLATED TO STBD AUTOMATIC VALVE													
	REF: N6200100 REQ. NO. 02 REPAIR													

CHANGE NO. 02 — JOB SPEC.-NO. 29 MM2 30. 001 — 35. 020 TOTAL ELAPSED TIME/REQ'MT — 37 020, 39 020 TOTAL MAN-MINUTES — 47 001, 49 001 TOTAL NO. MEN

SHEET 1 OF 1 — PREPARED BY B. BLANCHARD — DATE 28 AUGUST 1965

FORM 1366F 12/63 — 11:4.0

FIG. C-3 Maintainability-task Analysis (Example). *(General Dynamics Corporation.)*

more than one possibility exists, then the one which would occur most frequently would be the base for the factor assignment. The reliability failure-modes-and-effects-analysis (*FMEA*) is the best data source for reflecting dependent failure characteristics.

5. PROBABILITY OF OPERATOR ERROR—Overstress of equipment due to operator human error is a possibility and is very significant in determining frequency factors. Under this category consider such questions as:
   a. What failures can occur if controls are inadvertently operated at various stages of item application? What are these probabilities?
   b. What failures can occur if integrated circuit controls are inadvertently operated at various stages of item use? What are these probabilities?
   c. What is the effect of normal methods of operation regarding "wear and tear" of the equipment?
   d. What are the chances of physical interchangeability conflicting with functional differences, and what failures would result if this did happen?
6. PROBABLE ENVIRONMENT OF MAINTENANCE ACTIONS—The environment of the maintenance action does have direct bearing on the quality of the repair and also affects the consideration of induced faults. Therefore, item failures can result from these indirect influences. Of significance are:

   a. Effect of level of lighting
   b. Effect of presence of hazards or obstacles
   c. Effect of use of special safety items, etc.

   If any of the above would tend to degrade the quality of repair or burden the personnel in the performance of the job, then it would increase the likelihood of item failure through direct or indirect happenings. The probability of such an occurrence must be resolved into a time relationship.
7. PROBABILITY OF EQUIPMENT DAMAGE THROUGH HUMAN ERROR (i.e., dropping, bumping, etc.)—Another very significant consideration in determining the frequency factor is the probability of item damage through human error in handling. The first question to ask yourself is: how often is the item moved or handled? The next question is: what is the chance of damage to the item in movement (transit), and/or what is the chance of damage caused by traffic movement? The latter should include activity of people with and without motorized equipment. Since there is no common base for this analysis, it is necessary to project oneself to visualize the conditions that could contribute to damage of the equipment item. The factor selected from judgment will be retained for application with other factors as discussed herein.
8. PROBABILITY OF INCLUDING DAMAGE THROUGH HUMAN ERROR

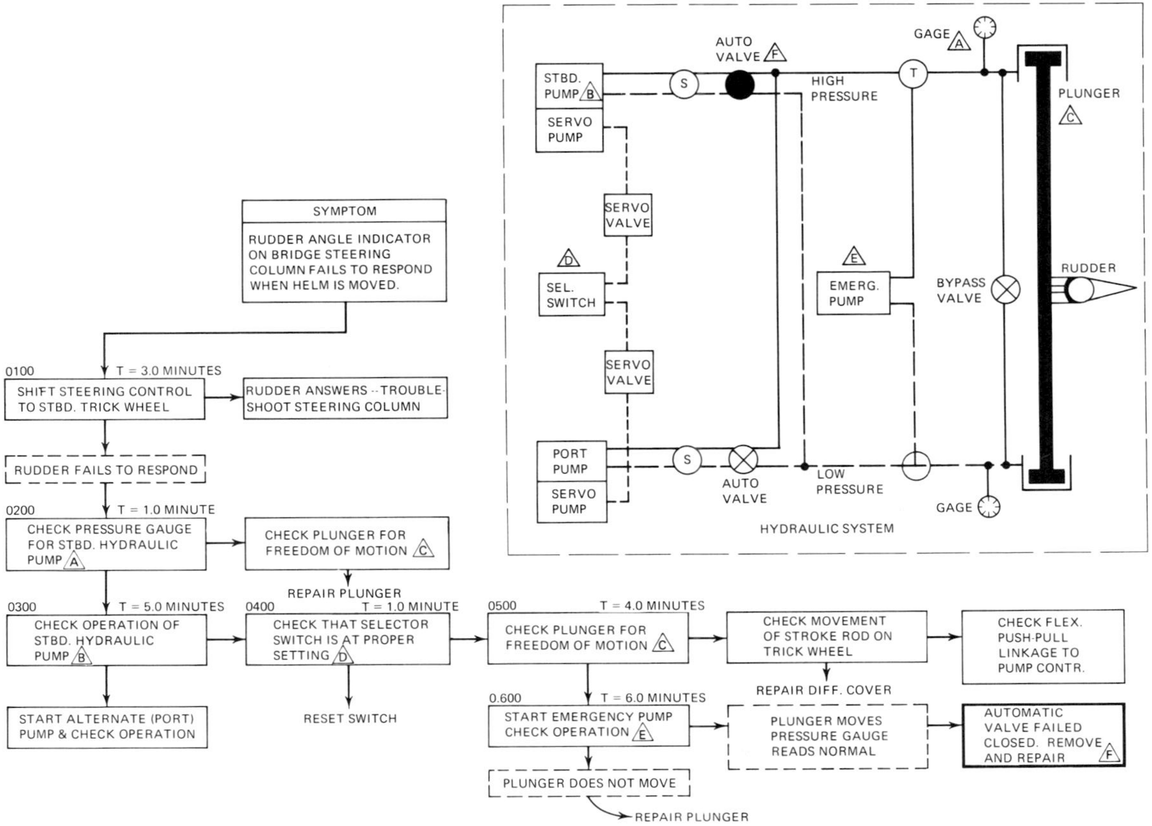

FIG. C-4 Logic Troubleshooting Flow for Ship Steering System (Example).

DURING MAINTENANCE ACTION—Damage to equipment caused by human error during maintenance actions and damage from faulty application of tools and test items must be considered jointly. To appraise this consideration it is necessary to define the following as a minimum:

a. Maintenance sequences and tool-test procedures
b. Equipment (part) layout and arrangement
c. Degree of equipment sensitivity and fragility
d. Environment of maintenance action
e. Maintenance actions performed on adjacent equipment as it affects the item in question

Once these are established, it is possible to evaluate the conditions and set a factor for induced faults based on judgment.

9. DEGREE OF EQUIPMENT SENSITIVITY, FRAGILITY, AND COMPLEXITY—Maintenance requirements are obviously dependent to a great degree on the equipment sensitivity, fragility, and complexity. This consideration overlaps with other reviews and becomes an inherent part of each analysis. It *cannot* be evaluated alone.
10. CALIBRATION AND ADJUSTMENT NEEDS OF THE EQUIPMENT TO MAINTAIN ACCEPTABLE PERFORMANCE LEVELS—Calibration and adjustment needs of the equipment are established on a scheduled basis. By virtue of this schedule, however, a measure is given to the probability of unscheduled needs. This affects performance-deterioration and should become a part of that analysis. The maintenance interval/frequency factor is determined for *each individual task* identified in blocks 12 and 13. The task frequency is based on consideration of the above factors (1 through 10). For example, the frequency of task no. 9040 (indicated in Fig. 13-4) may be calculated as indicated below.

	Consideration	Assumed Factor (Per Equip. Operating Hr)
a.	Inherent Failure Rate	0.00028
b.	Wearout	0.00000
c.	Performance Deterioration	0.00005
d.	Probability of Dependent Failures	0.00000
e.	Probability of Operator Error	0.00002
f.	Maintenance Environment	0.00002
g.	Equipment Damage (Handling)	0.00012
h.	Equipment Damage (Maintenance Induced)	0.00011
	Combined Factor	0.00060

The combined frequency factor for each individual task is entered in parentheses after the applicable task listed in block 13 (i.e., 0.00060 for task 9040).

The composite maintenance interval/frequency entered in block 10 for the overall requirement is the reciprocal of the sum of individual task frequencies as indicated below (refer to Fig. 13-4).

Task No.	Task/Description	Frequency
0010	Gain Access	*
0020	Prepare For Troubleshooting	*
9030	Isolate To Circuit Board 2A3	0.00050
9040	Isolate To Circuit Board 2A4	0.00060
9050	Isolate To Circuit Board 2A5	0.00030
9060	Isolate To Circuit Board 2A6	0.00033
Total Frequency	(Block 13)	0.00173

* Tasks 0010 and 0020 are applicable for each occurrence of Tasks 9030, 9040, 9050, and 9060; therefore, the task frequency is the combined frequency, or 0.00173.

The composite maintenance interval/frequency is

$$\text{Requirement interval/frequency} = \frac{1}{0.00173} = 578 \text{ hr}$$

Based on the above, the figure of 578 is entered in block 10 of the maintainability-task-analysis form. This indicates that corrective (unscheduled) maintenance on the computer assembly can be expected to occur once every 578 hr of unit *A* operation.

Block 11—References. Enter direct references to other requirements. For example, if the requirement being covered is requirement number 02 (block 3) for repair (block 4), reference related requirements such as:

A84220, REQ. NOS. 01 & 03

Requirement number 01 covers troubleshooting which leads to the repair action, and requirement number 03 covers functional test which is accomplished after the repair action is completed.

Block 12—Task Number. Insert a four-digit number, the first digit of which shall be 0 or 9. Tasks in a normal sequence of events (those required *each* time that the maintenance requirement is accomplished) shall be identified with a 0. Tasks which are optional depending on the applicability shall be identified with a 9. For instance, referring to Fig. 13-4, tasks 0010 and 0020 are accomplished each time that the maintenance requirement is necessary. On the other hand, either task 9030, 9040, 9050, or 9060 is accomplished for a single maintenance action depending on the failure (only one of these tasks will be applicable for a given failure).

The last three digits shall indicate the sequence of accomplishment (e.g., 0010, 0020, 9030, 9040, 9050, etc.). Subtasks may be identified further, if desired, through use of the fourth digit (e.g., 9041 9042, etc.).

Block 13—Task/Description. State in a concise technical manner

MAINTAINABILITY TASK ANALYSIS

GENERAL DYNAMICS | ELECTRONICS

SYSTEM SHIPS STEERING SYSTEM

1 PART NUMBER/NOMENCLATURE A11002/HYDRAULIC SUBSYSTEM — NEXT ASSEMBLY N6200000 — 2 MA CONTROL NUMBER N6200100

3 REQ. NO.	4 REQUIREMENT	5 CODE	6 TECHNICAL JUSTIFICATION	7 REMARKS
02	REPAIR	UM	A FAULTY STBD. AUTOMATIC VALVE IN THE HYDRAULIC SUBSYSTEM MUST BE REPAIRED TO RESTORE THE SHIP'S STEERING SYSTEM TO FULL OPERATIONAL CAPABILITY. REPAIR IS ACCOMPLISHED ON SHIPBOARD.	△1 ONE (1) GALLON OF SOLVENT REQUIRED. △2 0.000226 REPLACEMENTS/HOUR. REFER TO REQ. 03-FUNCTIONAL TEST
8 MTBF: 4424	9 OPERATING LIFE: 20000	10 REPL INT: 226×10^{-6}		
11 REFERENCES: N6200100 REQ. 01 & 03				

12 Maint Level and Task No	13 Task/Description	14 Elapsed Time - Minutes (crew size)	15 Elapsed Time Total/Task	16 Job Specialty or Rating	17 Basic Skill	18 Intermed Skill	19 Super Skill	20 Total	21 Qty/Assy	22 Part No. / 27 Part Name	23 Maint Qty Factor	24 Shelf Life	25 Tool No. / 28 Tool Name	26 Tool Use
0010	ACTUATE 3-WAY VALVES, SHUTOFF PRESSURE.	①	002	MM2		002		002						
0020	OPERATE HANDPUMP	③	015	FN	015			015						
0030	REMOVE 1½ INCH EXTERNAL TUBING	①	010	MM2		010		010					610012-1	010
		②		FN	010			010					WRENCH—1½"	
0040	REMOVE 3/8 INCH EXTERNAL TUBING	①	003	MM2		003		003					622881-2	003
													WRENCH--3/8"	
0050	INSTALL FLANGES. START PORT PUMP.	②	005	FN	005			005						
	STOP HANDPUMP. ACTUATE 3-WAY VALVES.	①		MM2		002		002					FLANGES	
0060	REMOVE CYLINDER ASSY.	①	011	MM2		011		011					633262-4	011
													WRENCH--3/4"	
0070	REMOVE VALVE ROD & PISTON	① (CYCLE 2)	015	MM2		015		015						
0080	REMOVE VALVE	①	005	MM2		005		005						
0090	REPAIR VALVE-MACHINING, CLEANING, ETC.	④	038	MMC			038	038					MACHINE SHOP	038
0100	CLEAN VALVE BODY, ROD, SPRING ASSY, PISTON.	① (CYCLE 3)	030	MM2		030		030					DUPONT 110 SOLVENT △1	
0110	INSTALL VALVE & SPRING ASSY.	① (CYCLE 4)	010	MM2		010		010						
0120	INSTALL VALVE ROD & PISTON	①	008	MM2		008		008	01	AN6526-10 GASKET	△2	NA		
									02	AN6857-17 "O" RING	△2	12		
0130	INSTALL CYLINDER ASSY.	①	010	MM2		010		010						
0140	STOP PORT PUMP. ACTUATE 3-WAY VALVES	①	002	MM2		002		002						
0150	OPERATE HANDPUMP	③ (CYCLE 5)	010	FN	010			010						
0160	REMOVE FLANGES. INSTALL VALVE ASSY.		010	MM2		010		010					610012-1	010
	CONNECT 1½ & 3/8 INCH TUBING.	①											WRENCH--1½"	
		②	010	FN	010			010					622881-2	003
													WRENCH--3/8"	
0170	STOP HANDPUMP. ACTUATE 3-WAY VALVES.	①	002	MM2		002		002						

CHANGE NO. 02 — JOB SPEC-NO: 29 MM2, 30 001, 31 MMC, 32 001, 33 FN, 34 002

35 Total Elapsed Time Req'mt		36	37	38	39
128	Total Man-Minutes	050	120	038	208
	Total No. Men	002	001	001	004

SHEET 1 OF 1 — PREPARED BY B. BLANCHARD — DATE 28 AUGUST 1965

FORM 1366E 12/63

FIG. C-5 Maintainability-task Analysis (Example). *(General Dynamics Corporation.)*

the task(s) which must be performed to fulfill the maintenance requirement. The task listings and descriptions are stated in sufficient detail to enable development of a complete maintenance checklist.

Block 16—Job Specialty or Rating. Opposite each task, enter the AFSC (Air Force Specialty Code), Rating Code (Navy), MOS (Army), or company job/skill rating of each man required for the task. The following documents are applicable guides for the Air Force, Navy, and Army, respectively:

1. AFM 39-1, personnel manual
2. NAVPERS 18068B, personnel information
3. AR 611-201, personnel selection and classification

Company job/skill-level ratings may be in terms of technician 1, technician 2, field-service representative, field-service engineer, engineering aide, senior engineering aide, etc. Such ratings are dependent on the individual company's job-classification structure.

Block 17–20—Man-minutes. Blocks 17 to 19 refer to the total man-minutes required per task for each of three basic skill levels. These skill levels may be classified as follows:

1. BASIC SKILL LEVEL (BLOCK 17)—This skill level applies to the qualifications of Air Force personnel with a 3-level AFSC, Navy service-pay grades E3 and E4, or civilian equivalent, who have completed the appropriate Air Force *basic* technical school, Navy class "A" service school, or its equivalent field or fleet experience. These personnel are assumed to have received some specialized training in order to perform routine checks and to assist more highly skilled personnel.
2. INTERMEDIATE SKILL LEVEL (BLOCK 18)—This skill level applies to Air Force personnel with 5-level AFSC, Navy service-pay grade E5, or civilian equivalent. These personnel are assumed to have OJT, and have received specialized training in the Air Force schools, Navy class "B" service schools, Navy maintenance trainers, and/or contractor's factory training programs.
3. SUPERVISORY SKILL LEVEL (BLOCK 19)—This skill level applies to the qualifications of Air Force 7- and 9-level AFSCs, Navy service-pay grades E6 or above, or civilian equivalent. These personnel are assumed to have an extensive amount of OJT and to have completed specialized service schools (i.e., Navy class "B" and/or class "C" schools). This classification implies qualification to perform the highest level of field or fleet maintenance and to instruct other less-skilled personnel in these functions.

   Task listings are in the natural order of occurrence and are limited to significant functions. Listings should describe or indicate *what* has to be done. *How* it is done will be covered in appropriate technical manuals. It is not necessary to include in the task listing minor steps which by their nature are self-evident in the prosecution of the task.

Task descriptions, entered on the same line as task numbers, shall include action words (e.g., GAIN ACCESS, PREPARE FOR, TROUBLESHOOT, ISOLATE TO, CLOSE-UP, SECURE, INSPECT, SERVICE, CLEAN, INSTALL, REMOVE, ADJUST).

After each "9" task, enter in parentheses the estimated frequency (per operating hour) of accomplishment. Task frequencies are based on the criteria presented in the instructions for the completion of block 10 above.

Insert in block 13 cautions or warnings necessary to ensure the safety of maintenance personnel and equipment during the performance of a task. These entries precede the listing of the task that presents a hazard and are included to alert the maintainability engineer of a potential design problem.

Block 14—Elasped Time (Minutes). Opposite each task entry (block 13) enter a line for the time involved for the task. Identify each man involved in the task by a separate line and a number (i.e., ① for man number 1, ② for man number 2, etc.)

Most tasks for electronic equipment can be accomplished in less than 30 min (the form in Fig. 13-4 was designed with this in mind). If additional time is recorded, it is indicated by extending the time line on the next line commencing at zero. This is accomplished for each man requirement, as applicable.

Relative to mechanical equipment, task-time spans often exceed 30 min. In such instances, the basic scale can readily be converted to hours.

Block 15—Elapsed Time Total/Task. The span of time required for each task (start to finish) is recorded in block 15. Do *not* total the times of the task segments. Record only the time between task start and task complete. For instance, if man no. 1 starts task XXXX 10 min from requirement start, and man no. 2 completes this task 14 min from requirement start, the total elapsed task time is 4 min.

Each task is analyzed from the standpoint of complexity, and the appropriate skill levels are assigned accordingly. Human-factors personnel should be consulted on an "as-required" basis when determining complexities. The total man-minutes for each task (the sum of blocks 17–19 for all skill levels) are entered in block 20.

- *Blocks 21–24, 27—Part Data.* The intent of recording part data is to identify the parts that are necessary to complete a maintenance requirement and/or the parts that are expended in the maintenance action. Spare/repair-part requirements are derived from this information.

1. BLOCK 21—QTY./ASSY.—Against each part listed in block 22, enter the quantity used (actually replaced or consumed) in accomplishing the task.
2. BLOCK 22—PART NO.—Enter the identifying number of the part which requires replacement or is consumed during the task.
3. BLOCK 23—MAINT. QTY. FACTOR—The maintenance quantity factor

MQF or replacement rate considers part replacements as a result of primary failures (castastrophic-failure rates) secondary or dependent failures operator- and maintenance-induced failures, quality-control or manufacturing defects, forced removals, etc. Such causes for equipment failure are discussed in the criteria presented for the completion of block 10.

Generally speaking, for each repair requirement where individual tasks indicate an item replacement, the task frequency and the MQF will be the same, since both factors consider the same causes for equipment failure.

The MQF for each part is entered in block 23 and reflects part replacements per 100 hr of equipment operation.

4. BLOCK 24—SHELF LIFE—Shelf life represents the time that an item can be stored on the shelf, removed, and put into operational use without requiring maintenance. Some items such as O rings and seals degrade after a period of time and must be replaced. Other items must be calibrated on a periodic basis (e.g., 180 days, 270 days). An applicable shelf life for these items is entered in block 24. Other items are not affected and are covered by NA.
5. BLOCK 27—PART NAME—Enter the identifying nomenclature of the part indicated in block 22. (Use abbreviations when space precludes entering the complete nomenclature.)

Blocks 25, 26, 28—Tool Data. The intent of recording this data is to identify all tools and support equipment required to complete a maintenance requirement.

1. BLOCK 25—TOOL NUMBER—List the military or manufacturer's part number assigned to each tool or support-equipment item required per task.
2. BLOCK 26—TOOL USE—For each task requiring a tool or support-equipment item, enter the total time that the tool is in use or the total time that the tool is not available for use elsewhere, whichever is greater.
3. BLOCK 28—TOOL NAME—Enter the identifying nomenclature of each tool or support-equipment item indicated in block 25.

Blocks 29–34, 40–45—Job Spec. No. On the last page of the requirement, insert each AFSC, rating code, MOS, or civilian specialty as identified in block 16. These entries utilize blocks 29, 31, 33, 40, 42, and 44 as necessary. After each job specialty or rating, enter the quantities required in the applicable three-digit block (blocks 30, 32, 34, 41, 43, 45).

Block 35—Total Elapsed Time/Requirement. In block 35, enter the total elapsed time for the requirement. All "9" tasks are averaged (by weighting) and added to the "0" tasks. The weighted average is based on the sum of the products of frequencies and times for each task divided by the sum of the frequencies. The weighted-time calculation is illustrated below (refer to Fig. 13-4).

Task No. (Block 12)	Frequency (Block 13)	Elapsed Time (Block 15)	Weighted Value
9030	0.00050	006	0.00300
9040	0.00060	011	0.00660
9050	0.00030	008	0.00240
9060	0.00033	013	0.00429
TOTAL	0.00173		0.01629

$$\text{Weighted average of ``9'' tasks} = \frac{0.01629}{0.00173} = 9.4 \text{ or } 9 \text{ min}$$

$$\text{Total elapsed time (block 35)} = 9 + 3 + 2 = 14 \text{ min}$$

- *Blocks 36–39—Total Man-minutes.* The sum of blocks 17, 18, 19, and 20 is entered in blocks 36 through 39 respectively.
- *Blocks 46–49—Total Number Men.* From block 14 determine the total number of men of each skill level needed to satisfy the maintenance requirement. Enter the number required in the appropriate space of the last page of the requirement.

4.0 Maintainability-task Analysis—Examples

Completion of the maintainability-task analysis follows the same basic principles regardless of the end item involved. Two examples are presented. Example 1 (Figs. C-1 and C-2) covers the troubleshooting and repair requirements for an electronic system, and Example 2 (Figs. C-3 to C-5) covers the troubleshooting and repair requirements for a ship's steering system.

APPENDIX D

Instructions for Maintenance-analysis Data Summary

1.0 GENERAL

The maintenance-analysis data summary (Fig. D-1) presents a composite coverage of major maintenance requirements for each significant repairable item. The summary sheet covers a brief functional description of the item, a description of the maintenance concept/policy for the item, and anticipated quantitative factors covering the various detailed scheduled and unscheduled maintenance requirements. This appendix is designed to cover instructions for completion of this form.

2.0 DATA COVERAGE

A maintenance-analysis data summary shall be completed for each repairable maintenance-significant item (i.e., end item, assembly, subassembly, etc.). The summary is supported by the individual maintenance requirements covered by the maintenance-task analysis presented in Appendix C.

3.0 INSTRUCTIONS FOR FORM COMPLETION

The following instructions should be employed when completing the maintenance-analysis data summary. (The data items discussed in Appendix C form the basis for the summary information outlined herein.)

System. Enter the proper nomenclature for the end item covered by the overall analysis.

Block 1—Identification Part Number/Nomenclature. Enter the manufacturer's part number and nomenclature for the item being covered by the maintenance-analysis data-summary sheet. The item may be a system, subsystem, unit, assembly, subassembly, etc.

Block 2—Identification Next Assembly. Enter the next higher assembly by part number and nomenclature. If the maintenance-analysis data summary covers the *subsystem* level, the *system* nomenclature is entered.

Block 3—MA Control Number. Enter the maintainability-analysis (MA) control number for the item being covered as identified in block 1. The maintainability-analysis control number is a numerical identification for maintenance-significant items. It may follow the work-unit code system (Air Force), the MEAR control-number system (Navy), or a contractor's designated numbering system. MA control numbers are assigned by indenture level of maintenance. MA control numbers are assigned by function and not necessarily by part number. If a modification of unit *A* is accompanied or an alternate configuration of unit *A* is employed, the manufacturer's part number changes, but the MA control number remains the same as long as the basic function of the applicable item is unchanged. MA control numbers are assigned in accordance with the basic instructions presented in Appendix C.

Block 4—Design Specification. Enter the applicable design or performance specification covering description and design requirements for the item identified in block 1. This entry serves as a point of reference relative to specific customer requirements, design criteria, etc., for the item being covered. The applicable specification may be oriented to the end-item level and not to the subsystem or assembly level. In such instances, enter the end-item specification number and date.

Block 5—Supporting Documents. List (in brief form) major supporting documents directly applicable to the information on the maintenance-analysis data summary. For instance, reference the applicable reliability report serving as the source for *MTBF* and failure-rate data, and reference major trade-off–study reports supporting the maintenance concept. In the event that space is inadequate, utilize the space available in block 21, or use an addendum sheet.

Block 6—Functional Description. Briefly describe the basic function of the item covered (as identified in block 1). The description should be complete enough to enable a reviewer to ascertain what the item is and what function it serves. In addition, the description should provide enough information to support the maintenance concept presented in block 7. The maintenance-analysis data summary should be able to stand on its own without requiring a lot of supplementary descriptive data.

Block 7—Maintenance-concept Description. Briefly describe the maintenance concept or policy for the item covered. Include reference to the levels of echelons of maintenance, depth of main-

GENERAL DYNAMICS
Electronics Division

MAINTENANCE ANALYSIS DATA SUMMARY

SYSTEM XYZ SYSTEM

1. IDENT PART NUMBER/NOMENCLATURE: A58180/COMPUTER ASSY.
2. IDENT NEXT ASSEMBLY: A37701/UNIT A
3. MA CONTROL NO: A84220
4. DESIGN SPEC: ZA88446 (9/4/65)
5. SUPPORTING DOC: RELIABILITY REPORT RA100-2; TRADE-OFF STUDY REPORT M-106
6. FUNCTIONAL DESCRIPTION: SYSTEM XYZ IS THE AIRBORNE ELECTRONICS PORTION OF THE SHORT RANGE TACTICAL-AIR-NAVIGATION SYSTEM INSTALLED IN THE F-XXX AIRCRAFT. THE COMPUTER ASSY (A58180), INCORPORATED IN UNIT A OF SYSTEM XYZ, PROVIDES FINITE RANGE AND BEARING INFORMATION FOR THE OVERALL SYSTEM. SUCH INFORMATION IS REQUIRED FOR ACCURATE RANGE AND BEARING MEASUREMENTS. THE COMPUTER ASSY CONTAINS FOUR MAJOR CIRCUIT BOARD SUBASSEMBLIES (2A3, 2A4, 2A5, AND 2A6).
7. MAINTENANCE CONCEPT DESCRIPTION: UNSCHEDULED MAINTENANCE -- UPON DETECTION OF A FAULT AT THE SYSTEM LEVEL, THE APPLICABLE FAULT ISOLATED TO UNIT A (A37701), UNIT B (A37702), OR UNIT C (A37703) THROUGH BUILT-IN SELF TEST. THE APPLICABLE UNIT IS SENT TO THE ELECTRONICS FIELD SHOP FOR CORRECTIVE MAINTENANCE. FAULT ISOLATION OF UNIT A TO THE COMPUTER ASSEMBLY LEVEL IS ACCOMPLISHED THROUGH USE OF THE SYSTEM TEST STATION (I.E. -- OSCILLOSCOPE, POWER SUPPLY, VOLTMETER) IN THE FIELD ELECTRONICS SHOP. FAULT ISOLATION IS ACCOMPLISHED TO THE CIRCUIT BOARD LEVEL; THE APPLICABLE CIRCUIT BOARD (I.E. -- 2A3, 2A4, 2A5, OR 2A6) IS REMOVED AND REPLACED; AND THE COMPUTER ASSEMBLY IS CHECKED OUT THROUGH A FUNCTIONAL TEST OF UNIT A. FAULTY CIRCUIT BOARDS ARE DISCARDED AS "THROW-AWAY" ITEMS. MAINTENANCE IS ACCOMPLISHED BY AFSC 31330 & 31350 PERSONNEL IN A NORMAL SHOP ENVIRONMENT.
SCHEDULED MAINTENANCE -- NONE REQUIRED FOR THE COMPUTER ASSEMBLY.

8. REQ. NO.	9. REQ. NOMEN.	10. MAINT. LEVEL O	I	D	11. MAINT. TYPE SCH.	UNSCH.	12. FREQUENCY (HOURS)	13. MTBM (HOURS)	14. MTBF (HOURS)	15. MMM/OH	16. T.B.O	17. ELAPSED TIME (HOURS)	18. PERSONNEL DATA JOB SPEC	B	I	S	19. SUPPORT EQUIP (1)	20. SPARES TYPE (2)
01	TROUBLESHOOT		X			X	0.00173	578	1163	0.1436	N/A	0.233	31330 31350	1 –	– 1		OSCILLOSCOPE POWER SUPPLY VOLTMETER	
02	REPAIR		X			X				0.0657		0.283	31330	1	–		SCREWDRIVER	C.B.2A3 C.B.2A4 C.B.2A5 C.B.2A6
03	FUNCT. TEST		X			X				0.0208		0.100	31330 31350	1 –	– 1		SPECIAL TEST SET	
TOTAL							0.00173	578	1163	0.2301		0.616		1	1			

21. NOTES:
(1) REFER TO ADDENDUM A (SHEET 1) FOR SPECIFIC QUANTITIES AND TYPE OF SUPPORT EQUIPMENT.
(2) REFER TO ADDENDUM A (SHEET 2) FOR SPECIFIC QUANTITIES AND TYPE OF SPARES.

FORM 2297B 8/67

PREPARED BY: J. JONES 27 APRIL 1966

FIG. D-1 Maintenance Data Summary (Example). (*General Dynamics Corporation.*)

tenance recommended for each echelon, maintenance environments, and any additional narrative-type data considered significant to the maintenance concept. References supporting the concept (e.g., trade-off–study reports) are included in block 5.

When preparing maintenance-analysis data-summary sheets for each maintenance-significant equipment indenture level (indicated by the MA control number), it should be realized that some overlap will exist in the narrative descriptions presented in blocks 6 and 7. However, for the sake of clarity and completeness, it is necessary to present a complete picture relative to maintenance of the item.

Block 8—Requirement Number. List all applicable maintenance requirements (e.g., troubleshooting, repair, servicing, calibration) by identification number. (Refer to Appendix C to obtain code numbers.)

Block 9—Requirement Nomenclature. The requirement name (for each requirement number listed in block 8) should be entered in this block.

Block 10—Maintenance Level. Check the applicable level or echelon at which the specific maintenance requirement is accomplished. Maintenance requirements listed in blocks 8 and 9 may all be accomplished at one level or echelon or a split concept may be recommended involving several levels or echelons. The levels are:

O: organizational maintenance
I: intermediate (field) maintenance
D: depot maintenance

Block 11—Maintenance Type. Check whether the maintenance action covered by each requirement is scheduled (SCH) or unscheduled (UNSCH).

Block 12—Frequency. The value entered in this block is the requirement frequency (in terms of maintenance actions per equipment-operating hour).

The total frequency (bottom of block 12) is the composite frequency covering all different maintenance actions for the item as identified in block 1. For instance, the example presented in Fig. D-1 reflects only one type of maintenance action, an unscheduled action requiring corrective maintenance. Since in this case the requirements of troubleshooting, repair, and functional test are necessary to complete the maintenance cycle each time that an unscheduled maintenance action is accomplished, the frequency for each of these requirements as indicated in block 12 is the same. On the other hand, the item covered may also require periodic (scheduled calibration at 6-month intervals. In this event, the calibration requirement is covered in blocks 8 and 9, and the frequency entered in block 12 is 0.000961. The total frequency is 0.00173 plus 0.000961, or 0.00269.

Block 13—MTBM. The mean-time-between-maintenance (*MTBM*) factor is entered against each requirement and as a total value (bottom of block 13). This value is the reciprocal of the frequency factor entered in block 12.

Block 14—MTBF. The mean-time-between-failures (*MTBF*) factor for unscheduled maintenance actions resulting from catastrophic (primary) equipment failures is entered in this block. The *MTBF* value is obtained from reliability-prediction data. The applicable reliability report is referenced in block 5.

Block 15—MMH/OH. The maintenance man-minutes per equipment-operating hour (*MMH/OH*) is determined by multiplying the frequency (block 12) by the total man-minutes. The result is entered in block 15 for each requirement. Total *MMM/OH* is the sum of the individual values for each requirement.

Block 16—TBO. Enter the time-between-overhauls *TBO* as required. This factor is more applicable to mechanical items (e.g., gear trains) than to electrical or electronic equipment.

Block 17—Elapsed Time. Enter the elapsed time for each requirement. In instances where different maintenance frequencies exist, the total elapsed time (bottom of block 17) for the item identified in block 1 is obtained in a similar manner as the total elapsed time for each individual requirement. A weighted average is derived by dividing the sum of the products of frequencies (block 12) and times (block 17) for each requirement by the sum of frequencies (block 12).

In the example presented in Fig. D-1, only one type of maintenance action is reflected, an unscheduled action requiring corrective maintenance. Since in this case the requirements of troubleshooting, repair, and functional test are necessary to complete the maintenance cycle each time that an unscheduled maintenance action is accomplished, the total elapsed time is the sum of the individual elapsed times. On the other hand, the item covered may require a 6-month periodic calibration with an elapsed time of 6 min per calibration action. In this event, the calibration requirement is entered in blocks 8 and 9, the frequency of 0.000961 is entered in block 12, and the elapsed time element of 60 min or 1 hr is entered in block 17. The total-elapsed-time calculation is illustrated below.

REQUIREMENT	FREQUENCY	ELAPSED TIME	WEIGHTED VALUE
Troubleshooting Repair Functional Test	0.00173	0.616	0.00107
Calibration	0.00096	1.000	0.00096
TOTAL	0.00269		0.00203

$$\text{Total elapsed time (all requirements)} = \frac{0.00203}{0.00269} = 0.754 \text{ hr}$$

Block 18—Personnel Data. For each requirement listed in blocks 8 and 9, enter the personnel types (job specialties) and indicate the quantities of each skill level (basic, intermediate, or supervisory).

Relative to the total value (bottom of block 18), an estimate is made of the quantity of each skill level required to support all maintenance actions associated with the item covered by the analysis sheet. The estimate is a gross figure derived through review of the various maintenance requirements. Although a gross indication of this type is not adequate for accurate logistics planning relative to determining total manning requirements, it is adequate for the maintainability engineer in providing a quick check on system/equipment complexity. If the personnel quantities are excessive, or if the skill level requirements are unusually high, the maintainability engineer is alerted to a potential problem area and the system/equipment design should be evaluated for possible improvement.

Referring to Fig. D-1, the example presented indicates that one 31330 and one 31350 will be required to support the Computer Assembly, A58180. These figures provide a general indication of maintenance-personnel requirements. On the other hand, an accurate prediction of personnel quantities (by skill level) requires an analysis of the individual maintenance tasks for each requirement. Personnel quantities depend on whether tasks are accomplished in series or on a parallel basis and on whether maintenance personnel can be assigned to several tasks by time sharing. Review of all requirements and specific tasks, and of their relationships in terms of the planned system/equipment life cycle, is necessary to derive the total personnel manning for maintenance support.

Personnel data in the maintenance analysis is compared with human-factors personnel-requirements data to ensure compatibility.

Block 19—Support Equipment. List the proposed support equipment (by type) against each requirement covered in blocks 8 and 9. Utilize an addendum sheet if space is inadequate.

Once again, the information as presented in this form reflects a summary of basic needs providing the maintainability engineer with a quick reference. An accurate listing is based on a review of all individual requirements (blocks 25, 26, and 28 of the form in Appendix C) relative to support-equipment types and utilization. Time sharing of support equipment for different requirements is a major consideration.

Support-equipment information in the maintenance analysis is used for provisioning purposes (i.e., formal support-equipment plan, recommendation data sheets, etc.).

Block 20—Spares Type. List the proposed spare-parts types against each applicable requirement involving a remove-and-replace action. Utilize an addendum sheet if space is inadequate.

The spares data listed in this block are sufficient to verify the maintenance concept relative to the type of item removed and replaced at each level or echelon of maintenance and the replacement frequency. Detailed information on replacement parts is derived through the review of individual requirements (blocks 21–24 and 27 of the form in Appendix C). This information in turn is utilized in the preparation of spare-parts provisioning data.

Block 21—Notes. Include additional information (notes) as required to amplify or support the entries in any of the other blocks. Utilize an addendum sheet if space is inadequate.

Index